The Highlands Controversy

Science and Its Conceptual Foundations

David L. Hull, Editor

DAVID R. OLDROYD

The Highlands
Controversy

Constructing Geological Knowledge through
Fieldwork in Nineteenth-Century Britain

THE UNIVERSITY OF CHICAGO PRESS

Chicago and London

David R. Oldroyd is associate professor in the School of Science and Technology Studies, University of New South Wales.

The University of Chicago Press, Chicago 60637
The University of Chicago Press, Ltd., London
© 1990 by The University of Chicago
All rights reserved. Published 1990
Printed in the United States of America
99 98 97 96 95 94 93 92 91 90   5 4 3 2 1

ISBN-13: 978-0-226-62634-5 (cloth)
ISBN-13: 978-0-226-62635-2 (paper)

Library of Congress Cataloging-in-Publication Data

Oldroyd, D. R. (David Roger)
    The Highlands controversy : constructing geological knowledge
  through fieldwork in nineteenth-century Britain / David R. Oldroyd.
        p.   cm. — (Science and its conceptual foundations)
     Includes bibliographical references (p.   ).
     ISBN 0-226-62634-2 (alk. paper). — ISBN 0-226-62635-0 (pbk. :
  alk. paper)
     1. Geology—Great Britain—History—19th century. 2. Geology—
  Scotland—Highlands. 3. Geology, Stratigraphic. 4. Geology—Great
  Britain—Field work. I. Title. II. Series.
  QE13.G7043   1990
  550′.941′09034—dc20   89-20610
                                                                              CIP

Contents

Acknowledgments

The following institutions have kindly granted me permission to refer to, or quote from, unpublished material in their care, and I wish to express my thanks for their courtesy in allowing me to do so: the British Geological Survey; Edinburgh University Library; Elgin Museum; the Geological Society, London; Haslemere Educational Museum; Imperial College, London; the National Library of Scotland; the Public Record Office, London; the Royal Museum of Scotland; St. Andrews University Library; the School of Earth Sciences, University of Birmingham; Wellcome Institute, London. I should also like to thank these bodies for allowing me to make use of their facilities during the course of my research.

The specific items of copyright material that I have utilized are acknowledged at the appropriate places in the text, but I should mention that Crown copyright material in the Public Record Office is reproduced with the permission of the Controller of Her Britannic Majesty's Stationery Office; material in the archives of the British Geological Survey is reproduced with the permission of the Director; and material in the archives of the Geological Society is reproduced by permission of the Library Committee. An aerial photograph of the Loch Assynt region (fig. 4.1) is reproduced by permission of the Controller of Her Britannic Majesty's Stationery Office.

I should like to express my appreciation of the generous assistance given me by the staffs of the libraries of the following institutions: the Aberdeen Public Library; the University of Aberdeen; the British Library; the British Geological Survey; the Geological Society, London; Imperial College, London; University College, London; the University of Edinburgh; the University of New South Wales; the University of Sydney. Particularly, I would mention the assistance rendered by Miss J. M. Fitch, formerly with the British Geological Survey in London; Miss W. Cawthorne of the Geological Society Library; Mr. C. Will at the Scottish Branch of the Survey in Edinburgh; and Mr. J. Howard, Keeper of Manuscripts at the Edinburgh University Library. Mr. C. A. McLaren, Archivist and Keeper of Manuscripts at the University of Aberdeen, responded very helpfully to several inquiries; and so too did Mr. G. McKenna, Chief Librarian and Archivist of the British Geological Survey. I should further like to express my thanks to Mr. A. Jewell, formerly Curator

of the Haslemere Educational Museum, for his assistance. At the University of
New South Wales, my friends in the Library's interlibrary loan service, Val
Wotton, Margaret Pratt, and Margot Zeggelink, have performed wonders on
my behalf, succeeding in tracking down every one of my numerous requests.

I am further greatly indebted to Peter Osborne of the School of Earth Sciences at the University of Birmingham, who allowed me to use the Lapworth
archive in his care, and who arranged for the photograph of the Lapworth
field map (fig. 8.8) to be prepared. Likewise, John Thackray, Archivist to the
Geological Society, allowed me to utilize the society's archives and obtained for
me the photograph reproduced in figure 4.2b, and I would like to thank him
for this.

I should mention that I benefited from discussions with Beryl Hamilton
about the life and work of Lapworth, and I wish to thank her for the generous
gift of a copy of her unpublished list of the papers in the Lapworth archive.

I enjoyed an extensive discussion with Jim Secord about issues in the history of British geology; and likewise with John Thackray, who was also kind
enough to read a draft of the book and provide me with a set of well-informed
comments on the work. Martin Rudwick and Rachel Laudan were also good
enough to read the draft and give the book their blessing. In Sydney, Tom
Vallance went through the text with his kindly and critical eye and, with his
probably unequalled knowledge of the history of mineralogy and petrology,
saved me from some errors that I was only too glad to correct. He also suggested a considerable number of improvements for the glossary, arranged for
the preparation of some thin sections of rocks that I had collected, and advised me on their interpretation; and Dr. F. I. Roberts prepared some photomicrographs for me. Other photographic work was done for me skillfully and
promptly at the University of New South Wales by Belinda Allen, for which
I specially thank her. Dr. M. B. Katz was able to produce some samples of
Saxon granulites for me to inspect. Professor B. K. Martin gave me some assistance with the rendering of Gaelic place names in English.

In my own department, I have benefited from discussions with David Miller about the content of chapter 11, and I have discussed other issues with
Randall Albury, whose advice is always judicious. Down the corridor in the
School of German Studies, Olaf Reinhardt has always been willing to find time
to provide me with translations from his language, and I am extremely grateful to him for this.

At the University of Chicago Press, the series editor, David Hull, has been
extremely supportive, generous in ideas, and careful in his reading of the
text, with numerous suggestions for improvement. The press's reader cautioned me against trying to generalize too much about the nature of either
geology or science on the basis of my investigation, and I accept his suggestion
for restraint and trust that the reader will think that I have heeded it sufficiently. The editorial staff in Chicago have been consistently helpful, encouraging and efficient, and I owe them a considerable debt, which I should like to
acknowledge publicly. My sincere thanks also to the production staff.

Finally, I may as well say thank you to the middle fingers of my left and right hands, which have painstakingly typed every word of the text! And a big bouquet is due to the person(s) who devised the program Microsoft Word 3.01, with its marvelous system for dealing with footnotes and numbering them automatically. Without that program, I doubt this book would ever have been completed. Moreover, it would never have been started but for the funding generously provided by the University of New South Wales for a period of study leave in the first half of 1987.

To all these people and organizations, then, I express my warm thanks. My thanks go also to my wife, Jane, for tolerating confusion and mess in our household until the project was completed. For any errors that may remain in the text, readers should hold me responsible—not the people mentioned above, who have been generous with their assistance in so many different ways.

1

Nineteenth-Century British Geology and Its Historiography: Some Themes, Goals, and Methods

The story of the origin, the publication and the extraordinary success of the Murchisonian hypothesis of the Highland sequence, in spite of the manly opposition of Nicol, and of the labours, the discoveries, and the conclusions of its opponents, from the date of the bold re-opening of the controversy by Dr. Hicks, down to the issue of the report of Messrs. Peach and Horne, will form a most interesting and instructive chapter in the history of British Geology. It is to be hoped that some geologist [or historian], who is familiar with all the facts, and has not identified himself with either of the contending parties, will write this story for the information and edification of our scientific public in general . . .

—Charles Lapworth, 1886

Geology is a branch of science that has been accorded relatively little attention by historians, yet it is an area of considerable interest, not only to the historian of science, but to the geologist and the general reader. It also raises important questions for those interested in philosophy of science and sociology of science, since the problems that the geologist has to deal with are somewhat different, both methodologically and epistemologically, from those encountered in other sciences.

So far as the social history of science is concerned, nineteenth-century British geology offers a particularly suitable object of inquiry. The Victorian geologists took pride in their communality and were indeed one of the most visible scientific communities in Britain at that time. Their ideas are reasonably easy of access, through their voluminous correspondences, many of which are preserved in public archives, and they were prolific publishers. However, their numbers were not so great as to make their activities impossible to comprehend as a group. In this book we shall examine one segment of their community in detail, looking at the social relationships and the ways in which social activity affected the production of geological knowledge.[1] In this way we

1. The book covers a time span of about a hundred years: that is, through the nineteenth century and up to 1907. In that year, a major memoir was published (Peach et al. 1907) that summarized all the earlier work and marked the culmination of years of bitter controversy about the rocks of the northwest. It also displayed the consensus that was eventually reached when the controversy was resolved. Thus 1907 marks a convenient cut-off point for the present study.

1

may hope to add to the growing understanding of the way in which scientific knowledge is generated and sustained within a scientific community.

Primarily, geology's domain of activity is the field, rather than the laboratory, and its empirical work is not based on experimentation to the same extent as in, say, physics. Geological problems are often ones of local interpretation rather than general principle, and the sheer mass of reported information concerning areas that may be very distant from one another and observation claims that often cannot be conveniently checked by co-workers have sometimes caused great difficulty for geologists, and also for historians of science who would seek understanding of what was done in the past. On the other hand, the rocks themselves don't change significantly over time, so the historian of geology has the opportunity to recreate the observations of former workers and to understand their thinking in a way that is not always feasible for other sciences.

My work builds on two important recent monographs, by Martin Rudwick (1985) and James Secord (1986a), which together have broadened and deepened our understanding of nineteenth-century British geology to a considerable degree. Both studies deal with controversies that stemmed from the work of the energetic and influential Victorian geologist, Sir Roderick Murchison (1792–1871), the second director general of the Geological Survey of Great Britain, patron of scientific exploration, founder of the Silurian and Permian systems, and intimately concerned with the establishment of the Cambrian and the Devonian systems. But there was a third major controversy concerning the interpretation of British rocks that developed from Murchison's work, which occurred during the closing stages of his career. It is with this that we shall be concerned in the present study.

Rudwick's definitive work examines in detail the bitter contest in the late 1830s and 1840s between Murchison and the first director of the Survey, Sir Henry De la Beche (1796–1855), concerning the establishment of the Devonian system. Rudwick shows how the two geologists became embroiled in controversy about certain sedimentary rocks of Devon, which De la Beche sought to understand and classify with the help of lithological criteria, while Murchison favored reliance on palaeontological criteria. The opposing views of the age of the rocks started off wide apart; but gradually a *via media* or consensus was established, with both geologists altering their views substantially during the course of the controversy and as it came to be recognized that there were sediments in Devon that were approximate temporal equivalents of the well-established Old Red Sandstone in other parts of Britain. Thus it was established that there could be rocks of different "facies" but similar ages;[2] and as a

---

2. "Facies" is a somewhat vague, but commonly used, geological term, referring to the total appearance and properties of a sedimentary deposit, reflecting the conditions and environment existing at the time of its deposition. The word is also sometimes used in relation to an igneous or metamorphic rock, particularly where there is a departure from the typical rock of the mass to which it belongs. For example, a mass of granite might be said to have a porphyritic facies near its margins.

result of Murchison's unflagging efforts, the Devonian came to be recognized as a "worldwide" system, definable in terms of characteristic fossils. But in achieving consensus and closure of the Devonian debate it was not simply the case that one side was right and the other side wrong. Both Murchison and De la Beche had to give ground significantly, and both of them had some "right" on their side (from the point of view of the consensus that was eventually established).

Rudwick does not simply give an account of the technical details of the Devonian controversy and the numerous shifts in theoretical viewpoint. With great virtuosity, he displays the ways in which geological debate was conducted in the nineteenth century between men with different social interests—the gentlemanly-amateurs on the one hand and the newly emerging bureaucratic professionals of the Survey on the other.[3] Rudwick further shows how alliances were formed in order to promote the various interests of the contending parties. In particular, Murchison had as ally the respected Cambridge geology professor, the Reverend Adam Sedgwick (1785–1873).

The manner in which the Devonian debate was conducted is of special interest. Observations were made in the field, and the collected data—and theoretical interpretations thereof—were brought before meetings of the Geological Society of London or the peripatetic British Association. In these venues, the issues were debated almost in parliamentary fashion, and the geological community thereby made up its collective mind on the issues under debate, with the men of highest standing in the geological community naturally having the greatest influence in the formulation of the theoretical views of the community as a whole.

Thus the emergence of geological knowledge, while based on empirical evidence and theoretical reasoning, was, on Rudwick's account, essentially a social activity. He sees scientific knowledge as being shaped or forged in the heat of scientific debate, with the empirical evidence constraining, not determining, the interpretations that are reached. His book is remarkable for its interesting diagrams, which display the social relationships of the geological community and depict in an ingenious manner the trajectories of the theoretical ideas of the several participants in the Devonian controversy. I have found it helpful to utilize a number of Rudwick's ideas in my own investigation, for it seems to me that he captures very successfully the character of British geology in the first half of the nineteenth century, showing in particular how scientific knowledge emerged from the intense controversies that developed within a tight-knit scientific community. I have not, however, sought to emulate his elaborate diagrams in order to depict the scientists' conceptual changes.

In the Devonian controversy, though a compromise was eventually reached, it is probably fair to say that De la Beche was largely vanquished by Murchison and Sedgwick (though it was acknowledged that some of the structures advocated by the director of the Survey were sound). Meanwhile, research was

3. "Bureaucratic" is not intended here as a term of opprobrium.

being pursued on the ancient rocks of Wales by both Sedgwick and Murchison, and it was from this work that the second of the three battles mentioned above grew. In 1831, Sedgwick started investigations in the mountainous regions of North Wales, working in rocks that were soon designated the Cambrian system.[4] Fossils were found, but many of them were ones that Murchison later claimed did not belong in the Cambrian but in his own system, the Silurian. The basis of Sedgwick's characterization of the Cambrian was chiefly structural, being based on his interpretation of numerous geological sections. Murchison started in the Welsh Border region and South Wales. He proposed a system which he named the Silurian and which was clearly identified by characteristic fossil assemblages.[5]

At first, it did not appear that there were going to be any difficulties. After a joint field trip in 1834—which took place before the names Cambrian and Silurian were formally proposed for two different geological systems—it seemed that a satisfactory boundary line could be established between the two "territories," but as time went on it became evident that this boundary had not been drawn satisfactorily, and each geologist sought to annex more and more of the other's ground. This led to a notorious and bitter dispute, with the two former friends falling apart completely in the 1850s. Secord analyzes the story in detail and with skill, showing the manifold influences of the controversy on British geology in the nineteenth century. He shows also how wider developments, such as the professional mapping of Wales by the Survey, gradually changed the character of the contest between Murchison and Sedgwick and strongly affected the eventual outcome.

In his own time, Murchison largely succeeded in defeating Sedgwick, for he was able to claim that the Cambrian system was not properly characterized or defined by palaeontological criteria.[6] By the mid-nineteenth century, following the outcome of the Devonian controversy, it was recognized that fossils should serve as the basis of stratigraphical classification. But while Murchison was largely successful in his contest with Sedgwick, and while the Murchisonian doctrine of the Silurian was adopted by the Survey, in time it came to be acknowledged that Murchison's claims were too strong.[7] The Cambrian was eventually characterized satisfactorily by fossils in Wales, and a new system (the Ordovician) was proposed in 1879 by Charles Lapworth (1842–1920), which included the rocks of the region of Wales between the Cambrian and the Silurian contested so keenly by Sedgwick and Murchison. However, though

4. Secord (1986a, 100) points out that although the term "Cambrian" is traditionally associated with Sedgwick, it was actually Murchison who, in the first instance, perceived the need for the postulation of such a system, below his Silurian. But it was Sedgwick who first suggested the name that eventually came into general usage.

5. Eventually, he regarded *his* system as having the first forms of life on earth.

6. This came about partly as a result of the Devonian debates, for what were at first believed to be characteristic Cambrian fossils in Devon were redesignated as Devonian, leaving Sedgwick's Cambrian bereft of palaeontological definition. Sedgwick thus found that fossils in "his" territory were described and figured as Silurian in Murchison's *Silurian System* (1839).

7. Sedgwick, indeed, retained significant support in his home base at Cambridge, and in time it became evident that some of his complaints against Murchison's claims were warranted.

the Ordovician was firmly based on palaeontological principles, the Survey did not accept it into its official publications until the beginning of the twentieth century—and this had as much to do with the social relationships within the British geological community as with the strengths and weaknesses of Lapworth's arguments. In his study, Secord emphasizes the importance of names for Victorian naturalists: the Cambrian and Silurian systems were seen as labels for the scientific "properties" of Sedgwick and Murchison. He also demonstrated that the intense debate yielded much scientific fruit, even if the personal estrangement of Sedgwick and Murchison has to be deplored.

The present study, then, examines in detail the third great battle of Murchison's career in relation to the strata of the United Kingdom: his debate with the Scottish geologist, James Nicol (1810–79), concerning the interpretation of the geological structure of the rocks of the Northwest Highlands of Scotland.[8] The central point in this debate was the question of whether there was or was not a regular upward ascending sequence toward the east, from the Fundamental Gneiss of the Hebridean islands and northwest coast. This sequence, Murchison believed, could serve as a firm basis for the understanding of the whole stratigraphical column in Britain. But on Nicol's view, Murchison's understanding of the structure was in error. The region did not offer a straightforward ascending sequence. It was, Nicol believed, ruptured by a huge line of fault running from the north coast down to Skye. This resulted in a repetition of the metamorphic rocks and there was, he supposed, an intrusion of igneous rock along the line of fault. The decision eventually reached was that neither Nicol nor Murchison was exactly right. Rather, it was held that there was an exceedingly complex structure, resulting from the exertion of lateral forces acting from the southeast leading to low-angle thrust faulting. The upper metamorphic rocks had very likely been formed during the processes of lateral earth movement. Such views were developed independently, first by the "amateurs," Charles Callaway and Charles Lapworth, and a little later by the professional officers in the Survey, notably John Horne and Benjamin Peach.

In chapters 2 to 6, I offer a close examination of the Murchison-Nicol debate. But this is only the beginning of the story, and by tracing it further we find ourselves led naturally and conveniently into a study of the details of the British geological community in the second half of the nineteenth century, and toward an examination of newly emerging ideas and research techniques and the further professionalization of geological work in Britain. In particular, I examine the work of Archibald Geikie (1835–1924), initially a young surveyor and disciple of Murchison, but a man who, attaining the position of director general of the Survey in 1882, grew to be the most powerful force in British geology in the 1880s and must naturally draw the attention of any historian interested in this period of British geology. The Highlands controversy, as I shall call it, was of central importance to Geikie's career, for by

---

8. There were other battles too, for example, about his theories on the origin of gold deposits worldwide, that were not specifically concerned with British geology.

supporting Murchison's views he created a stepping-stone for his own advancement. By studying the controversy in its earlier and later phases we find a convenient link between the work done in the era of the gentlemanly-amateurs such as Murchison and Sedgwick and the more professional activity of the later years of the nineteenth century.

It might, I suppose, be thought that there are other, more seemly, things for the historian of science to do than be constantly concerned with scientific controversies, and it could be argued that readers have already had enough of Murchison and his quarrels. But major scientific contests exert a natural attraction for the historian, and they are very much the nodal points of the history of science, as wars and revolutions and other kinds of human conflict are in the history of society. Moreover, scientific controversies are commonly a source of special interest to the philosopher of science, who typically is concerned with the processes of scientific change and the growth of knowledge. Controversies provide valuable information for the social historian of science or the sociologist of knowledge since it is in times of intellectual conflict that some of the more interesting aspects of the sociology of science are revealed. I make no apology, therefore, for attending to a third controversy stemming from the work of Murchison, and my study of features of the closing stages of his career rounds out our knowledge of this interesting and influential leader of the community of geologists in Victorian Britain. He was indeed a combative character, to the extent that in examining the several controversies in which he became engaged during the course of his career one is looking at a sizeable portion of the history of nineteenth-century British geology.

The Highlands controversy has, of course, long been known to professional geologists in Britain, and some of the ground in the north of Scotland where the contest was conducted continues to be visited every year by successive generations of students, who are regaled with accounts of the exploits of the early investigators. There are already several small-scale accounts of the debates written by geologists—either contemporary literature surveys such as those of Hudleston (1879), Bonney (1885), or Judd (1886), which we shall meet in later pages, or later historical accounts by writers such as Woodward (1908), Flett (1937), or Bailey (1952), whose work I have also utilized. However, the nineteenth-century descriptions were to some extent partial since they were written by geologists who were themselves participants in the debate, and the later writing of Flett and Bailey, though illuminating, bore the mark of the Survey itself, which these two authors served as directors. Apart from a chapter in a doctoral dissertation by Beryl Hamilton (1979), the Highlands controversy has not yet been the subject of research by modern historians.

I do not claim that my study has revealed anything that is startlingly unexpected in relation to the Highlands controversy. There is certainly no reason to suggest that the accounts we have to date are wrong because some vital piece of evidence, hitherto overlooked, has now been brought to light. But through examination of archival material that was not utilized by earlier writers, I have found it possible to display the motivations at work in the debate, and to demonstrate the empirical, theoretical, and social practices of its participants. Thus I have sought to bring into focus what has previously been a vague and incomplete outline, and the full magnitude of which has not been appreciated hitherto.

Fieldwork, of course, was and presumably always will be an essential element of geological research. It can, however, be discussed by historians in a rather abstract manner. I have tried to do just the opposite by tracing out the more important journeys and localities of work of some of the major protagonists in the Highlands controversy, where possible comparing the contents of the geologists' field notebooks with what the modern observer may see and record with the help of a camera. In this way, I have sought to engage directly with nineteenth-century fieldwork, and I hope, thereby, that some of its characteristics may be more clearly revealed.

In this way, one may notice a feature of early geological research that hitherto has not attracted much attention.[9] The geologists could not examine their subject matter visually as a unified whole. Rather, they climbed and walked over it, somewhat like ants over a log. So there was difficulty in forming a composite picture of the history of the "log" on the basis of the numerous isolated observations of its present parts. By the second half of the nineteenth century, techniques for forming an overall view of a region with the aid of maps and sections were quite well established. Indeed, that was the proper task of the officers of the Survey. But when dealing with reconnaissance work in more remote parts of the world, before full-scale mapping was undertaken, there was great difficulty in seeing the picture whole, and in part the Highlands controversy arose from the fact that the several observers were often looking at somewhat different parts of a whole. Thus the geologists often had problems in replicating their opponents' observations. Although the rocks did not change from one visit to the next, a geologist could never be sure that the rock exposures that he was examining were precisely the same as the ones that his opponent (or colleague) had seen and described. By my own work in the field, I have found evidence to suggest that this problem was not uncommon; and of course the historian of science has to face it too.

The task of the historian should be, as far as possible, to understand and depict the ideas and actions of people of the past as they actually were. It is, I believe, particularly valuable for the historian of geology to engage in fieldwork, for as Collingwood (1939) suggested, considerable insights may be achieved by, as it were, placing oneself in the shoes (boots!) of the persons about whom one is writing.[10] Collingwood urged historians to attempt to recreate the ideas of the people about whom they were writing—to rethink their thoughts, so that their intentions and motivations might be understood. I do

9. I am indebted to David Hull for drawing my attention to this point.
10. Collingwood's method is well illustrated in his description of his archaeological work and his studies of ancient British history (Collingwood 1939, chap. 11).

not have quite such an ambitious or hazardous program, for it is scarcely feasible to climb into the minds of long-deceased persons! But I firmly believe
that by trying to recreate the physical experiences of the earlier geologists—to
re-enact their experiences as Collingwood might have approved—one can
have a greatly enhanced prospect of understanding their thinking, and why it
was that they behaved as they did. Accordingly, in preparation for this study, I
have spent as much time in the field in the Northwest Highlands as in archives
(but more time in reading and writing!). I like to think that the approach I
have used may signal the expansion of historical fieldwork of this kind, which
may complement the interesting work done in recent years in so-called laboratory studies or ethnomethodology. However, while understanding may be
facilitated, there is undoubtedly the risk of imposing a modern theory-laden
vision on the recorded observations of the geologists that one is examining,
thus falling into the snares of Whiggery and anachronism.

There is no simple solution to this problem. All I can say is that I believe the
risks to be worth taking, and I have made a conscious effort to avoid becoming
Whiggish or anachronistic. Where a twentieth-century opinion or comparison
has intruded, I have sought to acknowledge it openly, my intention being
simply to clarify a point, rather than to make anachronistic judgments or interpretations of early geological work. There is in fact a two-sided problem:
on the one hand knowledge of later events may lead one to misunderstand or
misconstrue the ideas of earlier scientists;[11] on the other hand, lack of understanding of later theory or of the relevant field relationships may make it difficult or impossible to understand the problems faced by earlier scientists or the
ideas they put forward. The best that can be done is to attempt to recreate a
scientist's earlier thinking, while being fully conscious of the fact that it may all
too easily be distorted by modern perspective.

However, no matter how successful one may be in recreating the actions
and thoughts of earlier scientists, this is a somewhat superfluous exercise if it
is not linked to some wider program of inquiry. My task, then, has been to set
out the details of the Highlands controversy as it developed through fieldwork. But one must also examine the manner in which the information thus
collected was "processed" in the "agonistic field" of the geological community.
Accordingly, I have sought to examine the interplay of ideas and the way in
which they were shaped by power relationships within the geological community, thus affecting the constitution of scientific knowledge. Such relationships
need careful examination as much as the rocks themselves, for as is well
known, there is a great deal more to the establishment of knowledge than the
acquisition of empirical information, the establishment of theories therefrom,
and the publication and acceptance of the results. I am thus particularly concerned with the manner in which personal ambitions and fortunes were integrated with, indeed intrinsic to, the formulation of ideas and their acceptance

---

11. An interesting example of this has recently been reported by Perrin (1989), in relation to the
work of Lavoisier.

or rejection, with how geological "truth" was constantly modulated by the realities of power. In particular, the work of Murchison and his disciple Archibald Geikie illuminates the ways in which one may seek positions of authority within a scientific community in order to gain acceptance of one's ideas. Conversely, one may take up and promote ideas, or reject them, in order to advance one's social position. The careers of these two men show further how interests may be linked together for the advancement of scientific ideas. Geikie and Murchison were both highly adept in such negotiations. What became the received view of the geology of the Northwest Highlands, enshrined in maps and in textbooks, was a structure to which Geikie gave his official support for reasons that were partly personal in nature.

Although we are required to look closely at the social dimension of knowledge, I do not wish to argue that scientific knowledge is, as it were, merely a puppet on the strings of social forces, hegemonies, interests. However, I do hold that it is essentially a social product, not simply an outcome of "pure" observation and "pure" reason. I also believe that by and large scientific knowledge becomes more "truthful" in the course of time; and I believe this to be so because increasing empirical and theoretical "coherence" is achieved as scientific research proceeds, the social processing of knowledge claims notwithstanding. My reasons for this opinion will be canvassed in the concluding chapter. The text as a whole, of course, shows how the empirical basis of the knowledge altered and increased through the several phases of the inquiry.

Besides seeking to illustrate some general principles about the construction of scientific knowledge, on a less exalted plane this study is also intended to give an account of certain important aspects of the earlier history of the Geological Survey of Great Britain. There have been valuable studies of this organization by staff members (Flett 1937; Bailey 1952; Wilson 1985), and (for the Irish branch) by the historian Gordon Herries-Davies (1983). Flett's book, in particular, is a useful compendium of empirical information, and it provides valuable inside knowledge of the workings and traditions of the Survey that would otherwise be inaccessible today. But there are important aspects of the history of the Survey that are revealed by the Highlands controversy that hitherto have not been subjected to the historian's scrutiny. Indeed, the whole operation of the Survey was influenced by the controversy in the last two decades of the nineteenth century, and in reporting my results I endeavor to display the manner in which this occurred. I also seek to show how the information collected by the surveyors in the field was processed by the "system," and how the director general had to try to be (or sought to be) the repository of knowledge for the whole geological enterprise in Britain, and also the source of authority for stratigraphical classification and geological theory as a whole within the British geological community. This way of proceeding raised difficulties for the relationships between the "professional" surveyors on the one hand and "amateur" geologists on the other—a distinction that greatly exercised the minds of British geologists in the latter part of the nineteenth century. The term "amateur" was used in a way that we may not immediately recognize, extending as it did from persons who held university posts in geology to those who made occasional field excursions and collected rocks, minerals, or fossils in a desultory fashion. The troubled relationship between "professional" and "amateur" is, then, a further topic of inquiry here, and the effect of this relationship on the knowledge that was generated will be included in our concerns.

As mentioned above, another feature that must attract our attention is the gradually changing research techniques that were developed during the course of the nineteenth century. In Murchison's younger days, much could be achieved by rapid, but rather superficial, reconnaissance surveys of large tracts of land. Communication of theoretical understanding was typically made by means of hypothetical sections along lines of traverse, or by small-scale maps of whole regions. Certainly, Murchison learned early on that stratigraphical work was best carried out using palaeontological rather than lithological criteria. But it was not possible to use palaeontological methods very effectively in the Northwest Highlands, where initially only one rock unit was found to be fossiliferous. Nevertheless, the opening shots in the battle of the Highlands were fired as a result of the use of differing stratigraphical criteria: fossils or rocks. One of the major problems that emerged had to do with the conflict between the theory of the two "professionals," Murchison and Geikie, arrived at by means of broad reconnaissance survey work, and the much more detailed investigations of the "amateurs," Nicol, Charles Callaway, and Charles Lapworth.[12] Eventually, however, the professional officers of the Survey, notably Benjamin Peach and John Horne, proved themselves with meticulous map work in the Northwest Highlands, work that probably had no equal in the world in the nineteenth century. The ants, so to speak, learned to coordinate their work so as to see their subject matter whole. Inevitably, this took time—but less than a working lifetime. Part of the trouble in the Highlands controversy arose, then, from the attempt to sustain doctrines arrived at through old-style reconnaissance work in the face of the more detailed information made possible by systematic surveys.

It may be thought that geological research in the surveying of a country cannot be treated as a series of isolated investigations. In a general way, this is certainly true: everything interlocks with everything else. Nevertheless, the extent to which the Northwest Highlands work can be treated as a discrete enterprise is remarkable, and it is doubtless fortunate for this study that this is so, otherwise my inquiry could never have been completed. In fact, the only other place in the United Kingdom for which I have found it necessary to give a fairly detailed account of work is the Southern Uplands of Scotland, where another "great mistake" of Geikie's directorate occurred (over and above the

12. Murchison was, of course, at first the paradigmatic amateur geologist. But he joined the ranks of the professionals when he accepted the directorship of the Survey in 1855. While an able administrator, who did much to advance the interests of the Survey, his own personal "style" in geological research and fieldwork did not change significantly as a result of his turning professional.

misinterpretation of the rocks of the northwest).[13] The two cases parallel one another to a remarkable degree, for in both the Southern Uplands and the Northwest Highlands it was the "amateur" geologist Lapworth who demonstrated serious deficiencies in the official work of the Survey and showed how the problems might be solved by the application of meticulous map work and ideas concerning the repetition of strata due to "reverse" faulting. In both cases the surveyors Peach and Horne were the ones to pull Geikie's irons out of the fire, so to speak, and rescue the Survey's reputation by their own meticulous yet imaginative work. The only other excursion from the Northwest Highlands that we need make during our exposition (apart from some very brief journeys to Scandinavia, Saxony, and America) is to Switzerland, where ideas concerning faulting and folding, and metamorphism, were developed in advance of work in Britain.

We shall also consider the growing importance of petrography during the period under consideration, and the significance of work with the polarizing microscope for the processes of rock identification. British geologists tended to fall behind in this work during the second half of the nineteenth century (despite the initial lead provided by the investigations of Henry Clifton Sorby). But in the Scottish Highlands, where palaeontological techniques were largely unavailing, it was essential for stratigraphical work that reliable petrographical information should be available to the fieldworkers. Geikie came to realize this, and with his appointment of Jethro Teall (who eventually succeeded him as director) as Survey petrographer, he made every effort to see that the northern investigations were conducted with the best available resources. Finally, I should mention the considerable impulse given to studies of metamorphic rocks, and to experimental work on mountain building, as a result of the Highlands controversy. Studies of metamorphism in particular assumed a role of great and lasting significance. But here again the British geologists tended to follow where the Continental geologists were already leading.

The foregoing remarks set out some of the main themes that the reader will encounter in this book. I would again emphasize that a prime concern has been to demonstrate the manner in which fieldwork—the geologist's chief mode of experimentation—developed during the course of the nineteenth century, within the context of the social history of the British geological community. I am interested in the manner in which knowledge is constructed, empirically, theoretically, and socially, how it grows, and how the communication system works in science. Controversies have long been the focus of attention for historians of science, for the early scientist-historians, like scientists themselves, were inordinately fascinated by priority disputes. The question of priority of discovery in the Highlands controversy does not, however, excite me

13. Actually, Geikie was only director of the Scottish branch of the Survey at the time of debacle of the Southern Uplands.

greatly. Indeed, I am not even particularly concerned with who got things "right" and who got them "wrong," for such judgments may be transitory, and in a sense are more the concern of scientists than of historians of science. Rather, I am interested in the kinds of processes whereby "cohering" interpretations of phenomena were formulated, emerging out of the intense activity within the "agonistic field" of scientific endeavor. In this way I hope that the present study will be of interest to a wider audience than the handful of specialists who write on the history of nineteenth-century geology. In particular, I attend to the interesting question of how scientific controversies are concluded, and how agreement may be reached (or imposed) during the construction of knowledge. I attempt to throw some light on the question of whether the conclusion of a controversy may be said to entail chiefly "epistemic" or "nonepistemic" factors. On the evidence provided by the present study, I argue that although observational work was of paramount importance in the Highlands controversy, one cannot find *pure* epistemic factors (e.g., observational evidence) determining the outcome of the debate in an unequivocal manner or causing scientists to change their minds on that basis alone.

A further point of clarification is perhaps in order here. In my concluding chapter, I say something about the empiricism of nineteenth-century geology, with particular reference to the methodological views of Archibald Geikie; and I "complain" that his views may seem naive today with their lack of regard to (say) the role of theory in the making of observations. Yet the critic may say, perhaps, that my own methodology is likewise overly empiricist and inductivist. So at least to make clear what I did and did not do in the preparation of this book, and how things developed, the following indications may be of interest to the reader.

At one time I had in mind a life of Archibald Geikie, and I planned to collect material with this end in view during a period of leave in Britain in 1987. But I knew there was a likelihood that this task could not be accomplished in the time available to me, so I had an alternative plan, namely, to write a history of the Highlands controversy. With these possibilities in mind, before traveling to Britain I had collected copies of virtually all Geikie's printed work, conveniently listed by Cutter (1974), and all papers pertaining to the Highlands controversy (Horne 1907). Soon after arriving in London I decided that my work was to be on the Highlands controversy, not a biography of Geikie. A period of work in the London archives, and in Haslemere, Birmingham, and Edinburgh, revealed the localities where the main protagonists in the controversy had made their observations, and I was able in several cases to obtain copies of their field notes or field maps.[14] I also collected material with an eye to the social relationships between the geologists and their changing positions in the geological community, thinking in terms suggested by Rudwick's monograph on the Devonian controversy and the more general literature on the

14. Some of Geikie's field notebooks are preserved at the Haslemere Educational Museum. The Lapworth papers are held at the University of Birmingham.

sociology of science. After a visit to the Southern Uplands of Scotland and after visiting Aberdeen, Elgin, and Cromarty,[15] I then made a journey into the Northwest Highlands, and worked my way southward from Durness to Skye, visiting important localities such as Assynt, Ullapool, and Loch Maree. My field excursion was completed in Galloway, and then there was follow-up work in the London libraries. The account was "written up" in Australia, with further calls on printed sources and manuscripts that seemed to be required as the project progressed. As might be expected, it became apparent at times that some of my initially formed ideas were in error or in need of modification. For example, I did not at first appreciate the significance of the term "granulite" in Nicol's theoretical vocabulary, and only came to examine this issue quite late in the study, though it was, I now think, a point that was central to the story.[16] But on the whole, things proceeded smoothly enough.

Despite my empirical approach, I would hold that my story is not simply a distillation of the empirical material to be found in books and papers, the archives, and the field. It has been modulated by an interest in the social relations obtaining between scientists, and between scientists and nonscientists. I am interested in such issues as patronage, hegemony and legitimation, interests and the "translation" of interests, and in the replication of experiments and observations, as well as the practical and economic importance of scientific knowledge. As a result, I looked for any documents that might furnish evidence on such questions as patronage and social interests, in addition to details of particular field observations or theoretical conjectures. In consequence, my account necessarily bears the imprint of the current interest in the social dimensions of science. I do not think any apology need be made for this. Historians are children of their time, as are scientists and politicians. In this sense, the methodologies of scientists and historians of science are not unlike. What I would deny, of course, is that interesting and useful history can be written simply by the antlike collecting of data. Like the geologist, the historian needs to see his subject whole; but for this end the assistance of some guiding theories or principles of inquiry is manifestly required. Empiricism alone is insufficient to the task. On the other hand, the more effectively the archives are searched, the more accurate should be the history that is written. It is sometimes said, only half in jest, that the best geologist is the one who has seen the most rocks. Similarly, no doubt, the historian can do better work the more thoroughly he or she knows the archives.

However, while I hope that my inquiry moves well beyond old-style positiv-

15. I hoped to find some papers of Nicol in Aberdeen, but none are held at the university. The Elgin Museum has the papers of a local nineteenth-century geologist, the Reverend George Gordon, who corresponded with several of the disputants in the Highlands controversy, including Nicol. But again, no significant information was gleaned from this source. Cromarty was the birthplace of Hugh Miller, who played a part in the early stages of geological work in the Northwest Highlands.

16. Even having realized the significance of the "granulite problem," I did not at first appreciate the difference between the German, French, and British usages of the term. I am much indebted to Professor T. G. Vallance for his assistance in this matter.

ist or empiricist historiography, I am also endeavoring to retain the core of rationality in science (which some playful metascientists such as Paul Feyerabend [1975] seem to have been at pains to eradicate). Thus while it is undoubtedly the case that social interests are at work in the construction of scientific knowledge, I suggest there is an increasing "coherence" of scientific knowledge, achieved (for example) when logically independent methods are deployed to determine the same physical properties and consistent results are achieved. Thus scientific knowledge becomes more verisimilitudinous as time passes. This argument is not paraded explicitly to any great extent in the present study, but it may be mentioned here as a background assumption that has underpinned my investigation and the presentation of my results. In general terms, then, I believe that there is a *via media* between the Scylla of positivism and empiricism and the Charybdis of epistemological relativism and methodological anarchism. What success has been achieved in finding this middle way, I leave to the judgment of the reader.

One important issue that calls for consideration here is the extent to which my findings about geology in Britain in the second half of the nineteenth century can be regarded as typical of geology as a whole, or may even be extrapolated to science as a whole. It seems to me that the scientists I have studied were typical enough of the geologists of their day. They enjoyed their fieldwork. They regarded themselves as members of a kind of fraternity (Porter 1978). The surveyors had a clearly understood task—to discover what rocks and minerals were to be found in the United Kingdom, and to record this information in the form of maps and descriptive memoirs. They also wanted to understand the structures of the rock formations that they studied and the geological history of Britain, insofar as it could be discovered by examination of the strata. Also, as has been pointed out by Secord (1986b), there was a general program in the Survey, initiated by De la Beche, to reveal the former environmental conditions existing at the time of deposition of the different strata. All this was done in the recognition that the government supported geology for economic reasons; so the Survey had an obligation to meet the economic needs of the nation, especially in mining and agriculture, and it also had a certain educational role to play.

In all this, the main work was stratigraphy, to which mineralogy, petrology, and palaeontology were in a sense but handmaidens. The surveyors were not really required or expected to generate new theories, except insofar as they pertained to stratigraphical matters.[17] The kind of work they did seems to me to typify what geologists in general were doing in the second half of the nineteenth century. But the problems they encountered in the Highlands controversy and in the work in the Southern Uplands were an order of magnitude

17. An exception was James Croll (1821–90), in his remarkable theory of glaciation resulting from secular changes in the ellipticity of the earth's orbit (Croll 1875). But Croll was not a great success as a practical surveyor, and his work was quite atypical. Some geologists such as Alfred Harker, whom we shall encounter during the course of our narrative, chose to leave the Survey to take up university positions, where they could do work of a more fundamental kind.

more difficult than anything that had previously been tackled in Britain. For some time the surveyors were baffled; it might be said that they did not know that they did not know. Eventually, however, after considerable prodding by the "amateurs," they succeeded in solving their stratigraphical and structural problems by essentially standard methods; and this led on to more general questions about the ways in which mountains were formed, earth movements, and "theories of the earth." But such matters were the province of large-scale synthesizers such as Suess (1904–1909) or grand theorizers such as Wegener (1966), not the typical field surveyor.

So above the work of the surveyors there were the constructions of the grand theorizers; and below it there was the work of local amateurs and collectors, whose methods and interests were in some ways much the same as those of the surveyors, but usually more modest in scope. At the same level, so to speak, were university or museum specialists such as Bonney and Davies (who figure in chapter 7), but whose activities were in principle calculated to be supportive of survey work. There were still, in our period, truly amateur geologists (as we would understand the term), such as Hicks and Lapworth (see chapters 7 and 8), who could challenge the surveyors' findings and sometimes show them to be in error. But the period of ascendancy of the amateur was already in decline. As has been shown by Menard (1971), there is an almost inevitable tendency for scientific work to divide into specialties of ever-increasing narrowness, because of the ever-increasing quantity of information that has to be assimilated, and because of the career advantages that become available through the division of disciplines. Such changes were just beginning in the period with which we are concerned (O'Connor and Meadows 1976), with palaeontology already beginning to split off as an important subdiscipline, which, so far as the Survey was concerned, ministered to the needs of the stratigraphers. Mineralogy had, in a sense, preceded geology as a branch of science, but it was now to become a separate subdiscipline (marked in Britain by the foundation of the Mineralogical Society of Great Britain and Ireland in 1876), and in time a fully-fledged discipline. Petrology, glacial geology, geochemistry, palaeobotany, to name but a few, were soon to follow. In this sense, then, the state of geological research in Britain in the second half of the nineteenth century was certainly not characteristic of the whole of geological science. It marked a specific stage in an ongoing process of specialization. It was rather more concerned with delineating stratigraphical boundaries on maps, and less with matters of high theory, than was the case either earlier or later. It had its own preoccupations that were natural to it at that time and place. On the other hand, already geology was becoming an international enterprise, with strenuous efforts being made to link the stratigraphical columns in different parts of the globe and to establish uniform nomenclature.[18] From this view, the general nature of survey work was much the same in Britain as

18. This is shown by the establishment of the International Geological Congress, the first of which was held in Paris in 1878 (Ellenberger 1978), or by texts such as Kayser's (1893).

elsewhere (except that the British climatic and physical working conditions were not too difficult compared with those that many geologists had to endure in other parts of the world).

As to the question of whether the scientific work that is described in this book is characteristic of science as a whole in some very broad sense, one can only resort to somewhat vague generalizations. To understand science, the philosopher may try to devise general analytical schemes and see to what extent these are actually exemplified in the history of science. Or he may engage in a number of "microstudies," which, it may be hoped, collectively instruct about the nature of science. Obviously, what I have undertaken here is just one such microstudy, and it would be foolish to suppose that it can tell us about all science, everywhere and at all times. Nevertheless, I think it does offer interesting and generalizable information on the way in which scientific controversies are conducted, and the way in which they may eventually be resolved or closed, as I discuss in chapter 11. Also, it seems to lend support to a thesis recently advanced by David Hull (1988), who has argued (on the basis of his studies in the community of zoological taxonomists) that commitment to one's ideas, self-interest, and even bias, are quite normal in science, and that scientific knowledge grows because of such characteristics, rather than in spite of them.

My book was, in fact, written without knowledge of the ideas that Hull was developing;[19] yet it seems to me they are confirmed to a significant degree in my study. For example, self-interest was undoubtedly a major feature of Geikie's career, and he flourished exceedingly. Nicol was committed to his own ideas, but when they were rebuffed he tended to lose heart and did not fight strongly enough for them to prevail; he did not have enough allies and supporters whose interests were linked to his. Lapworth fought for his ideas, and although his initial position was not so very strong, as a schoolmaster amateur geologist, he gradually built up his allies, through correspondence and personal friendships, until he and his supporters became a force to be reckoned with. Of course, it was not just Lapworth's assiduity as a correspondent that ensured his eventual success. It was the remarkable ingenuity and perseverance that he displayed in his empirical work, his skill in presenting his ideas in wonderfully clear maps and sections, and his ability to convince by offering highly cogent explanations of the phenomena he described that enabled him to enlist supporters. Nevertheless, he really had to fight hard for his ideas, and he almost ruined his health in the process. If he had not engaged

---

19. I may mention also that I did not read McMullin's essay of 1987 until 1988. So although this essay is important to my concluding chapter, as it happens my empirical information was not collected with McMullin's arguments in mind. Thus one can see how ideas may develop and expand in response to ideas that one is constantly drawing in from all sorts of sources. On the other hand, while collecting my empirical data, I *did* have in mind ideas on coherence that I adumbrated in 1987, and that I was trying to substantiate in the present study; and my thinking on Geikie develops an account of him that I published in 1980.

the Survey over the Southern Uplands, it is likely that a muddled view of the structures of these rocks would have persisted for some considerable time, for all the maps and memoirs would have been published, setting in official concrete the erroneous ideas that Geikie and his surveyors initially entertained. By Lapworth's challenge, the Survey was forced to rethink and revise its work over large areas of Scotland, though it too fought hard to preserve its position. Thus I think that we can see significant similarities in fields so very different as the zoological systematics studied by Hull and the geological survey work that I have examined. In both cases, we can see scientific progress emerging from the heat of scientific battle.

Such issues are, I dare say, chiefly of interest to professional historians, philosophers, or sociologists of science, with their own ongoing debates, but my investigation is intended to have a wider appeal. The Highlands controversy was a major episode in the early history of British geology. Yet it was no mere local episode. Though in the period that we shall be examining the debate was chiefly focused on the geometrical arrangement of certain rocks and their correct placement in the stratigraphical column, rather than grand theories of (say) a cooling and shrinking earth, it had major implications worldwide for the understanding of the phenomena of mountain building and the processes of metamorphism. Indeed, the region where the controversy was centered is a focus of research to this day, especially in relation to metamorphism and the interpretation of the gneisses and schists of the northwest. Students still visit the Northwest Highlands as a kind of geological Mecca. It may be, then, that this book will be of interest to geologists of the present generation, to those now in training, and those still to come forward.

For such readers, I should perhaps make a word of apology that I have chosen to provide a fair amount of elementary geological information. But this is necessary, for the book is intended to be of use to those who are not geologists as well as those who are. In addition, the book is intended for those who may not live in Britain or who may never have visited its far northwestern corner. To assist these readers, I have been at pains to give a good deal of topographical information, much of which is summarized in figure 1.1, which locates nearly all the places in the Northwest Highlands mentioned in the book. Some of the details of Scottish topography may be tedious to those who already know the north of Scotland well. But I want to conduct readers to that rugged and beautiful part of the world, to create in their imaginations the most interesting rocks that may be found there, and not to lose them on the way! Also, we need to attend to just a few rock types in some detail. The Highlands controversy was quite finely focused on a small number of rock units, most with clearly recognizable lithological features. It will be necessary to get to know these if any sense is to be made of the debate.

Finally, considering the ever-growing interest in Scottish history, this book is submitted also to the judgment of the general reader—the person who is perhaps traveling through a (now decreasingly) remote corner of Britain, and

Fig. 1.1. Sketch map of the Northwest Highlands of Scotland, showing the localities referred to in this book.

who may wonder how scientific knowledge of that district came to be established. Such a reader may be surprised to find that the forging of geological knowledge was sometimes accomplished almost with the vigor of an ancient battle of the clans. The way the battle was conducted is set forth in the following pages.

2

Early Geological Investigations in the Northwest Highlands

The northwest of Scotland is a wild and sparsely inhabited region that was not explored by naturalists and scientists to any significant degree before the nineteenth century. It required considerable physical, mental, and financial resources to do serious work in those beautiful but challenging regions. Indeed, through much of the nineteenth century to visit Sutherland was, for southerners, not unlike exploring a foreign country, with an incomprehensible language, poor transport, an impoverished economy, and inadequate maps. Even today, if the weather is bad, it can be a considerable undertaking to venture far off the few tracks and to ascend the higher peaks. Few do so in winter.

Almost nothing was known—or recorded—about the geology of Sutherland before the beginning of the nineteenth century. In reviewing the history of the Highlands controversy, John Horne (1907) mentioned only two eighteenth-century observers: Thomas Pennant (1726–98) and John Williams (ca. 1730–95).[1] In July and August of 1772, Pennant traveled north from Ullapool on the west coast at Loch Broom toward the Assynt region and briefly recorded his impressions (Pennant 1775, 314–17). He noted seeing "mountains of stupendous height, and generally conoid forms" such as the "sugar-loaf hill of *Suil-bhein* [Suilven]." These were the famous red-brown mountains of what are now called Torridon Sandstone, and are classified as pre-Cambrian age—mountains such as Coigach, Cul Mor, Suilven, Canisp, and Quinag. They often have rounded, cone-shaped peaks, as Pennant described Suilven, and some of them have cappings of white quartzite. To the southwest, the range extends with mountains such as An Teallach, Slioch, Ben Eighe, Sgurr Mhor, and Ben Bhan, and continues into the southern part of Skye.

Besides the red-brown sandstones, Pennant also recorded the occurrence of white limestone on his journey from Ullapool to Loch Assynt. This would have been the Cambrian rock known as the Durness Limestone, named from the region on the north coast (at Loch Durness) where it is particularly well exposed. Pennant recorded its use for lime burning for fertilizing the land. A little to the northwest of Cam Loch, near the farm of Ledbeg, he recorded the

---

1. On Horne, see chapter 9.

occurrence of a "white marble, fine as *Parian*," used for building purposes.[2] Today, this is mapped as a Cambrian limestone, altered by a neighboring igneous intrusion. Pennant reported being hospitably received by the inhabitants, though he observed their condition as "most wretched." The region has long been denuded of vegetation by burning and felling to allow the grazing of sheep. But Pennant attributed the denuded appearance of the countryside to some great convulsion that "shook off all the vegetates."

In a general survey of the mineral deposits of Britain published in 1789, Williams recorded observing a "fine white statuary marble" accompanying a "prodigious rock of grey limestone of a granulated texture," which, he said, occurred in the river bed near a large house, about two miles south of the church at Assynt (Williams 1789, 2: 26). This, I suggest, may have been near the farm of Stronchrubie, which we shall meet again as a place visited by Sir Roderick Murchison.[3] There one finds good exposures of Durness Limestone, but the marble rock outcrops a few miles further to the south, at a place called Ledbeg. Williams stated that it was not then used for commercial purposes, being too inaccessible. This is still the case, though the "Assynt Marble" has been quarried at Ledbeg from time to time for the purposes of decorative building and furniture.

The first effort toward preparing a geological map of Scotland was that of Louis Albert Necker (1786–1861), a Genevan who came to Edinburgh to study at the university in 1806 (Eyles 1948). Necker came from a well-known Swiss family, several of whose members had scientific interests, and when he arrived in Scotland he already knew some mineralogy and geology. In the next three years he traveled extensively in Scotland, and by 1808 he had gathered together enough information to prepare a rough geological map of the country colored onto an eighteenth-century topographical map of Thomas Kitchen (n.d.). But this contained no information about the northwest of Scotland, and according to Eyles's reconstruction of his journeys, Necker did not visit that region before he compiled his map. After returning to Europe for a time, he eventually settled in Skye as a recluse, and although he made geological observations there, these played no part in the Highlands controversy.

The first major investigation of the rocks of northwest Scotland was made by John Macculloch (1773–1835), who traveled widely in the region in the years 1814 to 1824. Macculloch's work, which has been extensively studied by Cumming (1981, 1983, 1984, 1985), is of considerable importance for our narrative.[4] For on the one hand, Macculloch made a number of fundamental observations and inferences that were utilized profitably by all subsequent investigators, and he made the first significant geological map of our region.[5]

2. The island of Paros, one of the Cyclades, is famed for its white statuary marble.
3. See chapter 4.
4. See also Eyles (1937, 1939) and Flinn (1981).
5. A geological map of Scotland was published by Aimé Boué accompanying his *Éssai géologique sur l'Écosse* (Boué 1820). But he stated (ibid., 82–83) that his data from the northwest were derived from the work of Macculloch. Boué showed most of the Highlands as mica schist. He gave a

On the other hand, he made certain suggestions that we now think are erroneous and which, being adopted uncritically by some of his successors, were in some measure the direct source of many of the difficulties attending the Highlands controversy

Macculloch, who was born in the Channel Islands and spent much of his earlier life in Cornwall, studied medicine at Edinburgh, where he also received training in chemistry and gained an interest in geology. He enlisted in the army as a surgeon's mate, but changed careers on appointment as an assistant chemist to the Board of Ordnance. He was also employed in his early years to search for a supply of limestone suitable for use in the grinding of gunpowder (Cumming 1984). Thus it was that the years 1811 to 1813 saw him prospecting for limestone in Scotland, and no doubt taking a clue from Williams's text, he visited the Loch Assynt region in that period. He had joined the Geological Society of London in 1808, soon after its foundation, and had made an early reputation for himself, which led to his appointment as geologist to the Trigonometrical Survey in 1814, a position he occupied until 1821. He also lectured to the Woolwich Academy and the East India Military Seminary at Addiscombe. Macculloch rose to the position of president of the Geological Society in 1816 and was elected to the Royal Society in 1820, following the success of his treatise on the geology of the Western Islands of Scotland (Macculloch 1819). He published widely on many topics besides geology and chemistry.

During his period with the Trigonometrical Survey, Macculloch traveled over much of northwest Scotland as part of his project to prepare a geological (or mineralogical) map of the country. In 1821, however, the money for the survey was terminated and he had to carry on by himself for several years. Then, in 1825, by suitable lobbying, he succeeded in gaining Treasury support to continue his work, which he did until his untimely death as a result of a fall from his honeymoon carriage in 1835. Macculloch's later years were made difficult for him by criticisms of the expenses of his survey work (Flinn 1981), emanating chiefly from his rival, the Edinburgh professor of natural history Robert Jameson (1774–1854), who felt that he if anyone should have been engaged by the government to conduct a geological survey of Scotland.[6] In his later years, Macculloch also came in for criticism for undue reliance on the mineralogical classification of rocks as the basis of his survey work, at a time when geologists were turning increasingly toward the use of fossils to identify

---

"Primary Red Sandstone" in the northwest and a distinct red sandstone to the east, in Caithness. He showed a band of "roches quartzeuses et chloriteuses" running irregularly northwestward from Skye to Durness. The Outer Hebrides were shown as gneiss, as was a small region near Loch Assynt.

6. For the relationship between Macculloch and Jameson, see Cumming (1985). Jameson did some of the earliest survey work in north Scotland, but he confined his attention chiefly to the Hebridean islands and to the Orkneys and Shetlands. To the best of my knowledge, he did not make any significant published contributions to the geological understanding of the region in which we are interested.

stratigraphical horizons (Murchison 1833). Macculloch also showed ever-increasing enthusiasm for "scriptural geology" (Macculloch 1831), at a time when it was losing ground among the leaders of the geological community. His critics notwithstanding, Macculloch's map was eventually published post-humously in four sheets (Macculloch 1836). Four issues appeared, the last in 1843.[7] The portion of the map particularly relevant to our present inquiries is shown in plate 1.

For anyone familiar with the terrain of northern Scotland, Macculloch's single-handed achievement must seem most impressive.[8] His classificatory sys-tem was simple, and we cannot suppose that he walked over all the ground that he colored in on his map. Nevertheless, comparison with a modern map, or with ones published later in the nineteenth century, shows that Macculloch had successfully identified a number of the major features of the geology of northern Scotland. He recognized the red gneiss of the west coast, which forms the greater part of the Isle of Lewis in the Outer Hebrides and hence is customarily called the "Lewisian."[9] However, he did not differentiate this from the schists of the central portion of northern Scotland, and both were represented in yellow in the map of plate 1.[10] Macculloch delineated the out-crop of the red-brown sandstones already noted by Pennant, but conflated these with the Old Red Sandstone of the east coast, both being shown in pink on the map.[11] He showed the important band of "quartz rock" or quartzite (in green) and associated limestone (in red) running southwest from the regions of Loch Durness and Eriboll on the north coast.[12] But he judged the limestone by its lithological appearance to belong to the so-called Mountain Limestone,

7. The first issue merely bore the title of the topographical map of Samuel Arrowsmith, on which the geological information was entered. The second issue (1836) bore the legend: *A geological map of Scotland by Dr. MacCulloch, F.R.S. &C, &c, &c. Published by order of the Lords of the Treasury by S. Arrowsmith. Hydrographer to the King*. On Macculloch's mapwork, see Cumming (1981).

The version I have used is the 1843 issue. A copy, in one sheet, is held at the British Library. For discussion of the different issues of Macculloch's map, see Eyles (1937).

8. Eyles (1937, 121) states that Macculloch did have the services of an assistant, possibly his nephew, during some of his survey work.

9. Gneiss is an important type of coarse-grained metamorphic rock, consisting chiefly of quartz, feldspar, and mica, with bands containing granular minerals alternating with bands containing platy minerals, notably mica. Paragneisses are believed to be metamorphosed sediments, while orthogneisses are metamorphosed igneous rocks. The term gneiss originated from the mining regions of Saxony or Bohemia. It was derived from an old Slavonic word (*gnezdo*) meaning "nest" (Czech: *hnízdo*). The metalliferous veins of the *Erzegebirge* occurred in a body of gneiss, which might be thought of as a "nest" for the ores.

10. A schist is a metamorphic rock with a laminated or foliated appearance due to the occurrence of flaky mica crystals of various kinds of rod-shaped amphibole crystals. There may also be sub-sidiary minerals (e.g., garnets) that are characteristic of the intensity of the metamorphism that the rock has undergone. In the early days of geology, schists were sometimes regarded as sedimentary.

11. Doubtless, Macculloch was influenced in this by the paper of Sedgwick and Murchison (1828), which is discussed below.

12. Quartzite is a rock formed by the metamorphism of sandstone in such a way that the quartz grains recrystallize and become interlocking.

Fig. 2.1. View of *An Teallach* from east, showing capping of "quartz rock" on a mountain otherwise formed of "red-brown" (Torridon) sandstone.

well known in the English Pennines.[13] The labor involved in producing even Macculloch's very preliminary depiction of the disposition of the rocks of north Scotland can scarcely be overemphasized. It must have involved endless hours of walking and riding, sailing in dangerous waters, and a certain level of conceptual analysis in making and adhering to the discriminations of rock type in a region where there was virtually no previous work on which to build.

It is evident that Macculloch fully accepted the distinction between the red-brown sandstones of the west coast and the overlying white quartzites or quartz rocks, but it is not exactly clear when he made this distinction. In a paper published in 1814 (Macculloch 1814), he gave an extensive account of the quartz rock, describing a number of varieties but maintaining that they were all generically the same. It is not obvious that at that time he regarded the quartz rock and the underlying red-brown sandstone as distinct formations. However, his descriptions seem to apply chiefly to the quartz rock. This he described as forming white cappings on several of the red-brown sandstone (Torridonian) mountains, and the sort of thing he would have seen is shown in figure 2.1. It was Macculloch who recommended the use of the term

13. This was a reasonable but unsuccessful guess. The Mountain Limestone (or Carboniferous Limestone) is today placed in the Carboniferous. The Durness Limestone is today classified as Cambrian. The first fossils were found in it by C. W. Peach in 1853, and were claimed by Murchison for the Silurian system (see chap. 3).

Fig. 2.2. "Quartz rock" of Northwest Highlands, showing characteristic "pipes."

"quartz rock," and this usage was followed by subsequent nineteenth-century geologists.

An important observation recorded by Macculloch was "imbedded cylindrical bodies" (Macculloch 1814, 461) in the quartz rock, which pipe-like structures he thought might represent the remains of some kind of marine-worm burrowings, indicative of former animal life in the quartz rock.[14] (See figure 2.2 for a modern photograph of these.) But this left Macculloch in a quandary as to the proper place of the quartz rock in the stratigraphical column. The worm structures and the rock's layered appearance suggested that it belonged to the "*Floetz* Series" of the German mineralogist, Abraham Werner (1749–1817).[15] But in places schistose rocks appeared to alternate with the

14. It is supposed to this day that the remarkable, and characteristic, "pipes" of the quartz rock do in fact represent the remains of filled-in worm burrows. The "pipes" come in different forms, as may be seen by comparing figures 2.2 and 7.10.
15. Abraham Gottlob Werner, of Freiberg in Saxony, developed what was arguably the first paradigm in stratigraphical geology. This is not the place for a full exposition of his doctrines, but the following brief outline may be useful. Werner supposed that at the time the earth was first formed it consisted of a large, irregularly shaped core surrounded by a vast universal ocean, containing much matter in solution. This ocean slowly evaporated, depositing a layer of "Primitive" crystalline granite all over the core. The evaporation continued until the upper parts of the core appeared, covered in granite, above the surface of the ocean. These high regions would then become subject to erosion, so that subsequent material deposited from the ocean would be

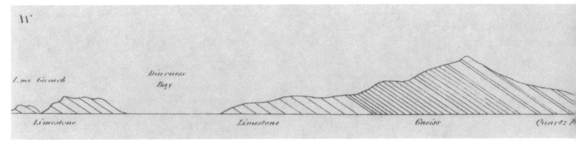

Fig. 2.3. Section from Durness to Loch Hope, from Macculloch's *Description of the Western Islands of Scotland* (1819), vol. 3, plate 32, fig. 1.

quartz rock, which suggested membership in Werner's "Transition Series." The Scottish rocks were not fitting into the general classificatory scheme to which Macculloch gave allegiance.

Macculloch's *Description of the Western Islands of Scotland* (1819) was the first major synthesis of his views. The treatise was well illustrated with line drawings, three of which should attract our particular attention. The first of these shows a sketch section from Loch a Chairn Bhain (near Kylestrome) southwest toward Ullapool, passing through the red-brown sandstone mountains of "Cuniach" (the Quinag group), Suilven, Cul Mor, and Cul Beag.[16] These were figured as consisting of approximately horizontal sediments, lying unconformably on the underlying complex of gneissic rocks. Thus Macculloch recog-

---

mingled with mechanically derived sediments giving "semicrystalline" rocks such as schists. These formed Werner's "Transition Series," which would form layers round the granitic mountains and lie banked up against them. In time, the ocean level sinking still further, the source of crystalline granite in the ocean would be exhausted and subsequent deposits would consist wholly of mechanically derived sediments. These would form approximately horizontal or gently sloping layers of rocks, banked up against the Transition Series. Werner called them *Floetz* (or layered) rocks and noted that they normally contained fossil remains. The *Floetz* rocks seemed to occur in a regular sequence that could be found in a number of parts of the world. The theory was able to account for the common observation that many mountain ranges consisted of a granitic core, round which could be found beds of schistose rocks, surrounded in turn by layers of sedimentary rocks in a generally recognizable order. There were, needless to say, enormous difficulties with Werner's theory, but it gained many adherents in the late eighteenth and early nineteenth centuries. The great attraction was the idea of a generally recognized and recognizable sequence of rocks, which could serve as a guide to researches in many parts of the world. Such studies were based chiefly on an examination of the mineralogical features of rocks, rather than their fossil contents. The main sources of information on Werner available in English are the writings of Ospovat (1967, 1969, 1971, 1980). See also Geikie ([1905] 1962) for a very negative but influential account of Werner's work. For a recent summary, see Laudan (1987, 87–112). Werner's ideas were actively promoted by Robert Jameson at Edinburgh, where he taught many generations of students and founded a "Wernerian Society," which published papers sympathetic to the German theory. As a result of Jameson's decision, specimens that might illustrate the rival Huttonian doctrines (see note 34) were kept from display at the university.

16. Vol. 3, plate 31, fig. 3 in Macculloch (1819). Macculloch inadvertently reversed the names Cul Mor and Cul Beag on his diagram.

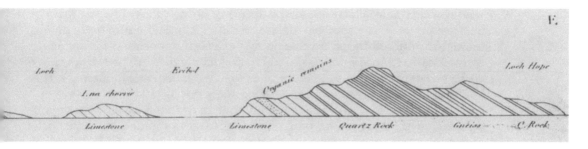

nized the fundamentally important unconformity between the "red sandstone" and the "gneiss."[17]

The second figure I should like to mention shows what Macculloch took to be a comfortable section along the northern shore of Loch Broom (at Ullapool), with the red-brown sandstone dipping under the quartz rock, which in turn was shown dipping *under* gneiss.[18] There is indeed some gneissic rock shown at this point on modern maps, though not much of it, and Macculloch, as we have seen, generally conflated gneiss with schist. Be this as it may, the area in question subsequently became a classic locality for geological observation and theoretical discussion. The point that gneiss or schist might sometimes *overlie* the quartz rock series in an apparently conformable sequence, as well as (elsewhere) unconformably *underlie* the red-brown sandstone series, was now graphically made and could thereby easily influence the thinking of subsequent investigators. It should be emphasized that what Macculloch recorded at Loch Broom represents a perfectly reasonable first interpretation of what may be observed there.

The third drawing we should examine is shown in figure 2.3. It shows an idealized section along the north coast from Loch Durness to Loch Hope. The great value of this region for unraveling the geological structure of northwest Scotland was fully recognized by subsequent investigators, even though they might disagree as to the correct interpretation. The important thing is that along the north coast one has the opportunity to see the whole section displayed east to west. Unfortunately, however, appearances can be deceptive. While the *sequence* proposed by Macculloch was more or less as we regard it today (if we are willing to follow a somewhat sinuous line of section), the total structure that he suggested has not been sustained by subsequent investigators. Many of the boundaries marked by him on this section are now regarded as fault surfaces, some high-angled, some of low-angle.[19] However, the

17. Both the red sandstone (Torridonian) and the gneiss (Lewisian) are pre-Cambrian.
18. Vol. 3, plate 31, fig. 2 in Macculloch (1819).
19. A fault is a fracture along which there has been displacement of rocks on the two sides relative to one another parallel to the fracture. A fault surface is the surface along which the dislocation has occurred. If the fault surface is plane, one may speak of the fault plane.

way in which Macculloch attempted to bring order to a confusing mass of
data such as may be assembled along this difficult coastline is entirely understandable, and he offered a valuable first effort to understand the structure
of the district. He showed that quartz rock could indeed be found either
above or below the gneiss, and above or below the (Durness) limestone. It is
interesting that his section shows more than one quartz rock layer—one
layer to the west of Loch Eriboll and another layer to the east, the two being
separated by limestone outcropping on the island in the middle of the loch.
Moreover, the section asserts the existence of two distinct gneisses. The validity of these claims became one of the most hotly debated issues in the
Highlands controversy. Murchison, who, as we have seen, poured scorn on
Macculloch's later work, was probably much influenced by this section of the
Durness-Eriboll region. At any rate, there is much similarity between Murchison's and Macculloch's sections. But Macculloch's notion of two distinct
quartz rocks was undoubtedly bold, given that the stratigraphical boundaries
of this section are obscured by the cold and uninviting waters of Loch
Eriboll. Furthermore, while Macculloch's section showed the various rocks as
consisting of flat, though inclined, layers, in a further publication he noted
that the quartz rock near Loch Eriboll exhibited "considerable curvatures"
(Macculloch 1822, 57).[20] It is evident, therefore, that his section was highly
idealized.

In his *Geological Classification of Rocks*, Macculloch formally identified the
"Red Primary Sandstone" (which I have referred to as the red-brown sandstone),[21] stating that "this is now for the first time introduced into a catalogue
of rocks" (Macculloch 1821, 331). He noted also that it overlay the underlying
gneiss with unconformity. In its basal layers, Macculloch noted that the sandstone actually appeared to be formed of fragments of underlying gneiss (ibid.,
332). However, the nature of the contact was not always clear. Sometimes
there appeared to be a "regular alternation," in other places an "indefinite
transition." In his *System of Geology*, Macculloch did distinguish between the
western (red-brown) and the eastern (red) sandstones (1831, 2: chap. 29); but
as we have seen, this distinction was not maintained in the posthumously published map.

We see, then, that Macculloch's contributions to establishing the first framework for the study of the geology of north Scotland were considerable—indeed fundamental. At least until the 1870s, all geologists working in this area
would have thought it worthwhile to consult his work before setting out on
their own investigations. This was true despite the fact that his field work was
sometimes severely criticized.[22] Thus Macculloch plays an important part in
our story, but we must leave him at this juncture and turn to consider the early

20. Macculloch's paper of 1822 described work carried out by him in the years 1817 to 1819.
21. This appears below the cap of quartzite in the mountain depicted to the right (west) of figure 2.1.
22. See, for example, Geikie (1858a). See also the comments of Cunningham, mentioned on page
38.

investigations of the redoubtable Roderick Impey Murchison (1792–1871) in the northwest.

The life and work of Murchison has attracted considerable attention from historians, so although he plays a role of fundamental importance in the present inquiry I shall not offer much biographical detail here.[23] What I want to say will emerge during the course of my narrative. Nevertheless, in order to introduce him, a few points may be mentioned at this point.

As is well known, Murchison was a wealthy amateur of Scottish blood, a one-time soldier who adopted a scientific career at the suggestion of his admirable wife, in preference to the gentlemanly pursuit of fox-hunting, which he enjoyed as a young man.[24] But Murchison was anything but the quiet, retiring amateur scholar. He had a burning ambition to see his views accepted by his peers and to stand in high repute, and the means by which he sought to attain these ends provide a textbook example that other nineteenth-century scientists with similar ends in view might well have sought to emulate.[25] Indeed, his protégé, Geikie, did to a considerable degree tread the same path.

Murchison epitomizes the British nineteenth-century gentlemanly-specialist—to use the term favored by Martin Rudwick (1985). Although Murchison did in a sense turn professional toward the end of his career by assuming the directorship of the Geological Survey, this advancement did not mark a wholly new direction for him, or require the adoption of a totally new way of participating in science. He had long been active in the Geological Society, the Royal Geographical Society, the Royal Society, and the British Association for the Advancement of Science. The assumption of direct responsibility for the Survey as a professional scientific bureaucrat inevitably made a difference to his day-to-day activities. But really it was all grist to Murchison's mill—an extension of his range of activities that allowed him to deploy to even better advantage his remarkable talents for promoting his views, especially the merits of his beloved Silurian system, on which his reputation chiefly rested.

A few words need to be said here about the Silurian system and the role it

23. Major relevant publications are Geikie (1875); Thackray (1972, 1978, 1979, 1981); Secord (1981–82, 1986a); Stafford (1984, 1988a, 1988b, 1989); Rudwick (1985).
24. He was the eldest son of a man who had amassed a fortune while working for seventeen years in Lucknow as a surgeon in the serivce of the East India Company. It may be suggested that this background impressed on Murchison the advantages to Britain of its imperial activities. Murchison did all in his power to advance Britain's colonial interests, and as will be shown, he tended to see geological research in terms of empire building.
25. He was described by T. G. Bonney, who knew him personally, as "tall, wiry, muscular, of a commanding presence and dignified manner" (1894, 320). He was also said to be "a hospitable host, a firm and generous friend, though perhaps, especially in his later years, somewhat too self-appreciative and intolerant of opposition." Bonney added that Murchison was "a man of indomitable energy and great powers of work, blessed with an excellent constitution, very methodical and punctual in his habits."

played in Murchison's self-image. As mentioned previously, on the basis of his investigations in Wales, Murchison successfully established one of the major subdivisions of the stratigraphical column, the Silurian system; and his first major announcement of this was in his volume of that name (Murchison 1839). In subsequent years, he successfully demonstrated that his proposed system was valid, in the sense that it was characterized by a particular assemblage of fossil animal life and could be recognized in many parts of the globe besides Wales and the Welsh Border region, where it had initially been proposed. The name was chosen because the ancient tribe that inhabited the region of Britain where Murchison carried out his investigations were called the "Silures." In time, the Silurian rocks came to be thought of by Murchison (1854) as the first part of the stratigraphical column where fossil remains of life might be found.

This claim did not go uncontested by any means, and it certainly has not survived into modern geology. But that is not the point that interests us here. What is intriguing is Murchison's increasing obsession with his system. As a result of his subsequent investigations in Europe and Russia, he was able to show that the Silurian fauna could be found in many parts of the world, and thus geologists could gradually bring order to the inadequately charted territory of Werner's former Transition series. But as time went on, Murchison began to identify himself personally with the expanding area of land recognized as belonging to the Silurian system. He began to regard it as a kind of kingdom. The more this domain expanded, the greater Murchison felt his geological accomplishment to be. Thus he became intensely preoccupied with the territorial expansion of what he regarded as his domains. Former military man that he was, he came to think of the expansion of "Siluria" as a process of colonization, to be achieved by a quasi-military campaign. Anything and everything that enhanced this campaign pleased Murchison mightily, and he became willing to sacrifice any amount of energy toward the defense of his kingdom. I suggest that the Highlands controversy only becomes comprehensible when understood in the light of Murchison's special delight in the acquisition of geological "territory."

It was not just Murchison who thought of himself as the ruler of a geological empire. His contemporaries also saw him in this light, as was made abundantly clear in 1849, when the British Association for the Advancement of Science met in Birmingham. An excursion was made to nearby Dudley, to the west of Birmingham, where Murchison gave a public discourse in Dudley Cavern on the local Silurian rocks, and later that day in the open air to a very large audience. The occasion was described by H. B. Woodward, the historian of the Geological Society of London:

> In proposing a vote of thanks . . . the Bishop of Oxford (Dr. Samuel Wilberforce) said that although Caractacus was an old king of part of the Silurian region, yet Sir Roderick Murchison had extended the Silurian domain almost illimitably, and it was only just and proper that

there, upon a Silurian rock, he should be acknowledged the modern king of Siluria. The Bishop, then taking a gigantic speaking-trumpet, which he had brought with him, called upon all present to repeat after him the words which are given below. He then spoke through the trumpet, giving one word at a time, to enable those present [several thousands, it is believed] to repeat it all together—Hail—King—of—Siluria! Then, after a pause, the words were repeated a second and a third time.

The vast assembly thrice responded with stentorian voices and most hearty hurrahs, and ever afterwards Sir Roderick was proud to be acknowledged "King of Siluria." (Woodward 1908, 168–69)

Indeed he was, and this fact must constantly be borne in mind in seeking to understand the events that unfolded in the Highlands controversy.[26]

It was a controversy that Murchison pursued with unremitting zeal just as long as he was physically able to do so. He saw it as a major campaign, first for the extension, and subsequently the defense, of his kingdom of Siluria. It was the last major controversy in which he was involved, and by that late stage of his career he knew all that there was to be known about how to muster support and gain credence for one's views in the scientific community. The effort that Murchison put into persuading others of the correctness of his interpretations of the Northwest Highlands rocks is in large measure the reason it became so difficult to achieve subsequent changes in the theoretical interpretations, for he tied together his theoretical views with an extensive network of interlocking social obligations. Indeed, chiefly through his patronage of Archibald Geikie, he succeeded in gaining the acceptance of his views by the officers of the Survey (and hence by the general public) for some twenty years or more, even though the views were very much open to question. As Francis Bacon ([1620] 1960, 39) put it, knowledge and power "meet in one." But the case also had to do with the peculiarities of Murchison's psychological makeup, and his special personal identification with the Silurian system.

Murchison had first decided to take up a geological career in 1824 or 1825, and by the end of 1825 he had already completed his first piece of geological research, in the south of England (Murchison 1829). This accomplishment had quickly led to his election as secretary of the Geological Society in 1826. That year, Murchison met William Smith (1769–1839)—one of the principal founders of British stratigraphy, celebrated for his production of the first geological map of England and Wales (1815)—in Yorkshire, and was shown by him how to reason about strata with the help of fossils (Geikie 1875, 1: 131–32).[27] Ever after, Murchison seems to have been a specially ardent supporter of Smith's principle that strata can, and wherever possible should, be identi-

26. For more on the imperial theme in Murchison's work, see Secord (1981–82) and Stafford (1984, 1988a, 1988b, 1989).

27. By reasoning about strata with fossils, I mean the use of Smith's principle that particular rock formations are characterized by, and may be identified by, their organic fossil contents. This prin-

fied by means of their fossil contents. Murchison then traveled up to Scotland, but did not do much detailed research in the northern Highlands that year.[28] He had learned enough, perhaps, to recognize that he needed some kind of guidance if he was to achieve any kind of understanding of the geology of the mountain regions of that country. Accordingly, he sought the assistance of Professor Adam Sedgwick (1785–1873) of Cambridge, already known for his researches in the ancient rocks of the Lake District and later to achieve acclaim for his researches in Wales.[29]

Thus it came about that Sedgwick and Murchison made a joint reconnaissance of parts of northern Scotland in 1827. Their investigation, although provisional in character, was important for Murchison's study of the ancient rocks of the north of Scotland, for he apparently identified certain important sections that were to figure frequently in subsequent debates and controversies. Unfortunately, we have little archival material pertaining to this investigation, and we must therefore rely largely on the published papers that resulted from the fieldwork (Sedgwick and Murchison 1827–28, 1829).[30]

As the titles of the collaborative papers indicate, the main focus of attention was the red sandstones of the far north, not the ancient gneisses, schists, or quartz rocks, which later became the main bones of contention. The "secondary" rocks of the Caithness peninsula, and the quite young fossiliferous rocks of Brora on the northeast coast, Skye, and elsewhere were what chiefly concerned Sedgwick and Murchison. The tour was largely accomplished by sea, but this allowed a visit to the important exposures at Loch Eriboll. The geologists also went to the important exposures at Loch Assynt, but being hampered by bad weather they achieved no significant results there (Murchison 1867, 168). On the west coast, they noted with approval Macculloch's recognition of an unconformity between the underlying gneiss and the overlying red-brown sandstone and conglomerate. But like Macculloch, they regarded the red-brown sandstone as equivalent to the Old Red Sandstone of Caithness

ciple was of paramount importance in stratigraphy in the nineteenth century and is, of course, still essential to this day. In the nineteenth century, however, the scope of the principle was a matter of considerable debate. For those trained in the school of Werner, a stratigraphical analysis based upon the study of lithologies was still deemed appropriate; and in poorly fossiliferous regions such as the northwest of Scotland, the use of lithologies was unavoidable. Smith himself worked out his palaeontological method for determining strata in the Cotswolds, where suitable criteria were needed to distinguish between different limestones of similar appearance. He certainly used the principle in the preparation of his celebrated geological map of England and Wales, but it has been shown by Rachel Laudan (1976) that in practice Smith quite often resorted to the use of lithologies to identify and map strata. Indeed, it appears that some of Smith's observations were quite superficial, made on occasion from a carriage rather than at the rock face proper. I shall want to argue in later chapters that Murchison also was sometimes willing to argue for his interpretations on the basis of rapid surveys and hasty observations of lithologies.

28. His chosen task was to try to identify the formation to which the Brora Coalfield, on the east coast of Sutherland, belonged (Geikie 1875, 1: 132–34).

29. For a sketch of Sedgwick's life and work see Speakman (1982).

30. The joint paper was read to the Geological Society on May 16 and June 6, 1828. The paper of 1827–28 was an abstract of the full version of 1829.

and the east coast, thus offering a kind of symmetry to the geology of the north of Scotland.[31] The correlation of the western and eastern sandstones was made, despite the fact that the former were entirely devoid of fossils.

It is interesting to speculate on the reasons that may have led to the supposition that the western and eastern sandstones were one and the same formation. It was probably chiefly the suggestion of Sedgwick, who was always more inclined than Murchison to try to make correlations by means of lithological criteria. There may also have been some latter-day Wernerian thinking at work, for the symmetrical identification would allow *Floetz*-type rocks to appear to lap a central core of older semicrystalline rocks. As the geologists sailed along the north coast, they may have recognized the granitic outcrops in the region of Melvich, which could have been construed as an ancient mountain core, though such thoughts were not recorded in the published papers. Sedgwick and Murchison did, however, examine exposures of quartz rock at Loch Eriboll and disallowed Macculloch's suggestion that the "pipes" of the quartz rock were organic in origin.[32] If one rejected this idea, the quartz rock might be taken to represent nonfossiliferous primary strata, and this suggestion was in fact tentatively proposed. The red or the red-brown sandstones could both be regarded as potentially fossiliferous secondary deposits—even if fossils had as yet only been found in the east coast exposures. The great difficulty with such a scheme, however, was that it implied that the sandstones of the west coast would—if a suitable contact could be found—be seen to *overlie* the quartz rock. In point of fact, it is just the reverse, as can be seen in figure 2.1. It is not surprising or reprehensible that Murchison and Sedgwick erred on this point. After all, they were only engaged in a very preliminary investigation, and it would appear that a great deal of the northern part of their journey was in fact carried out at sea. But it was this issue that was subsequently to lead to the breach that occurred between Murchison and his former friend and traveling companion, James Nicol. We shall therefore return to this important point later in our narrative.

In the outcome, Murchison did not cling to the "secondary" interpretation of the red-brown northwest sandstones very tenaciously. The tour of 1827 was important for opening his eyes to the possibility of geological work in the northwest, and for suggesting suitable sites for closer examination. But a claim about the identity of the eastern and western sandstones, not backed up by palaeontological evidence and possibly more due to Sedgwick than to Murchison himself, could be renounced without too much difficulty. It was another

31. Actually, in preparing the 1836 map, Macculloch would have had regard to the views of Sedgwick and Murchison, for the eastern and western sandstones were represented by the same color. But a few years earlier, he stated that the western sandstone alternated with and graduated into gneiss in some places (Macculloch 1831, 2: 182). This should have been sufficient to make a clear distinction between the eastern and western sandstones. I assume that the authority of Sedgwick and Murchison influenced Macculloch when it came to coloring his map.
32. This term was not introduced at this stage to refer to the possible annelid traces, but I use it here anachronistically for convenience. Murchison later came to regard the "pipes" as in-filled worm burrows.

matter altogether when it came to Nicol's claim that the eastern and western gneisses were one and the same—for, as we shall see, this claim had fundamental implications for the territorial range of Murchison's Silurian kingdom.[33]

In 1836, after his death, Macculloch's map was finally published, along with some accompanying memoirs. In them he stated that he had decided to color as one the "primary red sandstone" and the "quartz rocks" (Macculloch 1836, 80). Perhaps he had made the change under the influence of Sedgwick and Murchison. I do not know. However, while to us it might appear a retrograde step, to Macculloch it probably seemed prudent, given that the two formations not infrequently appeared to be conformable and that neither was characterized satisfactorily by any fossils known at that time.

Considerable advances in understanding the geology of the Northwest Highlands, so far as regional mapping was concerned, were made by Robert Hay Cunningham (1815–42), a young man of good family who studied chemistry under T. C. Hope, and natural history under Robert Jameson, at Edinburgh University (Waterston 1957). The former was an exponent of the so-called Vulcanist (or Plutonist) views of James Hutton;[34] the latter, as we have seen, was the chief Scottish representative of Wernerian (or Neptunist) thought and founder of the Wernerian Society, which served as a platform for the promulgation of the German's views in Britain. Cunningham made several communications to this society, but probably had more sympathy for Huttonian than Wernerian doctrines. Whatever his theoretical views, it appears

33. That is, gneisses to the east and west of the outcrops of quartz rock and limestone, et cetera, running from Durness and Eriboll down to Skye.
34. As is well known, Hutton (1726–97) was a remarkable geological theorist and natural philosopher, whose geological opinions attracted much attention in Edinburgh and elsewhere in the late eighteenth and early nineteenth centuries, in opposition to the Wernerian ideas favored by Jameson. Hutton supposed that the interior of the earth was a vast reservoir of heat. From time to time hot, molten material would well up in different places into the earth's crust, forcing it up so as to form mountain ranges. This initially molten material would slowly cool and solidify, forming crystalline rocks such as granite; and the intrusion would supposedly alter the surrounding rock to semicrystalline materials such as schists. The mountains would in time become subjected to the erosive powers of wind and rain, and eroded material would be washed down into the oceans to form sediments, which would become compacted by further deposits and the action of heat to form sedimentary rocks. Seemingly, the processes would be endlessly repeated in great cycles of mountain building and subsequent erosion and deposition, each cycle requiring an immensely long period of time. As in Werner's theory, one would expect to find a mountain range to have a central core of crystalline granite, a surrounding region of altered or semicrystalline rock, and beyond that layers of inclined, or fairly flat, layered sediments. But since the whole process was repeated endlessly, one might find discordant layerings, or "unconformities," separating sediments belonging to different cycles. A textbook example illustrating Hutton's theory with remarkable success was discovered by him in the northern part of the Island of Arran, where the granite of Goat Fell is surrounded by a ring of schist and a further ring of sedimentary rock; but the observations to be made on Arran were more or less compatible with Werner's doctrines also. The literature on Hutton is extensive. See, for example: Tomkeieff (1962), Bailey (1967), Gerstner (1968), Dott (1969), Ellenberger (1972), O'Rourke (1978), Jones (1985) and Laudan (1987).

that he was well trained in mapping and rock and mineral identification by Jameson.

Although Cunningham died at the tragically early age of twenty-seven, he accomplished much in his few years of fieldwork. He gave descriptions of Mull and Iona in the Inner Hebrides, and some of the lesser islands, and also worked in Skye. In addition, he did fieldwork in the Lothians. With the demise of the Wernerian Society, he had to look for alternative publishing outlets, and he found support from the Highland and Agricultural Society, which awarded him several prizes for his work. The essay with which we are particularly concerned was entitled "Geognostical Account of the County of Sutherland," and was published in 1841, accompanied by a colored geological map of Sutherland (Cunningham 1841).[35] The manuscript map, held at the Royal Scottish Museum, is dated 1838 (Waterston 1957, 263).[36]

Cunningham's map, probably the result of three seasons' work, is a truly remarkable accomplishment. Two portions of the manuscript version that are specially relevant to our present purpose are reproduced in plate 2.[37]

The relevant portions of the color code are as follows:

Dark blue-grey, in some places crossed by diagonal blue lines (as at west of Kyle of Durness and west of Loch Assynt)	Gneiss series with Hornblende slate
Grey (as round Loch Hope and to east of Loch Glencoul)	Mica-slate series
Dark brown, sometimes with brown dots (as to north of Loch Assynt)	Red Sandstone and conglomerate
Light brown (as round Loch Eriboll and to south of Loch Assynt)	Quartz rock
Blue-grey (as to east of Kyle of Durness and to east of Loch Assynt)	Limestone

From plate 2 it can be seen how Cunningham successfully mapped in the Durness Limestone on the north coast and in the Assynt region. He recognized the western gneiss, which was easy enough, but also he distinguished this rock

35. The term "geognosy," introduced by Werner, was coined to refer to his kind of research program: a systematic investigation of the strata and their relative dispositions and the determination of their mineralogical composition. It was intended to be an objective, almost positivistic, empirical investigation, not beholden to theory. This was possible up to a point, though as we have seen Werner's students were, in fact, trained according to the overarching Neptunist doctrine.

36. Royal Scottish Museum, ref. 1959.4.1.

37. It should be noted that the southern boundary of the map is the county border of Sutherland, not a coastline.

from the "mica-slates" of the central parts of the map.[38] The exposures of what I have called red-brown sandstone were mapped in on the west coast, but as in the work of Sedgwick and Murchison, this formation was not distinguished on the map from the eastern sandstones, that is, the Old Red Sandstone. A truly impressive feature was the delineation of the outcrops of the quartz rock for its whole extent in Sutherland. This rock (often with its characteristic "pipes" and brilliant white color) is not difficult to identify, but it was a fine achievement to mark it in so accurately in so much inaccessible country. The marked outcrops at Lochs Glen Dhu and Glencoul, for example, indicate sound first-hand experience, and they clearly were not simply derived from Macculloch's map. Perhaps even more important for its contribution to an understanding of Highland geology was Cunningham's recognition of the difference in strike between the foliations of the western gneiss (approximately northwest by southeast) and those of the eastern "slates" (approximately northeast by southwest).[39] Cunningham (1841, 87) noted that the dips and strikes, extending quite regularly over large distances, could scarcely be accounted for satisfactorily by means of deep-seated (Huttonian/Plutonist) intrusions of granite alone.

Cunningham's memoir was furnished with numerous sections, as well as its valuable colored map. A number of these sections attempted to depict the structural relationships between the gneiss, the red-brown sandstone, the quartz rock, the limestone, and the "slates" (schists). From these several efforts no clear and distinct structure emerged, but that is scarcely surprising given that in large measure *that* was what the Highlands controversy was all about. Cunningham referred to the red-brown sandstone (or "red conglomerate with sandstone"), the quartz rock, and the limestone as distinguishable entities, but he viewed them as a single (composite) formation, which he called the "Quartz series" (ibid., 90). The red conglomerate was seen to overlie gneiss unconformably, but quartz rock sometimes seemed to "alternate" with conglomerate. Cunningham also noted, in association with the limestone, the occurrence of a somewhat earthy rock of "wood-brown appearance" and "calcareous strata of brownish-yellow colour" (ibid., 92). These would have been what were later called the "Serpulite grit" and the "fucoid beds"—Lower Cambrian strata occurring between the Durness Limestone and the quartz rock that by their characteristic appearance were to prove of inestimable value in subsequent efforts toward field mapping and structural interpretation.

38. "Mica slates" are more commonly referred to as "schists"—that is, the well-known "Moine Schists" in modern terminology. The Moine region is a desolate moorland area, to the east of Loch Eriboll. Moine House stands as a lonely outpost of civilization on the road between Loch Hope and the Kyle of Tongue. The Moine Schists are almost useless for any form of agriculture in the severe climate of that very northern part of Sutherland. It may be noted that not all the "Moines" are highly schistose. In some areas they may reasonably be described as schistose flags, rather than schists.

39. A "strike" line is equivalent to a contour line on a bedding plane, running perpendicular to the direction of maximum inclination (dip) of the bed at any given point. This matter was soon to be treated in much greater detail by Daniel Sharpe (1852a).

Cunningham observed that both the red-brown sandstone and the white quartz rock could, in different places, be found lying unconformably on the western gneiss. But the quartz rock could also be discovered to lie unconformably on the red-brown sandstone, as it did, for example, a few miles to the west of Quinag and to the north of Loch Assynt (ibid., 96). Thus he drew attention to what later came to be known as the celebrated "double unconformity," which is noticed in modern geological guide books to the area as a feature of special interest (e.g., Johnson and Parsons 1979, 35). Further unconformable junctions were recorded between the quartz rock and the gneiss right up to the north coast.[40]

But at Heilam (on the eastern side of Loch Eriboll), at the inaccessible Whiten Head (to the northeast of the loch), and at some other places, Cunningham observed gneiss *overlying* the quartz rock, apparently conformably. Then, somewhat to the east of Ben Arkle (near Loch Stack, to the southwest of the head of Loch Durness), where there was a nice exposure of quartz rock lying unconformably on gneiss, he found gneiss overlain unconformably by "mica slate" (Cunningham 1841, 99).

There is no doubt that Cunningham was sorely perplexed by these observations. They meshed with no known theory. Taking a cue from Macculloch, Cunningham suggested that there were in fact two systems of gneissic rock, one above and one below the quartz rock series. And above the upper one he described yet another crystalline series—the "mica slates." This hypothesis

40. An unconformity may occur in a sequence of rocks if
    1. a series of strata (*A, B,* etc) is deposited;
    2. there is a break in the sequence of deposition and earth movements occur so as to incline the strata;
    3. the strata are exposed to weathering and erosion;
    4. after the erosion, subsidence occurs and further strata (*X, Y,* etc.) are deposited on the now inclined strata (*A, B,* etc.).
The rocks may then appear in section as follows:

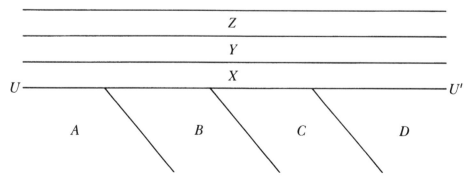

The plane *UU'* marks an *unconformity*. Its occurrence is indicative of cyclic processes of erosion and deposition, and it would presumably take a very long time to be brought into being. The notion of unconformity was introduced by Hutton in the eighteenth century, and the occurrence of unconformities was used by him as empirical evidence for the great age of the earth. In the foregoing figure, *A, B,* etc., are said to be conformable to one another, and so too are *X, Y,* etc.

implied that gneiss was not necessarily a primordial rock, for it might be found lying above a rock containing indications of fossils, that is, above the quartz rock series (ibid., 100–101). In point of fact, this was the theory, more or less, that was subsequently adopted by Murchison, though he never gave Cunningham or Macculloch much credit for it. It was a perfectly reasonable interpretation of the field relationships of the rocks, though it raised serious difficulties for anyone who might care to ask *how* the rocks attained their mineral characters and their relative dispositions.

Cunningham identified several valuable sections, which Murchison subsequently deployed in his debates with his fellow geologists, sometimes referring to them as "his" sections. Again, their prior "discovery" by Cunningham was conveniently overlooked. Ones that will particularly attract our attention were at Stronchrubie a few miles south of Loch Assynt, at Sangobeg on the north coast near Loch Eriboll, and at the aforementioned Heilam on the eastern margin of this loch.

Cunningham made a distinction (ibid., 102–3) between the eastern red sandstones of Caithness and the rocks of the "quartz series," in which, as we have seen, he included the western red-brown sandstones. Thus, by implication, if not explicitly, he was distinguishing between the eastern and western sandstones and offering a more sophisticated account than Sedgwick and Murchison on this issue. Cunningham censured Macculloch for failing to recognize clear evidence for unconformity along the north coast between the quartz rock system and the underlying gneiss (see fig. 2.3.) As he put it, this was "an error which the ascent of one hill, or the following of one stream, would assuredly have prevented" (ibid., 102). However, as we have seen, Macculloch clearly depicted such an unconformity on the west coast.

The achievements in fieldwork and structural interpretation made by Cunningham in just a few short years were remarkable, considering his relative inexperience and the nature of the terrain where he was working. It is highly regrettable that his early death has to a large degree led to a neglect of his attainments. But we must leave him here and turn our attention to the work of the better-known but perhaps less accomplished geologist, the charismatic autodidact Hugh Miller.

Miller (1802–56) was one of the most interesting figures in Scottish history in the nineteenth century.[41] He came from a seafaring family in the town of Cromarty, to the northeast of Inverness.[42] His father had died in a storm when Miller was still a child, and the boy, after apparently wild behavior at school, began life in the hard world of the itinerant stonemason. While work-

41. On Miller, see for example, Agassiz (1861), Miller (1854), Bayne (1871), Waterston (1966), and Rosie (1981).
42. Today, Cromarty is very much an economic backwater. Its main industry is arguably based on Miller's relics in the little cottage where he was brought up. There is an imposing statue of him behind the village, on a column of the same order of magnitude as that of Nelson in Trafalgar Square. It was not Miller's geology, however, that has given him such an important place in Scottish history, but his radical journalism, and the cause of religious emancipation in Scotland.

ing as a mason in the Old Red Sandstone deposits of Caithness, he became fascinated by the fossil fish that he discovered in the rocks and amassed a considerable collection of these strange remains. In due course, these were brought to the attention of the Swiss naturalist, Louis Agassiz, the leading authority on fossil fish, and Miller would achieve considerable renown for his accomplishments as an amateur palaeontologist with the publication of his highly successful work *The Old Red Sandstone* (Miller 1841, 1906). In addition to being an accomplished naturalist, he became quite a distinguished writer in a somewhat sentimental syle, with a wide knowledge of general literature.

In 1834, Miller managed to escape the drudgery of manual labor—which earlier he seems to have embraced in a romantic rebellion against the orthodoxies of middle-class life—and gained employment in a bank.[43] He continued his studies as an amateur geologist, but he also began writing of a more polemical character concerned with church politics and Scottish emancipation. This work led to his appointment as editor of the evangelical newspaper the *Witness*. Miller played, thereby, a significant role in the establishment of the Free Church of Scotland, and his paper, while being an important vehicle for Scottish nationalism, also served as a venue for many of his geological essays. As an ardent Christian, he was keen to demonstrate the harmony of the Bible with the latest geological knowledge. His chief effort in this direction was his work entitled *The Testimony of the Rocks* (1857), in which he argued that the several stages of creation described in Genesis could be identified with the chief geological epochs, if one regarded the single days of creation in an allegorical sense.[44] For example, the Coal Measures might be supposed to represent the stage of creation when plants were formed.

Even after taking up residence as a journalist in Edinburgh, Miller continued to make geological researches in the far north. In *The Old Red Sandstone,* he gave sanction to the picture of the structure of the Highlands that was emerging as a result of the work of Macculloch, Sedgwick and Murchison, and Cunningham. Miller sought to correlate the western red-brown sandstones with the eastern red sandstones. Indeed, he envisaged the whole of the north of Scotland as having once been covered with Old Red Sandstone, but with the central portions of the cover now being stripped away to reveal the more ancient underlying rocks (Miller 1906, 47–48). In a later paper, published initially in the *Witness* in 1854 and subsequently as an addendum to later editions of *The Old Red Sandstone,* Miller reported his conclusions based on fieldwork in the Assynt district in 1852 and 1853.[45] He recognized both an upper and a lower quartz rock, a point that was to become a central issue in the Highlands controversy.

---

43. He had earlier made some minor reputation for himself as a writer, which facilitated his change of station in life. Miller was undoubtedly stimulated to seek a higher position by his wish to marry a young lady of the bourgeoisie whom he met at Cromarty in 1831.
44. See also Miller (1861).
45. "On the Red Sandstone, Marble, and Quartz Deposits of Assynt; with their Supposed Organisms and Probable Analogues" (Miller 1906, 309–26).

Miller rejected Macculloch's suggestion as to the possible organic origin of the "pipes" of the quartz rock. But the chief interest of his 1854 paper was its attempt to provide a more exact correlation of the sequences of rocks on the west and east coasts. To this end, Miller suggested the following parallels (ibid., 322):

WEST COAST (e.g., a section at Ben More Assynt)	EAST COAST (e.g., Caithness)
Upper quartz rock	Red, white, and yellow sandstone
Limestone	Flagstones and associated limestones or calcareous nodular clays
Lower quartz rock	Thick arenaceous beds
Red-brown sandstone	Great conglomerate
Gneiss	[Gneiss]

He acknowledged that these correlations were not based on fossil evidence, but he thought the lithological resemblances of the various formations were too great to be coincidental.

It might be supposed that the hypothesis of a correlation between the fossiliferous eastern Old Red Sandstone deposits and the barren western red-brown sandstones—according with the aging Wernerian theory, but at odds with the prevailing southeasterly dip of the strata[46]—would have held little attraction to any knowledgeable geologist in the 1850s. However, Miller tried to muster support for this view in correspondence with Murchison, and in fact he seems to have had some success in this endeavor, for Murchison appears to have accepted the idea for a time.[47] Indeed, it may have been on this issue that disagreement first erupted in the Highlands controversy. Miller (1886, 97) liked the idea of the red sandstones forming, as it were, a picture frame round the ancient crystalline rocks.

We should now retrace our steps a little to introduce one of the leading figures in the Highlands controversy—James Nicol (1810–79)—an important nineteenth-century Scottish geologist, whose work has received little attention

46. The general dip of the stratified rocks of the northwest is to the southeast—that is, *toward* the central mass of Highland rocks. But the Old Red Sandstone of the eastern side of Scotland was figured by Miller (ibid., 24–25) as dipping *away from* the central mountains.

47. Miller to Murchison, undated (Edinburgh University Library, Gen. 523/4, letter 60). (There is an associated envelope, saying "Draft letter from Hugh Miller to Sir R. I. Murchison June(?) 1855.")

by historians of geology, despite its interest and significance.[48] No doubt, this has partly been so because he unsuccessfully opposed Murchison and Geikie in the Highlands controversy, and by his being on the "losing" side in this matter his reputation has been unjustly obscured. But there also seems to be very little surviving archival material on Nicol, which has no doubt made him an unpromising subject to study. I hope, nevertheless, to do something in this investigation to restore Nicol's name and reputation.[49]

Nicol was born at the Traquair Manse, near Innerleithen, Peebleshire, son of the local minister. James Nicol senior was a poet of minor reputation in his day, and also a writer on ecclesiastical history. He had gained some knowledge of medicine while at Edinburgh University and introduced vaccination in his parish. Thus there was some scientific background to the son's upbringing. James Nicol junior studied arts and divinity at Edinburgh, with a view to taking holy orders. But he was already much interested in geology and mineralogy, and he studied these subjects under Jameson. Subsequently Nicol attended lectures on geological subjects at the Universities of Berlin and Bonn.[50] On returning to Britain, he began extensive reconnaissance survey work in Scotland, presumably having sufficient private means to do so. Like Cunningham, he successfully submitted essays to the Highland Society, and by 1844 he had accumulated enough information to issue a valuable synoptic account of the geology of Scotland, with an accompanying colored geological map (Nicol 1844a). This production already bespoke a considerable familiarity with the rocks of the far northwest, though for this region the only works on which Nicol could usefully base his account were those of Macculloch, Sedgwick and Murchison, and Cunningham. Evidently Nicol had by then well and truly abandoned such Wernerism as he might have acquired in Germany or from Jameson, and he proclaimed Hutton's doctrines (ibid., 128). The year 1847 found Nicol in London, serving as secretary of the Geological Society and editor of its journal. Here he made the acquaintance of influential geologists such as Lyell, De la Beche, and Murchison, and with their help he was successful in gaining appointment in 1849 to the chair of geology at Queen's College, Cork, in Ireland. It appears that in this early period Nicol was soon on friendly terms with Murchison.

48. On Nicol, see Sorby (1880), Judd (1886), Bonney (1895a), Anderson (1899), and the anonymous "Nicol Memorial" in the *Geological Magazine* 57 (1920): 387–92. He should not be confused with William Nicol (1768–1851), the inventor of the polarizing prism.

49. It is perhaps worth mentioning here that when the Murchison-Geikie theory of the structure of the Northwest Highlands was finally overthrown, several commentators took considerable pleasure in the vindication of Nicol's ideas, even to the extent of representing them as more "correct" than in fact they were.

50. For Nicol's life as a student in Berlin, see Edinburgh University Archives (Gen. 713), "Journal of James Nicol (1810–1879) Edinburgh University student, from 10th October, 1840 to 21st March, 1841, during attendance at the Winter Session at Berlin University." Nicol attended lectures by the geologist Von Dechen, the meteorologist Dove, the palaeontologist/zoologist Ehrenberg, the chemist Schubert, and the mineralogist Weiss. He also attended lectures in theology and philosophy. His chief academic contact appears to have been with Ehrenberg.

It is worth noting that in 1848 Nicol published an important paper on the geology of the Scottish Southern Uplands, in which for the first time it was argued that these hills constituted an extension into Scotland of Murchison's beloved Silurian system.[51] The following year he furnished the first figures of graptolites from southern Scotland (Nicol 1850).[52] Murchison and Nicol worked together in the important Girvan area in 1850, and in 1852 Nicol publicly displayed a section through the Southern Uplands from the Pentlands to the Cheviots (Nicol 1853a, 55), which Murchison subsequently utilized in his magnum opus, *Siluria* (Murchison 1854, 152).[53] When Murchison (1851, 169) suggested tentatively that the rocks of the Highlands—or at least some of them—might be metamorphosed equivalents of the Silurian strata of the Southern Uplands, this idea was given support by Nicol (1852, 423).[54] Thus we can have no difficulty in understanding why Murchison should have been extremely well disposed toward his younger countryman. Indeed, in 1853 we find that Nicol succeeded, with Murchison's assistance, in gaining the chair of natural history at the University of Aberdeen, a position that would have been attractive to him by enabling him to return to his homeland; the Scottish chair also offered a better salary than that which Nicol received at Cork.

A few letters from Nicol to Murchison dating from the 1850s have been preserved in the archives of the Geological Society, and they give us some insight into the personal relations of the two geologists at that period.[55] We find Nicol taking Murchison's part in the latter's bitter controversy with Sedgwick concerning the Cambrian-Silurian border in Wales, and suggesting that the mica slates of the central parts of the north Scottish Highlands might be regarded as Silurian, as Murchison desired.[56] Though a "slow Huttonian," Nicol was willing to countenance a certain amount of sudden geological activity, in accordance with Murchison's views.[57] We also learn that Murchison and Nicol

51. Nicol was assisted in his fossil identifications by the Survey palaeontologist, J. W. Salter.
52. See figure 8.2.
53. According to Murchison, Nicol exhibited his section at a meeting of the British Association for the Advancement of Science at its meeting held in Belfast in 1852. However, only the titles, not the paper was published in the *Report* for the meeting.
54. It should be noted, however, that the support was not perhaps quite so strong as Murchison (1854, 163) would have had his readers believe. For one thing, Nicol (1852, 424) explicitly repudiated any idea that the greywackes (coarse grey grits) of the Southern Uplands, the crystalline rocks of the southern Highlands, *and* the "great gneiss formation" of northern Scotland might all be Silurian equivalents. He only had the Southern Uplands and the southern Highlands in mind. It is also worth noting that he proposed *this* equivalence prior to Murchison. It should be emphasized that the whole issue of the equivalence or otherwise of the rocks of northern and southern Scotland, upon which Nicol and Murchison would not be able to agree, was a major component of the whole Highlands controversy. Yet, as I say, some of the foundations of Murchison's position in the controversy were initially provided by Nicol.
55. Geological Society Archives, Murchison Papers, N8.
56. Murchison Papers, N8, Nicol to Murchison, Oct. 26, 1850. (It thus becomes apparent that the idea was put to Murchison in correspondence before it was published.) For Nicol's private support of Murchison at that time, see also Edinburgh University Archives, Gen. 1999/1/19, Nicol to Murchison, Oct. 11, 1852.
57. Murchison Papers, N8, Nicol to Murchison, April 30, 1851.

worked together in the field on Southern Uplands exposures for three seasons, in 1850, 1851, and 1852.[58] In 1854 there is evidence in the correspondence that Nicol was soliciting Murchison's support in a bid to obtain Forbes's chair of natural history at Edinburgh, but the attempt to secure advancement came to nothing.[59] Nicol and Murchison also corresponded on a geological map of Europe that they were preparing (Murchison and Nicol 1856).[60] But this was to be their last joint labor, for further cooperation between the two foundered as a result of the theoretical disagreements that developed between them concerning the Northwest Highlands.

So Nicol stayed in Aberdeen, and as we shall see, he never really recovered his progress in his career once he had fallen out with the geological establishment in the person of Murchison. The disagreement between the two friends first arose in 1855, during a joint field trip to the north. Thereafter, the breach rapidly widened, and Nicol developed his ideas independently of Murchison. Nicol made a number to attempts to press his views on the geological community but failed to make much headway. His last publication on the topic was his *Geology and Scenery of the North of Scotland* (1866, 29–47 and 87–96), which I discuss later in more detail. Nicol continued to make geological excursions into the north of Scotland until late in his career, but we know nothing about the development of his views after he ceased publication on the topic.[61]

It was only after Nicol's death that his achievements were sufficiently recognized by his peers, and at this stage we need not concern ourselves with his later work. I would, however, like to mention his early *Guide to the Geology of*

58. Ibid., Oct. 1, 1852.
59. Murchison papers, N8, Nicol to Murchison, Nov. 27 and Dec. 25, 1854.
60. Ibid., Feb. 15, March 8, April 29, July 9, July 17, Aug. 21, Sept. 4, Oct. 14, Nov. 18, Nov. 21, Dec. 25, 1854, Oct. 26, 1855.
61. Nicol also published a work on Iceland, Greenland, and the Faroe Islands (1840); two useful mineralogical texts (1849, 1858), one of which was reprinted from his article on mineralogy in the eighth edition of the *Encyclopaedia Britannica;* some popular geological texts (1842, 1844b); and an important early geological map of Scotland (1858), which is discussed later. He was known to his contemporaries as being "popular with his pupils and friends" (Bonney 1895a). He was described by Geikie (1904, 382–82) as being tall, of abundant sandy-colored hair, having a pronounced south-Scottish accent, and being a prominent personage at meetings of the British Association. By one of his former students, he was described as "somewhat large of bone, spare of flesh yet not lean, erect in figure and firm in gait, he looked a man in hard condition, unused to luxury and capable of physical endurance." He was also, it seems, a "kindly man . . . but something in the firm straight mouth told of a possible dourness it were better not to provoke" (Anderson 1899, 87).

Anderson also tells us that Nicol taught zoological anatomy as well as geology, and that his biological lectures were by no means as successful as those at which he spoke on geological matters. Only then did he seem to show any real enthusiasm. He was renowned for the fact that his students always passed—except on one notable occasion, as Anderson recounts (pp. 85–86): "Once, only once, tradition ran, had that pinnacle of academic distinction or depth of academic degradation [i.e., being failed] been achieved by a man otherwise graduand—in his case too with honors. King's College was amazed, the professoriate pleaded for pity on the unhappy wight, but Nicol stood firm: 'He said that the cow had no anal opening, and I cannot pass him.'"

Scotland (1844a). In this work, Nicol gave a very fair description of the range of the exposures in the northwest, without attempting much by way of structural interpretation (ibid., 210–16). He stated his indebtedness to Macculloch, Sedgwick and Murchison, and Cunningham, but I am confident that he had himself ventured round the north coast and down the west coast, by virtue of the clarity of his descriptions. The *Guide* was accompanied by a geological map, which, so far as Sutherland was concerned, did not offer any distinct improvement on that of Cunningham. At that early stage, Nicol subscribed to the prevailing idea that the western red-brown sandstones and the eastern red sandstones belonged to one and the same formation.

We now leave Nicol for a while, knowing that we shall meet him again on numerous future occasions. The only other geological text from this early period of the debate that we need to notice is a memoir (1852a) by Daniel Sharpe (1806–56), an energetic amateur geologist best known for his studies of the geology of Portugal (e.g., Sharpe 1850), his work on the cleavage of metamorphic rocks (1847 and 1849), the geological structure of Mont Blanc and its environs (1856), and some palaeontological researches.[62] He played quite a prominent part in the controversies between Murchison and Sedgwick on the question of the Cambrian-Silurian border, taking the part of Murchison (Secord 1986a, chap. 5). A businessman with interests in Portugal, Sharpe played an active role in the Geological Society, of which he was president in 1856. He was also a fellow of the Linnean, Zoological, and Royal Societies and had interests in philology and archaeology.

The feature of Sharpe's work that is of most interest to us relates to his investigations of the phenomena of cleavage. Here he made important and original theoretical contributions. Rocks may appear layered or banded in several ways and for several different reasons. First, there is the simple layering that arises from the sediments being deposited in layers over time, which is called bedding. Second, rocks may acquire joints owing to earth movements after the strata have been deposited and consolidated. It is often quite difficult to distinguish between bedding planes and joint planes with certainty. Third, rocks may be banded or foliated because of the intrusion of hot liquids and gases between or across the layers of preexisting rocks. This process may produce chemical and crystallographic changes in the original rocks, or it may force them apart, so that layers of crystalline material are deposited in the intervening spaces. Such changes, which are often accompanied by high pressures, can give rise to rocks such as gneisses, which, as has become apparent

---

62. On Sharpe, see Portlock (1857), Wrottesley (1856–57), Bonney (1897). See also Literary Gazette (1856) and Linnean Society (1857). Sharpe studied, among other things, the relationships between different types of strata and the damage caused in the Lisbon earthquake of 1755; the stratigraphy of the Vallongo Valley near Oporto; and of the region lying to the north of the River Tagus. He also gave original descriptions of certain brachiopods and gasteropods and made comparisons between the fossil remains of America and Europe.

already, are commonly found in northern Scotland. Gneisses are therefore often banded in a way that is distinct from the original bedding—which may, however, sometimes still be discerned with difficulty. Fourth, rocks may become layered by the application of pressure. Under intense pressure, some minerals may break down, to re-form as new laminar minerals (such as mica), the laminations being at right angles to the direction of the applied pressure. Moderate changes of this kind yield slates (the fissile nature of which is therefore due to metamorphism, not the original bedding). More intense pressure yields gneisses or schists, which are thought to contain different minerals according to the intensity of the pressure that they have undergone.

These broad interpretations were not all understood at the time of which we are speaking. But Sharpe made a major contribution to the understanding of metamorphic rocks by recognizing that cleavage was not necessarily an original feature of a rock, but one that could arise as a result of the application of pressure. He arrived at this conclusion by studying the distortions of fossils found in slates. The distorted shapes of the fossils indicated the direction of the applied pressure. It could then be seen that the direction of cleavage of the rock was always perpendicular to the applied pressure (Sharpe 1847). In his memoir of 1852, Sharpe sought to make a clear distinction between cleavage and foliation. Cleavage, he argued, was a splitting of a rock into parallel sheets independent of the stratification or bedding. Foliation, by contrast, involved division into layers of different constituent minerals, such as one might find in a gneiss or schist. He pointed out that this distinction between cleavage and foliation had often been disregarded—by Macculloch, for example. It is, however, not such a clear-cut distinction as Sharpe might have wished.

Nor is it clear from Sharpe's memoir just how much of Scotland he investigated personally. My impression is that his field researches were confined to the southern parts of the Highlands and that for the lonely tracts of Sutherland he relied chiefly on the data supplied by Macculloch and Cunningham, though he certainly went as far north as Loch Maree (Sharpe 1852b, 121). But this in no way detracts from a most important generalization he proposed, a generalization that came to be regarded as a major point of evidence, to be taken into account by all who would theorize about the structure of the Northwest Highlands. Sharpe believed that the foliated rocks of the Highlands were thrown up into a series of major archlike structures. He drew, for example, a hypothetical section through the Grampians (north-northwest/south-southeast) showing six of these structures, recognizable by the general dips and strikes of the rock foliations. The possible causes of such structures were not canvassed, but they were nevertheless an interesting feature so far as the huge masses of schists and gneisses were concerned.

Much more interesting for our present purpose was the geological sketch map of the northern parts of Scotland that Sharpe provided in illustration of his paper (1852a). There was nothing remarkably new in his broad outline of the outcrops. In Sutherland, Sharpe marked gneiss and schist as one color, and the sandstones of the east and west coasts as analogous "framing" formations. But drawing an imaginary line southwestward from Eriboll toward Skye, he marked the schists/gneisses to the west as having their foliations striking approximately northwest to southeast, while representing similar rocks to the east with their foliations striking northeast to southwest. This remarkable generalization suggested that the rocks to the east and the west of the Eriboll-Skye line belonged to two essentially different systems—or that they had undergone two quite distinct episodes of earth movement. This was regarded as an unassailable fact by most contestants in the Highlands controversy. In particular, it was an essential element in the theoretical armory of Murchison, to which he had frequent recourse in his debate with Nicol. What Nicol made of it will be recounted in chapter 6.

In considering the general stratigraphy of the Northwest Highlands, Sharpe maintained the old opinion that the eastern and western red sandstones were one and the same formation (the Old Red Sandstone). He found evidence at Loch Maree that suggested there were two quartz rock formations—an upper one, which he thought belonged to the western red-brown sandstone series, and a lower one, which contained some schistlike bands such that the whole had a gneissose appearance (Sharpe 1852b, 121–22). But this suggestion can scarcely have assisted a general understanding of the structural and stratigraphical problems of northwest Scotland. For if gneiss is the lowest rock of the region, then red-brown sandstone, then limestone/quartz rock, etc., and then schist, it is not an aid to understanding if one is told that the lower portion of the quartz rock is more gneisslike than the upper. Rather, Sharpe's interpretations underscore the great difficulties that were to be encountered in the northwest before understanding of the geology of the region was to be achieved.

We have now sketched the essential features of the geology of northwest Scotland as they were understood by the middle of the nineteenth century. The whole region had been surveyed in an approximate manner—at least, no regions were left uncolored in the published maps. A major structural feature was recognized running southwest from Eriboll to Skye. To the west of this line might be found gneisses, overlain unconformably by massive deposits of red-brown sandstone and conglomerate. Along much of the line itself, but particularly near Loch Assynt and at Durness and Eriboll, could be found a fairly narrow set of rocks—chiefly limestone and quartz rock, but with some flaggy or shalelike deposits also—dipping fairly gently to the southeast. To the east of this line there were further, but overlying, gneisses and huge exposures of schists and micaceous flagstones, with approximately the same dip and strike for their foliations. On the east coast, abundant outcrops of nearly horizontal red sandstones had been found, and attempts had been made to correlate these with the western sandstones. Apart from the eastern sandstones (which contained, incidentally, numerous fossil fish, the discovery and description of which formed the chief basis of Hugh Miller's reputation as a geologist), the rocks had proved unfossiliferous, though there were the curious "pipes" in the quartz rock, which were arguably of organic origin. The geological ages of all the rocks, apart from the eastern sandstones, were therefore unknown. Some granitic and syenitic rocks were known in the central portion of the far north, and so-called trap rocks had been discovered over much of the northern part of Skye.[63] But these need not concern us here. An important feature seemed to be the distinction between the western and eastern gneisses, based on their different foliations. No rocks or minerals of major economic importance had been discovered.[64] There appeared to be possibly two quartz rock formations, and possibly two gneisses, one overlying the presumed upper quartz rock.

Broadly speaking, such was the state of geological knowledge for the northwest of Scotland at the time when the Highlands controversy began to erupt. With these preliminaries, then, we may turn to a more detailed consideration of the major protagonists in the debate and the relationships that obtained between them.

63. Granites are a family of generally light-colored, coarse-grained, plutonic igneous rocks containing essentially quartz and alkali feldspars, with subsidiary minerals such as brown mica.

A syenite is a coarse-grained, plutonic igneous rock of intermediate acid/basic character. It consists chiefly of alkali feldspar such as orthoclase and some darker, more basic minerals such as hornblende and brown mica. There is usually a small amount of quartz.

The term "trap" rocks derives from the German word for step. Basaltic rocks, formed from successive lava flows, may assume a steplike appearance after weathering and erosion.

64. In addition to the limestones of Assynt and Durness, there is a small field of Jurassic coal near Brora on the east coast of Sutherland, which was investigated by Sedgwick and Murchison during their field trip of 1827, but which need not concern us here.

3

Geological Work of Murchison and Nicol in the Northwest Highlands: 1850s

This chapter examines the geological work carried out in the Northwest Highlands in the 1850s, particularly that of Murchison and Nicol. We shall see how Murchison reconstructed the geology of northern Scotland in such a way as to produce a very substantial addition to his Silurian "kingdom." The challenge to his arguments by Nicol and the protracted dispute that resulted we shall examine in some detail. No matter how unfortunate the dispute may have been, it did have one valuable effect at least, namely, it drew geologists' attention to this remote but most interesting region. It is also important in that it allows an entrée to the study of the geological community in Britain in the second half of the nineteenth century and tells us much about the manner in which fieldwork was conducted at that period.

Our first task in all this is to recount, as far as we are able from the surviving documentary evidence, just what was done as Nicol and Murchison formed their opinions and began to take up opposing positions. We shall endeavor to understand their lines of reasoning, based upon the field evidence that they encountered. Here we find opening up the major issues in the Highlands controversy, which were to be hotly debated for more than three decades.

As has been indicated, the early interpretations of the Northwest Highlands were rendered difficult by the lack of fossils. It was anticipated, however, that the most likely places to find fossils would be in the several exposures of the Durness Limestone. But at first nothing was found. Eventually, success came as a result of the observations of Charles W. Peach (1800–86), a coast guard from Wick in Caithness, who was inspecting a wreck in Loch Durness in 1854 and happened on fossil remains in limestones exposed near the "Auld Kirk" on the shores of the loch—one of the most northerly inhabited places on the British mainland (C. Peach 1858; J. Miller 1859).[1]

Charles Peach was an enthusiastic amateur naturalist who established quite a name for himself in the nineteenth century as a keen and reliable observer. He was a regular correspondent of Murchison and numerous other leading scientists, who were pleased to utilize his empirical discoveries. But Peach

1. On Peach see Smiles (1878–58) and Bonney (1895b), and the anonymous memorials in *Nature* 33 (1885–86): 446–47, and the *Athenaeum*, no. 3046 (1886): 362–63. By the time Peach was working at Wick, he had the status of comptroller of customs.

went somewhat further than being a mere observer and collector. He was a regular participant at the meetings of the British Association and presented papers there from time to time, reporting, for example, the first discoveries of fossils in Cornwall. On occasions, Murchison thought him an eminently suitable person to accompany him on his field trips, and he received on Peach's behalf some prize money presented by the Geological Society in recognition of his fossil discoveries. On this occasion, Murchison (1859a) eulogized Peach, describing him as "ingenious, modest and highly deserving."

Peach came from the Midlands but resided at a number of coastal towns, particularly in Cornwall and in Caithness, the most northerly mainland county. In Caithness, he formed an active partnership with Robert Dick (1811–66), the "Baker of Thurso" (Smiles 1878), who like Hugh Miller achieved renown for his fine collections of fossil fishes from the Old Red Sandstone. Peach and Dick took great pleasure in each other's company and made numerous field excursions into the countryside and along the coastlines. Peach is of special interest to the present inquiry by reason of the fact that he was father of the great Benjamin Peach, who will figure so largely in the later chapters of this book.

The fossils first discovered at Durness by Charles Peach were poorly preserved and hard to identify. But the specimens were packed up and sent down to London for the benefit of the opinion of Murchison and the Survey officers. At a first guess, Murchison supposed that they might be Devonian in age. He wrote to Sedgwick:

> You have no doubt heard of . . . the discoveries of fossils in the Durness limestone of Sutherland, by Peach. He has corresponded with me on the point, and has sent me some of the fossils. I have had them polished. The forms (rude and ill preserved as they are) look more like *Clymeniae* and *Goniatites* than anything else (with corals);[2] and if so, the calcareous masses which we saw from Assynt to Durness, interstratified in the quartz rock, are high in the Devonian! I would like to hear what you say to this *éclaircissement*. I see great difficulty in understanding it.[3]

Clearly, Murchison recognized that the preliminary identification was uncertain and unsatisfactory, and that the site where the fossils had been found required further investigation. Were a Devonian age accepted for the *limestone* it would make huge problems for the received interpretation of the northwest stratigraphy. The limestone belonged to a series *above* the western red-brown sandstone, which was itself thought to be equivalent to the eastern Old Red Sandstone of Caithness, containing Dick's and Miller's fossil fish. But the eastern Old Red rocks were, according to Murchison, Devonian in age and were definitely at a higher stratigraphical horizon than the limestone.

In May 1855, Lord Palmerston appointed Murchison director of the Geo-

2. *Clymeniae* are types of fossil resembling the modern *Nautilus*, typically found in the Upper Devonian. *Goniatites* is an ancient genus akin to the ammonites, with species ranging from Silurian to Devonian.
3. Murchison to Sedgwick, May 30, 1855, quoted in Geikie (1875, 2: 195).

logical Survey, in succession to his former adversary, Sir Henry De la Beche. Obviously, even a man of Murchison's energy and zeal would take a little time to seat himself comfortably in the saddle and learn what was required of him in his new role as professional scientific bureaucrat. But by August he was ready to take to the field once more, and it should come as no surprise that he determined to travel to the north of Scotland to examine the spot where Peach had discovered fossils, to see if more could be found, suitable for a more secure identification, and to explore the implications of the new information for the general stratigraphy of Scotland. He took with him as traveling companion James Nicol—a very suitable choice considering their ongoing cooperative work on the geology of the Southern Uplands and their work together on a geological map of Europe.

A number of Murchison's field notebooks have survived and are held in the Geological Society's archives. Others are to be found in the Survey's archives.[4] But they are difficult to use, partly because of Murchison's execrable handwriting and partly because they have been cut up in many places so that sketches and diagrams from them might be incorporated into the twenty-four volumes of Murchison's autobiographical journal. These bound volumes, held at the Geological Society, are in the clear hand of an amanuensis, who, however, seems to have had as much difficulty as the modern historian in deciphering Murchison's scrawl, for in a number of places the text contains gaps where the master's hand became unreadable. The journal largely follows the field notebooks, so that it is not unreasonable to trace the development of Murchison's fieldwork and his thinking on the basis of the fairly well-polished journal. Obviously, it was Murchison's intention that the journal should be published after his death, and he arranged for Archibald Geikie to carry out this task. However, Geikie went far beyond what Murchison might reasonably have expected, producing a fine account of Murchison's life and times in a beautifully written biography (Geikie 1875) that tells us a great deal about the history of geological research in Britain in the nineteenth century. The book was a splendid testimony to Murchison, to Geikie's literary abilities, and to his regard for and devotion to his former director.

Murchison's journal for the mid 1850s bears witness to its having passed through Geikie's hands, for there are a few penciled marginalia on the manuscript, sometimes initialed in Geikie's hand.[5] At one spot I notice that Geikie has written: "These sections have been touched up by Murchison since they were made in the field."[6] There is evidence, therefore, that Murchison occasionally modified his field notes to make them appear more consonant with his subsequent thinking. Nevertheless, one can, I think, regard the journal as

4. For details of Murchison's surviving papers and their locations, see Thackray (1972).
5. Murchison Journal, 1853–56, "Scotland Germany Ireland," Geological Survey Archives, Murchison papers, M/J23.
6. Ibid., 165.

a generally reliable guide to his thinking "on the spot." Comparisons between the field notes and the journal are reassuring; and in cases such as the one commented on by Geikie, it is easy enough to recognize that modifications have been made.

Murchison and Nicol seem to have had a very hurried journey, and Geikie states that they encountered bad weather for much of the time. Murchison's journal records the travelers as being at Inverness on the east coast on August 11, traveling northward by Bonar Bridge and Glen Oykel to Loch Assynt and being at Loch Inver on the gneiss terrain of the west coast on the thirteenth. On the fifteenth they are reported as traveling from Scourie to Laxford, still in gneiss country, and then on to Durness in the land of quartz rock and limestone. There they were met by Peach, who no doubt took pride in displaying his new fossil finds.[7] But the journal makes no mention of any new fossil discoveries. The travelers apparently stayed with the owner of Eriboll House, a farmstead on the east shore of Loch Eriboll (see fig. 7.11)—which we shall encounter several times in our later discussions as the location of a number of sections that gave rise to numerous disputes. The geologists then crossed the dreary Moine district in an easterly direction to the beautiful Kyle of Tongue on the north coast, before rapidly heading south to Murchison's old site of inquiry at Brora. Then they went back across the country and down to Glasgow, passing the region of Loch Torridon, where the western red-brown sandstones can be seen to the best advantage, to report on their recent observations at the meeting of the British Association in September.

Murchison's journal for this excursion is not very full, and considering the haste with which the geologists traveled it is scarcely surprising that he did not record a great deal in the way of observations. There is the first hint of disagreement with Nicol, for we find Murchison writing: "In this tour I made many fresh observations in Assynt, Durness, &c., and in the relations of the rocks, and had plenty of disputes with Nicol."[8] But what Murchison was actually making of his new observations in 1855 is by no means simple to determine. In his paper on the topic at the British Association (Murchison 1856), he held to the opinion, formed back in 1827, that the red-brown sandstones of the Torridon district belonged to the Old Red Sandstone—which by 1855 was well established as Devonian in age. Murchison also recorded his observation of a clear unconformity between the quartz rock series and the old gneisses near Durness and at Rispond at the western side of the entrance to Loch Eriboll. At Inchnadamph (the settlement at Loch Assynt) he reported, with Nicol's concurrence, the observation of a "hard red conglomeratic grit" lying unconformably on the lowly gneiss. This grit, he claimed, was *not* Old Red Sandstone—though why this was asserted is not so very clear. The assertion would, if taken seriously, have had the unfortunate effect of forcing a division

7. Ibid., 169.
8. Ibid., 163. Murchison continued: "for which see my publications as well as my note books." Unfortunately, the notebooks that I have examined do not furnish information on this head. What Murchison had to say in his publications we shall see in due course.

of the western red-brown sandstones. However, it is not clear exactly what rock Murchison was talking about at Inchnadamph. It is not obvious from his text that he was referring to the massive deposits of red-brown sandstone such as make up the mass of Quinag. Whatever it was, he likened it to a similar rock to be found at Ullapool, and it is just possible that he was referring to the controversial rock that later became known as "Logan Rock," which we shall have occasion to mention again on numerous occasions.

As we have seen, there are two major unconformities to be discovered in the northwest: first, the break between the gneiss and the red-brown sandstone; and second, the break between the red-brown sandstone and the quartz rock series. The second of these unconformities is not always obvious to the casual observer, and it is interesting that Murchison appears to have missed it during his 1855 field trip.[9] Again, what he actually saw is not clear. His journal shows two pages of sketches of sections in the Loch Assynt region, pasted in directly from his field notebooks.[10] The exact localities are not identified, but one is marked "when viewed from the N. side of Loch Assynt" and appears to be a view looking north toward Quinag. It shows a definite unconformity between "white quartzite" and "red conglomerate." But as noted by Geikie in his marginalia, the sketch was touched up later by Murchison, and what the original drawing showed cannot now be precisely determined. The sketch bears a note in Murchison's hand (in different ink and clearly added later): "These diagrams should be published to shew that I had made out the order in 1855." There is, I believe, some reason to doubt this claim, for another sketch on the preceding page shows a clearly conformable sequence between the red conglomerate and the quartz rock. On the other hand, it should be noted that in his 1855 paper Murchison recorded seeing western red-brown sandstone dipping to the northwest at the head of Loch Kishorn (near Lochcarron, on the mainland opposite Skye), while the quartz rock series and limestone dipped east-southeast.[11] This suggests the existence of an unconformity, but there is no suggestion of one in Murchison's paper, and the junction between the two rock units at Kishorn was obscured by alluvium. Murchison specifically referred to the red-brown sandstones of the Applecross-Kishorn district as Old Red Sandstone, and described the series as being wrapped round the ancient rocks. It is scarcely possible to make sense of all this, and Geikie's remark (1875, 2: 205) that "the results of the tour were not very convincing" seems just.

It would be interesting to know whether this was the issue about which Murchison and Nicol began to differ. Even if it was, I doubt whether it could have been the whole of the problem. There was no a priori reason why the white quartzites and the red-brown sandstones should be either conformable or unconformable. Nevertheless, one suspects that it was at least one of the

9. For example, the western red-brown mountains sometimes have sloping white cappings of quartz rock, which from a distance may seem to lie comfortably on the sandstone (see fig. 2.1).
10. Murchison Journal, 1853–56, 164–65.
11. This observation agrees with data of modern maps. But there are several places in that district where red-brown sandstone may be found dipping to the southeast also. The region is in fact structurally very complex, being the locale of the important Kishorn Thrust.

issues that began to divide the two geologists. Unfortunately, we do not have Nicol's field notes or his correspondence from Murchison. But a letter from Nicol to Murchison written in 1856 is revealing:

> The fact that the Quartzite series appears to overlie the Red Sandstone near Assynt you will find stated in my own notes unless I am mistaken. We had a long discussion on the matter at Loch Inver and on our return up the loch side made it beyond doubt. Then in the north side of Queenaig [*sic*] we saw the same thing—though there you thought a slip might have caused the appearance.
>
> At Durness I also made out that the Quartzite series was above the red sandstone and conglomerate and mentioned it to you on my return from Cape Wrath. I do not remember whether you noted it or not, but have no doubt of mentioning it to you.
>
> In regard to Queenaig I am almost certain you drew a section of it. When we were coming up from Loch Inver *You told me I would be the Magnus Appollo of Scottish geology if I could make it out.* And I think you came to admit it before we were done, in regard to that place.[12]

All this may appear rather puzzling. Nicol's letter does not state that he was thinking of a quartz rock–red-brown sandstone *unconformity* in 1855 (though he did propose one the following year [Nicol 1857, 25]). In the letter quoted, he is merely saying that the quartz rock lay *above* the sandstone—a point that might, I suppose, be in dispute as a result of the Loch Kishorn observations. But it should be remembered that according to Hugh Miller red sandstone (supposedly Devonian) "framed" the Highlands of Scotland. This would have required the sandstone to be *above* the quartz rock, and Murchison was, in fact, adhering to such a view at this early stage of the controversy, despite the fact that it is clearly at odds with what may be seen on the west coast of Scotland. It would appear, then, that this was the issue on which he and Nicol first parted company. If this interpretation is correct, it may appear that Murchison's views were influenced as much by Hugh Miller's texts as by the observations that he himself made in the northwest in 1855. As we have seen, back in 1827 Murchison had, in conjunction with Sedgwick, regarded the eastern and western red sandstones of northern Scotland as essentially one and the same secondary formation, and at that time he provisionally regarded the quartz rock as primary. Hence the quartz rock would be expected to be *below* the red-brown sandstone. Now, however, Nicol was beginning to suggest that the correct order was the reverse. It seems, therefore, that Murchison was showing a characteristic reluctance to renounce a theoretical viewpoint, once he had taken it up and once his name had become associated with that doctrine. We find this apparent reluctance even where a piece of Murchison's "juvenilia" was at stake. But to be fair, it was not long before he changed his view on this point.

12. Geological Society Archives, Murchison Papers, N8, Nicol to Murchison, Aug. 9, 1856 (emphasis in original).

Murchison does seem to have been clear about one thing in 1855. He reported that the limestone−quartz rock sequence of Durness and Eriboll passed conformably into an overlying sequence of schists and gneisses, which occupied the terrain between Eriboll and Tongue. And so he supposed that "the band of limestone containing the fossil shells discovered by Mr. Peach is a low member of this great crystalline series of stratified rocks of such diversified characters" (Murchison 1856, 86). Further, he tentatively suggested that the Durness fossils might in fact be Lower Silurian (not Devonian, as had first been supposed). If this suggestion was correct, and if there was no conformity between the fossiliferous rocks and the overlying schists and gneisses, then there was a case to be made that the crystalline rocks were also Silurian, since they definitely lay below the well-established "Devonian" Old Red rocks of the northeast coast. Thus one could see opening up the pleasing prospect of a notable extension of the Silurian territory in Scotland.

I must stress, however, that such territorial claims were not advanced in 1855. I am merely suggesting that Murchison had made observations that, subsequently elaborated, came to serve as the empirical basis for such claims. Actually, one can perhaps see some hint of the views of Nicol in Murchison's 1855 British Association report. For Murchison (1856, 88) referred to the *repetition* of metamorphic rocks that may be discerned in Sutherland and Ross, which was a point that subsequently became central to Nicol's argument.[13]

Another issue that attracted Murchison's attention in his British Association paper of 1855 was the question of foliation and cleavage, previously discussed, as we have seen, by Daniel Sharpe. Murchison asserted that layerings of different colored materials that might be discerned in the ancient crystalline rocks represented original stratifications, not superimposed foliations or cleavages. The views of Sharpe (not identified as such) were countered by reference to the ideas of much earlier geologists such as Hutton, Macculloch, and Boué. The question of what was and what was not bedding, foliation, or cleavage was to be an issue that would recur during the course of the Highlands controversy, as our later discussion will show.

We thus have no clear understanding of just what went wrong between Murchison and Nicol in their 1855 field excursion. Nicol made no independent statement of his views to the British Association that year, the paper that he presented to the Glasgow meeting being concerned with the glacial phenomena that he had observed in the north of Scotland (Nicol 1856). It may be, in fact, that relations did not become strained until the following year, when Nicol returned to the Highlands by himself to look over the ground more carefully and at greater leisure, began to develop his own thinking on the matter, and presented an independent account of his views to the Geological Society (Nicol 1857).[14]

In May 1856 Nicol wrote to Murchison: "You and I seem now gradually dropping out of acquaintance." But a few days later, writing of his plans to

13. Such a repetition had already been remarked on by Cunningham.
14. The paper was presented on Nov. 19, 1856.

travel north, he wrote requesting a copy of Murchison's Highland notes, which clearly indicates that the two geologists were still on good terms. And Nicol was evidently siding with Murchison in his dispute with Sedgwick.[15] There is no reason to believe, therefore, that the relationship had deteriorated to any significant degree at that stage. Nicol sent a draft of his paper to Murchison in October 1856, about a month before it was read at the Geological Society, soliciting his comments. After the paper was read, he wrote thanking Murchison for recommending that it be printed.[16] Indeed, Nicol maintained an unfailing courtesy in all his subsequent correspondence with Murchison. So what was it in Nicol's paper of 1856 that began to raise Murchison's ire to such a degree? And why should they become so deeply divided on the geology of the northwest?

Much of Nicol's paper was concerned with field descriptions and sections of several important areas of the northwest. He described in turn the north and south shores of Loch Broom near Ullapool, the Assynt region, the Kyles of Durness and Eriboll, Gairloch and Loch Maree, Loch Kishorn, and the southern part of Skye. I have traced his descriptions in all these regions and can testify that they all "make sense." That is to say, in all cases one can discern what exposures he was referring to and one can well understand why he said what he did. In fact—according to our present understanding of the matter— he was tracing the long line of the thrust planes from Durness-Eriboll down to Skye. Or, from his perspective, he was tracing out the long line of exposures of quartz rock and limestone, taking every available opportunity to make theoretical cross-sections of the strata. So let me say something about what Nicol recorded. Like him, we may begin with the Ullapool region.

Ullapool is situated on the north shore of Loch Broom near a junction of red-brown (Torridonian) sandstone (to the west) and quartz rock (to the east). Fairly narrow bands of quartz rock and limestone run northwest toward Assynt and Durness-Eriboll. They dip approximately to the southeast. Over these, a little further to the east, one may see some fairly small exposures of gneiss, and above these may be found a vast tract of schistose rock extending eastward— the so-called Moine Schists. The limestone and quartz rock are particularly well exposed in the valley running up to Loch Achall, just to the north of Ullapool, where the limestone is extensively quarried today, as it was in Nicol's day. There, good exposures may be found of the contact of limestone with *overlying* gneiss, and essentially similar sets of exposures may be found on the south side of Loch Broom, opposite Ullapool. The main mass of (underlying) western gneiss is not found at Ullapool, but only outcrops about fifteen miles to the north.

Like everyone since Macculloch, Nicol easily discerned the unconformity between the red-brown sandstones and the western gneiss. But unlike Mur-

15. See Geological Society Archives, Murchison Papers, N8, Nicol to Murchison, May 24, 1856 and June 3, 1856. On the Sedgwick dispute, see Nicol to Murchison, Oct. 14, 1854; Oct. 26, 1855; July 28, 1856.
16. Nicol to Murchison, Oct. 22, 1856; Dec. 13, 1856.

chison in 1855, Nicol recorded a significant difference in dip between the quartz rock and the red-brown sandstones, and he showed them as unconformable to one another in his sections. However, he was not the first to notice this. Colonel Henry James (1803–77) of the Ordnance Survey, and formerly on the staff of the Irish branch of the Geological Survey, had visited the northwest of Scotland in the early summer of 1856 to select a site suitable for the attempted measurement of the density of the earth, and at Murchison's request he examined the junction between the quartz rock and the red-brown sandstones (Murchison 1859b, 357). In a letter to Murchison, dated July 26, 1856, recorded by Geikie in his biography of Murchison (Geikie 1875, 2: 213), but which I have been unable to trace, James reported the existence of the important unconformity. Evidently Murchison passed on this information to Nicol, for in a letter dated August 9, posted in Oban, we find Nicol writing to Murchison and acknowledging the enclosure of James's letter.[17] Nicol did not seem to think that James's observations added very significantly to the understanding of the northwest geology. However, it seems reasonable to accord to James the credit for the discovery of the quartz rock–red-brown sandstone unconformity. John Horne (1907, 14) thought it sufficiently important to refer to it in his synopsis of geological research in the 1907 memoir on the Northwest Highlands.

Nicol acknowledged—with Macculloch, Cunningham, and Murchison— that the eastern gneiss *overlay* the quartz rock–limestone series. But what was particularly interesting, and essentially new to the discussion, was the observation of a somewhat peculiar reddish rock at the base of the eastern gneiss in the exposures on both the north and the east shores of Loch Broom. Nicol described it as a "massive intrusive rock, in some places a kind of syenite, or a felspar-porphyry with hornblende, in others rather an impure serpentine."[18] The exact nature of this rock was to prove a source of major contention in the Highlands Controversy. Later it came to be called "Logan Rock" after Glen Logan, where it is well exposed (see figs. 5.1 and 5.2 and plate 6).

17. Nicol to Murchison, Aug. 9, 1856.
18. See chapter 2, note 63. Murchison had previously reported the occurrence of syenite at Ben Loyal, a few miles south of the Kyle of Tongue on the north coast.
    A porphyritic rock is one made up of relatively large crystals in a finer-grained groundmass. Feldspars (or felspars) are a family of minerals such as orthoclase and plagioclase. Chemically, they are sodium, potassium, or calcium alumino-silicates. Often, they form well-shaped white or pink crystals and are one of the main constituents of granites. Hornblende is the commonest member of the family of minerals called amphiboles. It is a complex alumino-silicate containing, among other elements, iron and magnesium. The crystals are usually quite dark in color and are found typically in basic (i.e., low silica content) rocks.
    The name serpentine refers to a group of hydrous magnesium-silicate minerals. It is also used to refer to a rock made up chiefly of serpentine minerals, and is believed to be formed by the alteration of rocks rich in ferro-magnesian minerals such as augite and olivine. (It sometimes has the appearance of a serpent's skin.) Serpentines, which are common, for example, in the Lizard area of Cornwall, were not well characterized in the 1850s and 1860s, before the widespread use of the polarizing microscope, and it was not unusual to refer any greenish rock of doubtful nature as "serpentine."

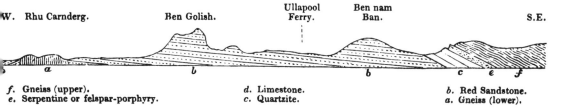

f. Gneiss (upper).   *d.* Limestone.   *b.* Red Sandstone.
e. Serpentine or felspar-porphyry.   *c.* Quartzite.   *a.* Gneiss (lower).

Fig. 3.1. Section of strata along the south side of Loch Broom, according to Nicol (1857, 22).

This course of action was adopted since the rock's petrographic character was so uncertain, and confusion about it became so widespread, that it was thought better to give it a theory-free name until such time as its nature could be determined satisfactorily. In fact, the rock's importance for Highlands geology was not apparent in 1855, and I shall not, therefore, attempt to explore its significance further at this juncture.[19]

Nicol's section of the strata at Loch Broom (Ullapool) is shown in fig. 3.1. North of Ullapool, Nicol examined the strata in the Loch Assynt region, as several geologists had already done, and as a large number were to do in the future. The general sequence appeared to be more or less the same as at Loch Broom. At the western end of Loch Assynt, he found red gneiss with western red-brown sandstone lying over this unconformably and rising to form the large mountain of Quinag. Then lying uncomfortably over the sandstone Nicol recorded the quartz rock, with its characteristic "pipes," which he called "cylindrical bodies."

There is a road leading over a watershed between Loch Assynt and Kylescu on Loch Glencoul—a sea loch to the north of Assynt. On this road Nicol (1857, 25) recorded "brown shale-like beds, with what appeared to be fucoid impressions" at a horizon above the quartz rock. A "fucoid" is a seaweed, and Nicol believed that some traces in these rocks represented fossilized plant remains. In this he was mistaken, it is believed today. But the rocks he was noticing (which are usually of a characteristic yellow-brown, somewhat earthy, appearance) are important in the stratigraphical sequence in the Northwest Highlands and are to this day commonly referred to as the "fucoid beds," which is the name I shall use. They are important marker bands and have proved very helpful for mapping in the northwest.[20]

At the lakeside, near the place called Skiag Bridge, Nicol did not notice the fucoid beds, and the quartz rock appeared to be succeeded immediately by limestone. This could be traced southward on the Ullapool road for a few miles to the farm named Stronchrubie, where it formed a scarp to the east of the road. It appeared to Nicol that there was a repetition of the sequence due to faulting. Also, at Stronchrubie, Nicol followed Cunningham's suggestion that there was further quartz rock lying *above* the limestone, and some igne-

19. Nowadays, the rock is regarded as a variant of Lewisian Gneiss.
20. It should be noted that they had previously been mentioned by Cunningham.

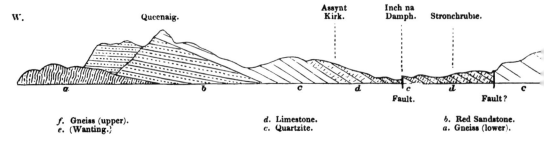

f. Gneiss (upper).   d. Limestone.   b. Red Sandstone.
e. (Wanting.)   c. Quartzite.   a. Gneiss (lower).

Fig. 3.2. Section of strata at Loch Assynt, according to Nicol (1857, 23).

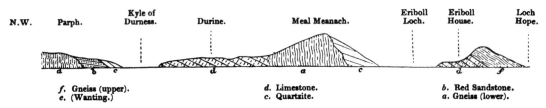

f. Gneiss (upper).   d. Limestone.   b. Red Sandstone.
e. (Wanting.)   c. Quartzite.   a. Gneiss (lower).

Fig. 3.3. Section of strata from Durness to Eriboll, according to Nicol (1857, 23).

ous rock (a "greenstone") appeared to outcrop along the base of the cliffs.[21] (Nicol's Assynt section is shown in fig. 3.2.)

Nicol discerned a somewhat similar sequence on the north coast in the Durness-Eriboll region: but the whole appeared twice in that area. To the west of Durness, on the Cape Wrath Lighthouse road, Nicol recorded quartz rock over red-brown sandstone over gneiss. Limestone was abundant to the east of the loch. But a little further to the east, the gneiss reappeared, to be overlain once again by quartz rock, which formed the western shore of Loch Eriboll. Then to the east of this loch, the limestone showed up again, but it was over-lain by a further deposit of gneiss, or in some places some more quartz rock and then the gneiss. Nicol noted particularly that there was no red-brown sandstone to be found at Eriboll. (His section for Durness-Eriboll is shown in fig. 3.3.)

Nicol then turned south and visited the beautiful region of Loch Maree and Gairloch. He noted, and placed particular emphasis upon, the unconformity between red-brown sandstone and the gneiss beneath, with fragments of the latter to be found in the former. However, it would appear from his description that his time was spent chiefly at the western end of the loch, rather than the interior, where the more interesting and controversial exposures were

21. Modern maps show many lamprophyre (basaltic) dykes in this district. Nicol was probably looking at one of these. In the 1880s, this rock was identified as a "diorite" (Teall 1886). This diagnosis would not hold today, since diorites are defined as coarse-grained igneous rocks of intermediate acid/base composition. But in the nineteenth century, it was customary to regard diorites as a subdivision of "greenstones," the latter being a very broad classificatory unit, used to refer to altered, dark, generally greenish, eruptive rocks.

subsequently discovered (at "Glen Logan"). Further south, at Loch Kishorn, Nicol described the rocks more or less as had Murchison in his 1855 British Association paper. However, Nicol saw that the disposition of the rocks there was problematical. The red-brown sandstone seemed to be deposited on the upturned ends of the (Durness) limestone. But this was scarcely compatible with the sequence established further north. He suggested that there might be a fault in the Kishorn district, bringing down the limestone.

Nicol then followed the limestones and quartzites further south into Skye, where the relationships appeared to be "exceedingly complex." He traced a section from Isleornsay on the southeast coast of the island approximately northward along the road to Broadford, and provided a section of his line of traverse. He recorded seeing first gneiss, then a "mass of claystone, or felspar-porphyry," which must, I suppose, be the "hornblende-chlorite-schist" shown on modern maps. Further north, and somewhat to the west, there appeared to Nicol to be red-brown sandstone. But then he supposed he found quartzite (quartz rock?), and then more sandstone. A little further north a "syenite or felspar-porphyry" was encountered as an apparently intrusive mass within the sandstone. Near Broadford there appeared low-lying rocks, easily identifiable as modern by their fossil contents.[22]

Nicol noted, rightly, that his sequence on Skye was substantially different from that which his investigations had revealed further north on the mainland. On Skye, along the chosen line of section, the dip appeared to be in general to the northwest, rather than the southeast. The limestone was apparently missing (though Nicol might have found it if he had looked a little further west), and some of the rocks did not seem to have quite the same appearance as those further north. Inspection of a modern map reveals all too easily the problem that Nicol had run into. He was, we now think, walking over a region of severe and multiple thrusting, with the red-brown sandstone series inverted. Thus what he thought was quartzite or quartz rock was probably no more than a variant portion of the red-brown sandstone.

What did Nicol make of all this? The Durness fossils, which as we have seen were at first very tentatively suggested to be Devonian in age, had by this time been investigated with more care by Murchison and the Survey palaeontologist, John Salter (1820–69), from which it appeared that they might be Lower Silurian, with forms such as *Orthoceras* and *Euomphalus*.[23] But the identifications were uncertain, and Nicol pointed out that the fossils could be Silurian, Devonian, or even Carboniferous. This gave him ample scope for theorizing in his discussion of the rocks of the Northwest Highlands.

22. These were oysters and ammonites of the Lias (Jurassic), which were unproblematical and which need not concern us further here.
23. On Salter, see Secord (1985). The genus *Orthoceras* has chambered shells, either coiled or straight, and is related to the modern *Nautilus*. It is a typical Palaeozoic fossil, but it has a very wide range and is not ideal for diagnostic purposes. *Euomphalus* is a smooth-shelled, flat-coiled gasteropod, also found widely in the Palaeozoic and again not very suitable for defining a stratigraphical horizon.

The hypothesis that Nicol put forward, then, was based on the old assumption that the red-brown sandstone was in fact Old Red (Devonian), as Hugh Miller had earlier supposed. If this was correct, then the unconformably overlying rocks of Durness and Eriboll (the quartz rock, the fucoid beds, and the limestone) could conceivably be Carboniferous—which would be feasible if the fossils were in fact those of the so-called Mountain Limestone, an important member of the Carboniferous sequence in the northern parts of England, and through very large areas of Ireland.[24] The fossil evidence at that stage was certainly not sufficiently strong to invalidate this hypothesis. And Nicol's suggestion, were it to prove true, would have the advantage of linking the Scottish rocks to two of the best-known formations in the British stratigraphical column. It would, however, do nothing for Murchison's territorial ambitions for his beloved Silurian system. Nicol emphasized strongly that he was only putting forward his suggestion in the most provisional manner.

As to the structural relationships, Nicol was already deeply concerned about the presence of metamorphic gneiss and schist above and to the east of the quartz rock/fucoid beds/limestone sequence. What were such metamorphic rocks doing lying *above* unmetamorphosed rocks, especially if these were only relatively young, say, Carboniferous? As we have seen, he thought he had detected igneous or altered rocks (syenitic porphyries or serpentines) at several places at the line of junction of the metamorphic rocks with the underlying sediments, and there was a good deal of field evidence to suggest faulting. How could all these considerations be united in a single theory?

Nicol did not attempt to settle the matter once and for all. He did not disallow the idea of the eastern rocks being metamorphosed *in situ*, as Murchison's reading of the stratigraphical relationships required. But it is evident from Nicol's paper that he leaned more toward an alternative hypothesis. This involved the following supposed sequence of events. The western gneiss was formed. Then, following the erosion of its surface into hummocks and hollows, the western red-brown sandstone was deposited on its surface, often with fragments of the older rock incorporated in the younger. This unit did not extend very far to the east. It came to be overlain unconformably by the quartz rock/fucoid beds/limestone sequence and possibly by an upper quartz rock (as favored by Murchison and Miller). Then came the eastern gneiss and overlying (Moine) schist. These might be sediments metamorphosed *in situ*, or they might represent the western metamorphic rocks brought to the surface again by large-scale faulting and "forced over the quartzite" (Nicol 1857, 38). These four words were, I suggest, the first hint in the literature of thrust faulting in northwest Scotland. (But one must not make too much of this. The general notion of thrust faulting had not been firmly established in the literature at that time, and Nicol never really developed the notion of lateral fault movement.) The serpentinous-feldspathic and porphyritic rock, which Nicol thought he had discerned in a number of places along the possible line of

24. The modern name for the Mountain Limestone is Carboniferous Limestone.

faulting between the eastern metamorphics and the underlying sedimentary series, was in his view a point in favor of the idea of large-scale earth movements or "convulsions."

It is a nice question as to exactly what it was that set Nicol's thinking in a different direction from that of Murchison. To subsequent writers on the topic such as J. W. Judd, Nicol was, as it were, the David taking on the great Goliath Murchison.[25] But as we shall see, Judd had some axes of his own to grind in relation to the Highlands controversy, and his interpretation should not be taken entirely at its face value. From Nicol's own later arguments, it would appear that he simply could make no sense of the notion of gneiss lying conformably on top of largely unmetamorphosed sediments.[26] He could only conclude, therefore, that the rock must have been emplaced by some kind of faulting process. Or perhaps Nicol, by his geological reading or by his Continental education, was somehow enabled to "see" the Scottish rocks in a different theoretical light than did Murchison, whose theoretical views were really quite unsophisticated—being dependent largely on the principle of superposition, the Smithian principle of identifying strata by their fossil contents, and the Huttonian notion of unconformity.

There may be a small clue as to what led Nicol to diverge from Murchison in his interpretations in a footnote to his 1857 paper, where he referred to the ideas of the French theorist of mountain formation, Élie de Beaumont (1798–1874). Nicol noted that the important structural line, running roughly southwest from Durness toward Skye, was approximately the same as one depicted by Élie de Beaumont in his *Notice sur les systèmes de montagnes* (1852).[27] But this scrap of evidence does not help us very much. Moreover, when we look at Nicol's philosophical position, as expressed in his inaugural lecture at the University of Aberdeen (Nicol 1853b), there does not seem to be anything about his opinions that marked him off from his contemporaries. The address was a typical piece of natural theology, with claims about the ability of science to effect moral improvement, such as Murchison or Geikie would have been happy to applaud.

I am inclined to think, therefore, that all that happened in the opening stages of the Highlands controversy was that Nicol reasoned about the rocks on a different basis from Murchison, giving less emphasis to palaeontological evidence. As we have seen, Nicol supposed that the western red-brown sandstone was Devonian in age, as was claimed by Hugh Miller and was envisaged by Murchison in 1855. Then, accepting and confirming by his own observations James's discovery of the unconformity between the quartz rock and underlying red-brown sandstone, and using the analogy with known Carbon-

25. See page 261.
26. See page 135.
27. The pattern of mountain ranges in Europe, as construed by Élie de Beaumont according to his attempts to arrange mountains into regular geometrical patterns, may be seen in his large maps at the end of volume 3. On Élie de Beaumont's theories, see Greene (1982) and Gohau (1983).

iferous sequences in the south of Scotland and northern England, Nicol found it plausible that the Durness Limestone was Carboniferous in age, since it appeared to be similar in lithology to the better-known Mountain (Carboniferous) Limestone. The fossil evidence for the Silurian age of the Durness Limestone was, he thought, insufficient.

In fact, Nicol supposed that the lithological parallels were stronger than those just indicated. In the Northwest Highlands, the following sequence had been established:

Limestone
Shale rocks, with carbonaceous plantlike impressions (fucoid beds)
Quartz rock
  (*unconformity*)
Red-brown sandstone

Nicol asserted that a similar sequence could be discerned in the coal-bearing regions of Fife and the Lothians in the Midland Valley of Scotland, and so he urged a Carboniferous age for the Durness Limestone: "It seems improbable that in such a limited region as Scotland there should have occurred in the palaeozoic age two such complex series of deposits, so nearly identical in mineral character and order, and yet that they should not be contemporaneous" (Nicol 1857, 35). He likewise sought to press this point of view on Murchison in their private correspondence.[28]

Having reached this view as to the status of the Durness Limestone, Nicol was induced to reject the upper gneiss and schists as Silurian. Then he began to look more closely at the line of contact between the overlying eastern gneiss and the underlying sediments. In several localities, he encountered a curious rock of quasi-igneous appearance. Thus he was led to the notion of the eastern gneisses and schists having been emplaced by some kind of process involving faulting, with the igneous rock supposedly having welled up along the line of fault. There does not seem to be any evidence that Nicol simply objected to Murchison extending yet further his burgeoning Silurian kingdom.

But Murchison was exceedingly unimpressed by Nicol's mode of reasoning. Indeed, according to Murchison's standpoint, it was a reactionary way of doing geology, reverting to the use of lithological analogies and criteria, even when there was some prospect of success in getting information from the fossils in the Durness Limestone. Murchison wanted to argue from this fossiliferous limestone to an age for the western red-brown sandstones. Nicol, by contrast, thought that the older view of the western red-brown sandstones as being Devonian could be retained if the Durness Limestone was located higher in the

28. Geological Society Archives, Murchison Papers, N8, Nicol to Murchison, Dec. 31, 1857. It is, unfortunately, not exactly clear what lithological correlations Nicol had in mind. In his published paper, he simply stated that the "close analogy of the [Durness Limestone] and associated sandstones, has forced itself on every observer, and in describing it comparisons with the coalformation in the central districts of Scotland constantly occur" (1857, p. 34).

stratigraphical column—a move that seemed to be warranted on the basis of lithological analogies and that was not, at that stage of their examination, contradicted by the fossils.

It is noteworthy, however, that the *sandstones* formed the main subject in the title of Nicol's Geological Society paper of 1857. We may also note that in a letter from Nicol to Murchison, dated a few days earlier than that cited above, Nicol agreed with Murchison that the (eastern) "crystalline strata" could be altered Silurian "no matter how we may dispute about the Red Sandstones."[29] Thus I am inclined to think that it was over the question of the interpretation of the western red-brown sandstones that Nicol and Murchison first fell into disagreement. But from the point of view of basic geological methodology, it was the old question for Murchison, which he had fought successfully with both Henry De la Beche and Adam Sedgwick: should rocks or fossils serve as the basis of stratigraphical analysis?

At this stage of the proceedings, matters might not have got any further out of hand. But not surprisingly, matters did not stop there. As we have seen, Nicol only identified the limestones of the northwest as Carboniferous in a very provisional manner, while Murchison was regarding them as Lower Silurian. Clearly, there was an urgent need for better palaeontological data. Peach returned to his collecting.[30] Attempts were also made to effect comparisons with Lower Silurian fossils collected in North America, some specimens of which had been brought to England in 1852 by the director of the Canadian Survey, Sir William Logan (1798–1858), for the benefit of British geologists.[31] On the basis of comparison with Canadian specimens, an announcement was made at the British Association meeting in Dublin in 1857 (Murchison 1858a), and at the American Association for the Advancement of Science meeting in Montreal (Murchison 1858b), that the Scottish fossils were indeed Lower Silurian in age. Murchison (1857–58) also dashed off a short letter, and a sample section of the Scottish Highlands according with his views, to Élie de Beaumont in Paris, to make sure that the French geologists were kept abreast of his thinking. A more detailed paper, with figures of the best

29. Nicol to Murchison, Dec. 26, 1857.
30. It appears from Nicol's correspondence with Murchison that Peach prepared some theoretical sections of the Eriboll district some time in 1857 and forwarded them to Murchison, who in turn sent them to Nicol for his information. Nicol had difficulty in understanding the sections. He wrote (Dec. 31, 1857): "If the facts be as [Peach] represents them I fear we are all in a mistake, and the whole fact of the overlying . . . of gneiss &c to the limestone, which we have been talking about these thirty years, will evaporate in a crush up of the crystalline strata, along a fault, over the quartzite. In his section No 1 on sheet 2 the gneiss on both sides of Loch Eriboll is all underlying and troughs the limestone and quartzites, which dip from it on both sides. In truth I do not see what can be made of them. They are quite different from any I have seen, but Mr. Peach had plenty of time and seems to have gone carefully to work." Unfortunately, I have not been able to locate Peach's sections, but the foregoing suggests that they may have hinted at the idea of a "crush up" of rocks at Eriboll to Nicol.
31. Logan read a paper to the Geological Society in March of that year, which referred to some of the relevant fossils. (Logan 1852). He was a fairly regular visitor to Britain since he had family connections in Edinburgh. The Canadian fossils were also described by Salter.

specimens from the Durness Limestone, was published in 1859 (Murchison 1859b; Salter 1859). At the Dublin meeting, Murchison argued that if the Durness Limestone and associated sediments were of proven Lower Silurian age, then the unconformably underlying, unfossiliferous red-brown sandstones might reasonably be allotted to the Cambrian of Adam Sedgwick. The overlying gneiss and schists (the eastern gneiss and the "Moines," that is) might also with reasonable safety be located in the Silurian, though there was, of course, no firm palaeontological warrant for this. Then, on the eastern coasts of Caithness and elsewhere, one found the firmly identified Old Red Sandstone of Devonian age. The oldest rocks of the whole sequence would, of course, be the western gneiss, found on the west coast and making up most of the Outer Hebrides, notably the Isle of Lewis.

All this must have seemed very satisfactory to Murchison. It would locate vast areas of northern Scotland in the Silurian. It might, perhaps, be thought surprising that he would accord any new territory to his foe, Adam Sedgwick, with whom he had been locked in bitter debate for several years about the correct placement of the Cambrian-Silurian border in Wales. But really it was not such a very generous gift. The red-brown sandstones were apparently quite destitute of fossils. So the rocks' Cambrian status was not fully authenticated, and might be rescinded if new palaeontological evidence were brought to light.[32] In any case, in writing to Logan, Murchison (1858b, 59) merely suggested that the red-brown sandstones might "pass for" Cambrian. It was a grudging concession.

It would not appear that Nicol thus far had stated his case in terms that might rile Murchison (though Murchison would have disapproved of Nicol's overreliance on lithological analogies), and their correspondence appears to have been quite conciliatory, to judge from Nicol's side of it. But Murchison was a man who could not tolerate any sort of opposition—and certainly none of the kind that might suggest trouble in his Silurian kingdom. It comes as no surprise, then, that we find him setting off to the Highlands once again in 1858. And it may also be no surprise that on this occasion he determined not to go with Nicol, but preferred Charles Peach as his traveling companion.

The two geologists started at Thurso, where they met the redoubtable baker, Robert Dick, and did some hammering among the fossil fishes of the Old Red Sandstone.[33] Then, arrangements no doubt facilitated by Peach's connections with the coastguard service, they proceeded by the steamer *Pharos* to the Orkneys, where the extension of the Old Red Sandstone beyond the

32. In fact, at that stage Murchison was wanting to claim that the Silurian had the very first forms of life, so it was quite acceptable to push the unfossiliferous red-brown sandstones down into the "Cambrian."

33. An attractive account of the meeting between Murchison, Dick, and Peach has been given by Smiles (1878, 272–76).

confines of the mainland was confirmed and several lighthouses were visited. After visiting the Shetland Islands also, the steamer turned south, and Murchison and Peach were dropped off near Cape Wrath, where Murchison would have been able to have a thorough look at the western gneiss, which, infiltrated by innumerable granitelike veins, makes up the massive cliffs of that awe-inspiring corner of Britain. In fact, following this visit, he came to recognize the gneiss as the very lowest basement rock of the whole stratigraphical succession in Britain. Hence he termed it the "Fundamental Gneiss." It seemed to be the British equivalent of the so-called Laurentian system of the Canadian Geological Survey.

I am uncertain as to the exact route that Murchison took during his 1858 expedition.[34] From Cape Wrath he would undoubtedly have gone to Durness. In his published memoir, he reports going to Suilven (the red-brown sandstone mountain to the southeast of Assynt), and it is evident that he was looking with care for the unconformity between the quartz rock and the supposed Cambrian sandstone, reported to him in Colonel James's letter of 1856, and also seen by Nicol that year and subsequently published by him. Peach apparently discovered the so-called Canisp porphyry in the depression between the red-brown sandstone mountains, Canisp and Suilven.[35]

It is further evident that the travelers visited the Assynt region, and Murchison (1859b, 367) reported Peach as tracing the exposures of the fucoid beds all the way from Ledbeg, south of Assynt, northward as far as Kylescu at Loch Glencoul. He noted, as others had done before him, evidence for unconformity between the quartz rock and the "Cambrian" sandstone on the flanks of Quinag, on the road from Kylescu to Assynt. An exposure of extreme importance was now described in detail (for the first time, to the best of my knowledge, though it had almost certainly been hammered previously) at a place called Knockanrock, some distance to the southwest of Assynt.[36] Here one may find a modest scarp where, conveniently near the road, there occur in ascending succession quartz rock, fucoid beds, grit, limestone, and overlying schist.[37] Murchison probably did not examine that precise spot, for he referred only to the contact between the limestone and the schist, noting that the uppermost band of the limestone was of a "peculiar aspect," "intermingled with schist and shale—some courses resembling volcanic ash or grit" (ibid., 369). This was an interesting remark when we consider that he was recording

34. The Geological Society archives appear not to have the journal for this year.
35. This igneous rock is marked as forming a number of sill-like structures on the eastern face of Canisp. Murchison thought it might have something to do with the porphyritic rock reported by Nicol at Loch Broom (Ullapool) and elsewhere, but in fact the two are quite distinct.
36. This place is today regarded as being of such geological importance that a nature trail has been established there for the convenience of tourists who wish to view the Moine Thrust and associated geological phenomena in comfort. See Nature Conservancy Council (n.d.[a], n.d.[b], 1986).
37. The grit rock, another useful marker band, is today referred to as the Salterella (or Serpulite) grit, by reason of its fossil contents.

an observation of what is today called "mylonite"—a rock ground, as if in a mill, by the action of the thrusting processes in orogenic movements.[38]

There is plenty of evidence in Murchison's paper that he and Peach made a close examination of the Durness-Eriboll area during their 1858 excursion, and it was in that region that Murchison's observations seemed to be particularly at variance with those of Nicol. However, Murchison confirmed that at Durness the "Cambrian" (red-brown) sandstones were absent, so that the quartz rock lay directly on the western gneiss.[39] A little further to the east— that is, in the vicinity of Durness village—Murchison described an area of great complexity, seemingly without being aware of all the difficulties that lay in the way of structural interpretation.

Just to the north of the Durness settlement there is a peninsula called Faraid (or Far-out) Head, which has mylonite below gneiss below schist. It is today regarded as an outlying fragment of the great mass of schistose metamorphic strata that have been thrust over the gneiss/red-brown sandstone/ quartz rock/Durness Limestone sequence, and it lies in faulted contact with the main mass of Durness Limestone lying immediately to the south (see figure 7.64). Murchison stated that the contact was obscured by wind-blown sand; nevertheless, he supposed that there was an uninterrupted sequence between the Durness Limestone, the upper quartz rock, and the overlying metamorphic rocks, which he termed "quartzose and micaceous flagstones." In fact, he treated the gneiss and the schist as essentially variants of the same unit. Thus what is today regarded as a rather complex structural relationship, which (rightly or wrongly) is construed in terms of faults of different ages and types, Murchison described and figured as an essentially simple conformable sequence (ibid., 371).

Further problems in Murchison's interpretation are suggested by the following passage:

> Previous to my last visit, I had supposed (and Professor Nicol had published a section to show it [Nicol 1857, 23, fig. 4]) that the Durness limestone was abruptly cut off by the old gneiss upon the east, between Durness and Loch Eriboll. But such is not the case; for Mr. Peach had observed, and I confirmed his observation, that between the interior ridge of old gneiss which extends from the western foot of Ben Keannabin[40] and the limestone of Durness, the dip is reversed, and the underlying quartz-rock is again brought up, lying between the gneiss and the limestone . . . (ibid., 372)

Now here both Nicol and Murchison were in fact right. There is indeed a long faulted contact to the east of Durness, where gneiss is brought up in such a way that it lies in contact with the Durness Limestone. But in addition

38. The term "mylonite" was introduced by Charles Lapworth (see chap. 8).
39. In further discussions, I shall now call the red-brown rock the Cambrian sandstone. At the due time I shall refer to it by its modern name of Torridon Sandstone (pre-Cambrian).
40. Ben Ceannabeinne on modern maps—the main ridge lying between Durness and Eriboll.

Murchison was right in his reference to an exposure of quartz rock between the gneiss and the limestone, for there is a pocket of this rock faulted in at the coast, at a little hamlet called Sangobeg. We see, then, emerging one of the chief problems that was to dog the Highlands controversy continuously. Geologists would go to an area of great structural complexity, make limited reconnaissance inspections, make mental sections of what they considered to be the best illustrative traverses, and describe the area back at the Geological Society or some other forum, publishing their descriptions and exhibiting and publishing their sections. What they did not do, however, was map the ground thoroughly. Or such maps as they may have prepared and exhibited at meetings of the Geological Society or the British Association were not published. So the overall complexity of the area did not always become apparent. Often the different investigators walked over slightly different areas, which could reasonably lead to differing structural interpretations. Often they did not seem to realize (or acknowledge) that they were examining different exposures, and so one frequently finds assertions that the other observer or observers were simply mistaken—as Murchison accused Nicol (and himself!) in the foregoing quotation. The problem was compounded by the fact that the published sections usually showed no scale. Often they lacked even some compass indications. And most of the descriptions, though quite detailed, would have been largely unintelligible without first-hand experience of the terrain under discussion.

The geologists were obviously not unaware of such problems. In well-known regions such as the south of England, easy of access and already well mapped, the technique of publication adopted did not lead to particular difficulties. But in remote regions, it was virtually impossible to verify empirical claims immediately. For less remote regions, such as the north of Scotland, the real difficulties soon became apparent. Regions of controversy emerged, and geologists crowded in to examine the disputed sites. But until detailed maps were published, it remained difficult to know whether one was in fact looking at what a previous observer had described. Naturally, the geologists did the best they could to ensure a flow of information, by private conversation, exchange of correspondence, offprints, and unpublished notes, maps and sections. Nevertheless, the problem remained and was, I believe, an important factor in the Highlands controversy.

Such problems were eventually to be dissipated by the careful mapping of the whole of the British Isles by the official Geological Survey. But this could only be accomplished slowly, and as the preparation of good topographical maps allowed. Meanwhile, there was ample room for fieldwork in areas not adequately mapped, by amateurs such as Charles Peach, by university geologists such as Nicol, or by members of the Survey such as Murchison still behaving in some measure as amateurs with rapid field reconnaissance work. And all the time the professional surveyors were methodically but slowly preparing their detailed maps. The tension between these groups is one of the most interesting features of the Highlands controversy. One thing that particularly annoyed the amateurs was that the Survey felt that it had a right to maintain firm views on the structural geology of regions that it had not yet properly mapped. This was indeed a messy situation, but it is hard to see how things might have been otherwise—other than that different geologists with different individual interests or personalities might have been involved.

To return to Murchison and Peach at Durness, Murchison acknowledged that the evidences at Sangobeg did suggest "a great upheaval of the old gneiss," and he was right to suspect some inversion of the strata in the district. He was also right to recognize that there was a general trough structure to the limestones of Durness.

Moving from Durness eastward to Eriboll, Murchison described a section across Loch Eriboll from west to east. To the west there was the gneiss hill of Ben Spionnach flanked by quartz rock dipping east-southeast on its eastern flank and extending down to the western shore of the loch.[41] On an island in the middle of the loch and on its eastern shore, there were good exposures of Durness Limestone, with an apparently similar dip. Murchison explored northward from the (now defunct) ferry crossing, which landed a little south of the hamlet of Heilam, to the bleak Whiten Head on the north coast of Sutherland. Here he was passing over some of the most complex geological structures in the British Isles, but he confidently informed his readers that he was able to confirm observations made back in 1827, during his excursion with Sedgwick. It seemed to Murchison that he could see quartz rock both above and below the limestone, and in this he was perfectly correct. So, he argued, there were in fact *two* quartz rocks: an upper and a lower, one above and one below the Durness Limestone.[42]

Here, Murchison's interpretations were in real difficulty, though he was apparently unaware of it. Most of his "lower" quartz rock was exposed on the western side of the loch, on the flanks of Spionnach. There is, to be sure, ample quartz rock on the eastern shore above the limestone and suggestive of an "upper" quartz rock. But whether in fact it is younger than the limestone on the island in the middle of the loch and on the eastern side is quite another matter. Murchison had no means of telling whether the rocks he was observing were lying the right way up or inverted, and he seems not to have explored the area with sufficient care to have observed the fine evidence for folding and overturning of the quartz rock that may be seen, for example, on Ben Arnaboll, a little north of Heilam. I also feel fairly sure that he did not on this occasion visit the southern head of the loch, but crossed by the ferry. For had he gone to the loch head, he would surely have seen the extraordinary hill, Creag na Faoilinn (rock of seagulls), where thrusts and overfolding may be observed even from a distance. Of course, it is unreasonable to expect Murchison to have viewed the rocks from a theoretical perspective that had

41. It is Ben Spionnaidh on modern maps.
42. This suggestion did not originate with Murchison. As previously mentioned, the existence of two distinct quartz rocks had earlier been made by Cunningham, and Nicol at first accepted the suggestion. As the Highlands controversy developed, the question of whether there were one or two quartz rocks became a central issue, to which Murchison attached the greatest significance.

not been developed at the time he was working. One may speculate, however, as to what he might have done had he observed that hillside, which is only a few miles to the south. For the most part, the structures at Heilam and Arnaboll only become apparent when subjected to a thorough survey and mapping, whereas at Creag na Faoilinn even the superficial appearance of the rocks suggests that their structures have been radically distorted by earth movements. However, we may be wiser to stick to our historical narrative than to contemplate what Murchison might have done had he traveled to the head of the loch. For our purposes, the point to be emphasized is that the question of whether there was or was not an upper quartz rock was to become almost an *experimentum crucis* in the Highlands controversy.

To the east of Eriboll, on the road to Tongue, Murchison found, as on his previous visit, a series of overlying "micaceous flagstones, schists and younger gneiss," dipping southeast and characterized by a high mica content. These he regarded as stratigraphically superposed to the quartz rock(s) and limestone. He referred to them as "the younger micaceous flagstones of Inverhope and the Moin [*sic*]," that is, the Moine Schists.

Murchison was quite right in his understanding of the relative geometrical relationships of the Moine rocks and the underlying Durness sediments. But he added to the problems in understanding the structures by failing to accept the arguments of Daniel Sharpe in the matter of the foliation and cleavage of metamorphic rocks. For Murchison the laminations in the schist were taken to represent "true stratification," the rocks themselves being regarded as "truly stratified aqueous deposits" (1859b, 391). Thus it seemed to him that the Moines represented a suite of rocks lying approximately conformably on the underlying sediments and metamorphosed *in situ*, rather than being, as modern opinion would have it, a series emplaced by large-scale thrust faulting. It must be acknowledged, however, that in some places in the middle of Scotland, where the metamorphism has not been very intense, it is a nice question as to whether one is looking at planes of original stratification or schistosity due to metamorphic pressure.[43] Nonetheless, Murchison was injudicious in supposing that the laminations of the Moine Schists represented their original stratification or bedding.

In his paper, Murchison referred to a claim by Nicol (who again visited the Durness-Eriboll region in the spring of 1858) that a porphyry was to be found at the base of the Moine Schist in the region between Loch Eriboll and Loch Hope.[44] This, Murchison reported Nicol as arguing, was an equivalent of that seemingly igneous rock that had been observed at Loch Maree. Modern maps show no such rock near Loch Hope, and Murchison asserted that he could find no such thing. I suggest that what in fact Nicol had been looking at was gneiss, two bands of which are mapped today to the east of Eriboll. But without the examination of the rocks in thin section with the assistance of a polarizing microscope, it is not always easy to make this identification. Just looking

43. See fig. 5.6 below, which may suggest how such an opinion might be formed.
44. Loch Hope lies inland, running north to south, between Loch Eriboll and the Kyle of Tongue.

at hand specimens, therefore, it is not remarkable that Nicol construed the rock as some kind of igneous material, a porphyry being the best approximation available to him.

Murchison did not actually deny the existence of Nicol's igneous rock at Eriboll. He simply said that he hadn't found it. In any case, even if there were igneous material to be found there, it wouldn't really prove things one way or the other so far as concerned the structural hypothesis that Nicol was beginning to develop. Murchison freely acknowledged that porphyries might be found in northwest Scotland. As we have seen, he and Peach had recorded one at Canisp, in the Assynt area. What he wanted to deny was that the existence of such a rock was in any way a proof of Nicol's emerging structural ideas. At the most, if the Durness sediments dipped approximately to the southeast and were overlain by gneiss and schist dipping similarly, the presence of an intercalated igneous band with approximately similar alignment could mark, perhaps, a sill-like intrusion—it would not prove a line of *upheaval* of the eastern metamorphics.[45] Thus Murchison felt fully justified in rejecting Nicol's suggestions and preferring his own fairly simple structural picture. It must have seemed very pleasing to him to be able to bring some kind of order to the apparent chaos of rocks in Sutherland, particularly since it enlarged the Silurian kingdom at the same time.

Murchison and Peach gave a considerable amount of time in their 1858 field trip to the problems of the Old Red Sandstone and some of the more southerly parts of the Highlands, but the details of this work need not concern us here. To conclude my discussion of the work of 1858, I need only mention that the travelers also passed Loch Stack and Loch More, which lie across the interesting line between Loch Assynt and Loch Eriboll. Loch More cuts right across the stratigraphical sequence, which appears deceptively simple at that point: western (Fundamental) gneiss, quartz rock, limestone, quartz rock, eastern gneiss, Moine Schist.[46] Thus Murchison could suppose that an understanding of the main structural relationships of the Northwest Highlands was in his grasp. There might be difficulties at Durness and Eriboll, but these were to be regarded as exceptions to an essentially simple and comprehensible structure.

In his report to the British Association held in Leeds in 1858, Murchison (1859c) gave the greater part of his attention to the Old Red Sandstone deposits of the east coast, rather than to the northwest rocks discussed above, and it may be that he avoided discussion that year of issues that were becoming contentious. It appears, however, that he did display a colored geological map of Sutherland at the meeting, which he had prepared that year. Nicol (1859) also spoke at the Leeds meeting, having made a lone excursion to the northwest that summer. Unfortunately, we only have a very brief account of his work in the Leeds *Report*, but it offers some idea of what he accomplished and the nature of his thinking. The *Report* states:

45. A sill is an igneous intrusion that has been emplaced between the layers of the sedimentary strata into which it has been intruded, rather than cutting across the layers.
46. There is no "Cambrian" sandstone in the section at Loch More. Quartz rock lies directly and uncomformably on the western gneiss.

[Nicol] described a section from Gairloch to the Moray Firth, and showed that the red sandstone and quartzite resting on the gneiss of the west, were cut off by a mass of felspsar-porphyry and serpentine from the supposed overlying rocks on the east. He had traced a band of similar igneous rocks at intervals of a hundred miles, from Loch Eriboll to Skye, and had observed other indications of fracture and convulsion along this line, as shown in the section exhibited. He therefore concluded that the overlap of gneiss on quartzite might be occasioned by a slip or convulsion of the strata, and not mark the true order of superposition. He also stated, that the unconformability of the quartzite to the red sandstone (Cambrian) which he described at Assynt and Loch Broom, did not seem to occur at Gairloch, where the two deposits, as seen in great sections on the escarpments of the mountains, appear quite conformable; the red sandstone, however, extending far beyond the quartzite on the west.

This synopsis gives us some indications of Nicol's itinerary. I suggest he traveled on foot or horseback from Eriboll down the long line of exposure of quartz rock, fucoid beds, Durness Limestone, and so on, through much wild and lonely country—some of it with, some without, pathways—to Skye.[47] Retracing his steps somewhat, then, he would have made a traverse of the Highlands from west to east from Loch Maree (Gairloch) to Inverness (Moray Firth). This was to become a fairly standard itinerary for the many geologists that were to interest themselves in the Highlands controversy as it gradually unfolded.

It will be noted from the British Association report of the Leeds meeting that the point initially at issue—the age of the western red-brown sandstones—was beginning to recede, to be replaced by the more fundamental question of whether there was a conformable sequence from the Durness Limestone to the overlying eastern gneiss and schist or whether there was some kind of faulted contact. Nicol was now of the opinion that he was dealing with a major region of faulting or "slip or convulsion." As we shall see, Murchison had good empirical grounds for arguing against Nicol's views and for preferring a structurally simple solution to the problem. But at this stage the battle had not been fully joined, and the full range of arguments that were to be deployed by Murchison and Nicol had not yet been developed. Further, the historical record appears not to be complete for these early rounds of the battle. We shall therefore have to leave matters here for the moment, to see in later chapters how things developed. It may be noted, however, that the amateur geologist, John Miller of Thurso, friend of Robert Dick and Charles Peach, visited the north-coast sites in August 1858; and though he commended Nicol's observations, he put his weight behind the theoretical explanation offered by Murchison, ascribing to him "astonishing sagacity in predicting from premises so slender such a clear and comprehensive, and . . . such correct views on

47. We shall be examining this terrain in more detail in chapter 5 when we recount the travels in the same district of Geikie and Murchison.

the geology of so wide a range of country" (J. Miller 1859, 579–80).[48] Thus, insofar as there was a "public opinion" on the controverted questions concerning the rocks of the Northwest Highlands, it might be said to be favoring Murchison.

There was one further contribution from Nicol in 1858 that I should like to consider in this chapter. In that year he published an important geological map of Scotland, part of which is reproduced in plate 3. In his correspondence to Murchison, Nicol explained that he planned to base the coloring of his map on lithogical principles, as Macculloch had done before him, since correlations of the large areas of unfossiliferous rocks in northern Scotland with fossil-bearing ancient rocks to the south could not be carried out satisfactorily.[49] To assist interpretation of the map, I list below extracts from its key, naming the rocks in the portion of the map reproduced in the plate.

		In all formations
Limestone	Blue	
Lias and oolite	Green	*l*
Devonian or Old Red [Sandstone]	Dull green	*h*
Quartzite	Yellow	*g*
Red sandstone of west coast	Dull orange	*f*
Clay slate, primary of north	Blue-green	*d*
Chlorite slate	Grey-green	*c*
Mica slate	Grey-blue	*b*
Granite	Red	*x*
Syenite	Red	*y*
Porphyry	Dark red	*p*
Serpentine	Dark red	*s*
Trap	Dark red	*t*

Looking at Nicol's map, we may readily see the band of quartz rock and limestone running southwest from Eriboll, as Macculloch and Cunningham had previously depicted them. To the west of this line will be seen the large areas occupied by the red-brown sandstone (e.g., to the immediate south of Cape Wrath, at Quinag, the mouth of Loch Broom, at Loch Torridon). To the

48. Miller's paper was read, by *Murchison*, on Dec. 15, 1858, Miller being out of town at the time. According to H. B. Woodward, Miller's supportive weight, if not decisive in the debate, was very considerable in other respects: "John Miller was in form perhaps the stoutest Fellow ever elected into the [Geological] Society; his medial circumference was so great that he could only sit on one end of any one of the benches in the meeting room, and when on his departure he attempted to occupy a four-wheeled cab, and finally obtained entry with a rush, it became an unsolved problem how he ever made exit again" (1908, 194). Accordingly, he was called Huge Miller (as opposed to Hugh Miller). John Miller was apparently a man of independent means, and he lived in London for most of the time, rather than his native Caithness. He purchased some of Robert Dick's fossils when the Baker of Thurso was in financial difficulties as a result of loss of flour in a shipwreck. See Smiles (1878, 335).
49. Geological Society Archives, Murchison Papers, N8, Dec. 26, 1857.

east, there are large areas marked as Old Red Sandstone (e.g., round Dornoch and Moray Firths). But what is specially interesting for our present purposes is that Nicol made no distinction between the western and eastern gneiss—for example between the gneiss of Cape Wrath and that of Whiten Head. Thus he was giving cartographic expression to the idea that he was beginning to develop, namely, that the eastern gneiss was merely the western gneiss brought to the surface by faulting. We may also note that Nicol mapped all the eastern gneiss and the Moine Schists as one.

As we have seen, Nicol believed that he could identify some kind of igneous rock (a porphyry, syenite, or whatever) at the base of the eastern gneiss, which marked a line of faulting, where molten material had supposedly welled up as some great convulsion occurred. Localities where Nicol thought he had discovered such rock were on Loch Broom, just to the east of Ullapool; at Loch Borolan, at the southern end of the Assynt limestones; and at the head of Loch Eriboll. (He had also noted the vast numbers of acidic and basic bands that are to be found in the western gneiss, to the north of Loch Inver, for example. But these need not concern us here.)

It will be seen that Nicol had enough information to color in the whole of the north of Scotland. But much of his information must have been derived from earlier investigators such as Macculloch and Cunningham. There is clear evidence that Nicol had *not* walked over some of the terrain that he mapped. For example, it will be seen that a large area to the east of the Assynt limestone (to the west of Ben More Assynt) is marked as quartzite. Yet much of that area is in fact gneiss. Comparison with the map of Macculloch (plate 1) suggests that Nicol relied on him to a considerable extent in the Assynt region at least. It should be remembered, however, that the region to the east of Assynt is one of the most remote and inhospitable in the British Isles, with weather patterns that are exceptionally bad even for northwest Scotland. It is not surprising that this area was not mapped accurately by the reconnaissance surveyors.

It is instructive to compare Nicol's map (1858) with that published three years later by Geikie and Murchison in 1861 (see plate 5). Also, we should note the title of Nicol's map, and its dedication. It was entitled *Geological Map of Scotland from the Most Recent Authorities & Personal Observations,* and it was dedicated to Murchison with the words: "In grateful acknowledgement of his unfailing friendship and in remembrance of wanderings together in the south, and in the north, of our native land." This was poignant, considering the bitter things that Murchison was to write and say privately about Nicol in the next few years. But these matters will be better addressed in a later chapter. We shall now turn to the events of 1859, when things really came to a head between Murchison and Nicol at the Aberdeen meeting of the British Association.

4

The Fieldwork of 1859 and the Aberdeen Meeting of the British Association

Murchison was not a man to accept any kind of opposition, especially in anything pertaining to his beloved Silurian kingdom. So although it could hardly be said that Nicol was maintaining a high-profile opposition to Murchison's interpretation of the Northwest Highlands, there were clearly some rumblings of opposition within the Silurian domains. We find, then, that Murchison traveled north once again in 1859 to try to provide a further vindication of his views. It was, of course, easy enough for a nineteenth-century geologist to assert that a particular sequence of rocks was this way or that way; to provide sections for a Geological Society audience and speak to them with great eloquence; to publish the most minute descriptions suitably illustrated; or to gain credence for his views by mustering support within the geological community by all sorts of social pressures, patronage, and reviews. But unless geologists could actually visit the places under consideration and form their own judgments, the observations would be at best vicarious, the witnessing "virtual," and the assent of the geological community hesitant or incomplete if the interpretations offered were contested.[1] Fossils, of course, might be brought home to testify to the age of the rocks whence they came. But structural relationships could not be demonstrated in a similar tangible fashion. A good map was perhaps the most persuasive document, but Murchison rarely chose to study an area in such detail as to be able to produce a map that would convince. He was one for rapid sweeps of terrain, with a quick eye for structural relationships. He liked to make traverses and construe a region and its structure with the help of sections. What, then, might he do in order to compel assent for his views when these were queried?

The best thing, surely, was to engage the services of the most reputable geologist possible to accompany him to the northwest and agree to the observations and interpretations that he was advancing. This is precisely what Murchison did. The Survey's second-in-command, Andrew Crosbie Ramsay (1814–91), was invited to accompany Murchison to the Northwest Highlands in the summer of 1859, and to attend the meeting of the British Association at Aberdeen

---

1. The term "virtual" has been suggested by Shapin and Schaffer (1985, 60) in relation to early work at the Royal Society.

in September, to make a public statement of their joint findings in Nicol's home territory.[2]

Ramsay was born in Glasgow, son of a manufacturing chemist. He started work as a clerk in a cotton-broker's office but soon acquired a keen amateur interest in geology. He made a careful examination of one of the classic areas for Scottish geology, the Isle of Arran near Glasgow. He mapped and modeled the island, presenting his findings to an appreciative audience at a meeting of the British Association in Glasgow in 1840. His work on Arran was published in book form the following year (Ramsay 1841). Murchison was greatly impressed and invited Ramsay to accompany him on his forthcoming geological explorations in Russia, but he also recommended Ramsay to Sir Henry De la Beche for a position at the Survey, and it was this latter position that Ramsay decided to accept.[3] (It was just as well that he had some employment in view, since he lost his clerical job in Glasgow as a result of his excessive attention to geology!) It came about, then, that Ramsay joined the Survey as an assistant geologist in 1841. He was an outstanding success, being appointed local director in 1845, professor at University College in 1848 (to 1851), professor at the Royal School of Mines (1851–76), and president of the Geological Society (1862–64). He began the work of the Survey in Scotland in 1854, starting in Dunbar. Ramsay became particularly well known for his researches on glacial phenomena (Ramsay 1860) and his interpretation of the geology of North Wales (Ramsay 1866). He also wrote one of the standard nineteenth-century texts on British geology (Ramsay 1863), which ran through a number of editions. After Murchison's retirement from the Survey in 1867, it was Ramsay who assumed the directorship, but he was not an unqualified success in this role, as the distinction came too late in his career, and he suffered grievous ill health in his later years, being forced to have an eye removed by operation in 1878. Even in 1859, Ramsay was not in good health, and he may therefore not have been able to give his best attention or support to Murchison's theories the way the old director anticipated. In his excellent biography of Ramsay, Geikie wrote of this period:

> In the summer of 1859 [Ramsay] accompanied Murchison into the North-West Highlands of Scotland, and assisted him in the preparation of his discourse for the British Association at Aberdeen. He seemed tolerably well and merry at the meeting, but afterwards, when among the hills in the south of Scotland, he complained of weariness. The symptoms of mental exhaustion increased during the autumn. (Geikie 1895, 260–61)

Ramsay thus may not have been the perfect supporting witness, but he could certainly be expected to lend credence to the Murchisonian view, and he

2. On Ramsay, see the Geological Magazine's Eminent Living Geologists series (1882), Young (1893), Geikie (1892, 1895).
3. Murchison to De la Beche, March 3, 1841, Wellcome Institute, London, Archives, file 63700.

carried sufficient weight in the geological community to enhance Murchison's position considerably in his debate with Nicol. Ramsay was a Scotsman, with a fine reputation for fieldwork. He owed the beginning of his geological career in part to Murchison; and in 1859 he would have been thinking of Murchison's future retirement and the possibility of being promoted to the position of director general. Murchison's choice of traveling companion was undoubtedly judicious.

We are favored with extensive archival material concerning the Murchison-Ramsay tour of 1859. Again, we may draw on the Murchison journal, held at the Geological Society.[4] Ramsay's field notes are available in the archives of the Imperial College, London (Pingree 1986). The two geologists met in Glasgow on July 28 and took a ship to Oban and on to Portree in Skye, where the "philosopho-misanthropist" Necker (as Murchison described him) was still residing. The boat steamed on to Ullapool on Loch Broom and Murchison recorded with evident pleasure a small exposure of gneiss on the southern headland to the loch (at Carn Dearg), in an area that Nicol had marked wholly as "Red Sandstone of West Coast" (172).

The exposures at Loch Broom and Ullapool provided some of the most important pieces of evidence that Nicol had been able to muster. One might suppose, therefore, that Murchison and Ramsay would have examined the area in some detail, both to the north and south of the loch. But on the contrary, they only spent about a day there, or as long as their ship allowed them to remain, and detailed investigations were not recorded. Murchison drew a section in his notebook roughly northwest to southeast, to the north of Ullapool, showing western gneiss under red-brown ("Cambrian") sandstone, unconformably under quartz rock, under limestone, under eastern gneiss; but he cannot have walked over the full extent of the traverse (174). He acknowledged that Nicol's sections agreed with what might be seen in the district. Nevertheless, Murchison's description and section were given so as to comport with his own theory. He made no apparent effort to look for Nicol's "igneous" rock (his "syenite" or "felspar-porphyry")—the rock at Loch Broom that had set Nicol wondering about the whole northwest sequence—and indeed the whole visit to Ullapool seems to have been rather perfunctory.

The travelers were transported by ship to Lochinver, another fishing village in gneiss terrain, about twenty miles north of Ullapool. There is a track from there running inland toward the red-brown sandstone mountains of Suilven and Canisp, and the two geologists examined this, Ramsay ascending Suilven alone in bad weather. "Greenstone" dykes in the gneiss were recorded by Murchison. After sheltering from the weather for a couple of days in the inn at Lochinver, the two geologists traveled by the mail coach to Inchnadamph to study the exceptionally difficult rocks of the Loch Assynt region.

On their first day, after Ramsay had been dropped off the barouche on the

---

4. Murchison Journal, 1857–59, Geological Society Archives, Murchison Papers, M/J 24. In this chapter, I cite the pages of the journal parenthetically in the text.

way to examine the "Cambrian" sandstone mountain of Quinag to the north of Loch Assynt, Murchison examined the more interesting, but exceedingly complex, set of rocks to the east and southeast of the loch, where limestone, quartz rock, fucoid beds, and Serpulite grit are found in an extraordinary intermingling. The grand mountain of Ben More Assynt, consisting largely but by no means exclusively of "eastern" gneiss, sits to the east of the loch, forming the main mountain in a range running approximately north-south. To the east of this again, we find the vast expanse of Moine Schists. It took later geologists many years of intense and arduous work to unravel the secrets of this region, the only saving geological grace being that the rocks, particularly the limestones, are often unusually well exposed. But Murchison and Ramsay only spent a week in the district on this occasion, and Murchison seems to have supposed that his quick and experienced eye could readily comprehend the general structure of the district.

There is a fishing inn at Inchnadamph ("water meadow of the stag, or ox"), at the head of Loch Assynt, well patronized then and now by reason of the fine trout that inhabit the waters of Gillaroo Loch (Loch Mhaolach), which sits in a corrie of limestone a few miles to the southeast of the inn. A short distance from the inn, there is a hillside of imbricated limestone, fucoid beds, Serpulite beds, and quartz rock, and also some igneous intrusive "lamprophyres."[5] This suite of rocks rises to form the hill, Cnoc an Droighinn ("hill of thorns or brambles"), which Murchison called Cnoc an Drein.[6] Some "eastern" gneiss may also be found near the summit of this hill. Beyond it, to the southeast, there is a mountain called Ben an Fhurain ("mountain of the spring") on modern maps, but which Murchison referred to in his journal as Benn y Urran. Part of the Ben More Assynt range, it consists of quartz rock, but there is some more gneiss, a bit further east, and another patch to the northwest of Ben an Fhurain. The hillside section of Cnoc an Droighinn became one of Murchison's favorite pieces of evidence; and being conveniently situated to the inn, it was later frequently visited by geologists and much discussed. An aerial view of Inchnadamph and Loch Assynt, showing the hillside of Cnoc an Droighinn, is shown in figure 4.1 (which may usefully be compared with plate 8).

While Ramsay was battling up Quinag, the older Murchison examined the lower and geologically more interesting terrain. He was rewarded in a double sense, as the following perhaps unexpected extract from his journal reveals.

> Walking up the glen, I met with a sprightly, handsome girl, Isabella
> Fraser, the daughter of the shepherd at Stronchrubie,[7] eighteen years
> of age; and she, on persuasion, offered to be my guide to the Gillaroo
> Lake. . . . She had been to the post to look for letters, as I told her,

5. The term "imbricate structure" refers to a series of beds faulted so as to lie one on top of another in a manner similar to the tiles on a roof.
6. The hillside is pretty bare today.
7. A farm on the road a few miles south of Inchnadamph. (The name probably means the "point with a tree.")

Fig. 4.1. Aerial view looking south over Inchnadamph, showing Loch Assynt (lower right); Inchnadamph Hotel (white house at center); Stronchrubie Farm (middle distance); Cnoc an Droighinn (hillside to left of hotel, northeast, toward foreground); Breabag (mountain at rear left of picture); Stronchrubie Limestones (center left); imbricate structure (lower left). Photograph DF 77, Cambridge University Collection. British Crown Copyright/RAF photograph.

from her sweetheart. She was one of the handsomest Highland lasses I have seen, and if [the artist] Edwin Landseer had been with me he would have had her in one of his foregrounds. (179)

Murchison and young Isabella were, then, walking along a path where modern geologists describe thrust planes exposed repeatedly in the limestone. This is how Murchison saw them:

I was all the time in the limestone which I came over the summit of in traversing from the Gillaroo Loch to Inchnadampff. On the whole it undulated in beds slightly inclined to the E. and E. by N., the strata being powerfully affected by transverse joints having a S.W., N.E. direction, and usually highly inclined or vertical. Swallow holes continually occur, and also many cavernous openings, through some of which waters flow when the country is wet. Isabella crept into one of them, and bounded out like a kitten. (179–80)

The cynic might suppose that Murchison's eyes were so captivated by this pleasing sight that he was for a while unable to attend to his rocks with the care that was warranted! On the Gillaroo track, one may find in places fucoid rock *above* the limestone instead of below it, as is usual. And the low-angle thrust planes, which might indeed be mistaken for bedding planes, are almost at right angles to the bedding planes expected for that district according to Murchison's theory.[8] So what modern geologists call thrust planes, Murchison called "joints." I would suggest that he erred in describing them as "highly inclined or vertical." Geikie's account of this Assynt excursion is therefore somewhat curious, when we have reason to suspect that for an afternoon, at least, Murchison's observations were less exact than usual. However, his biographer Geikie would have none of this. He wrote:

> The journal of these rambles, however, wholly devoted to quartz-rock, limestone, Cambrian sandstone, fundamental gneiss, and other geological matters, affords us no glimpses of its writer in any other capacity than as an enthusiastic hammerman, up early and out late, very pleased and elated to find his main proposition so completely sustained by his friend and companion's critical examination. (Geikie 1875, 2:226)

But I jest somewhat. To be fair, Murchison spent several more days in the district, going over the ground with Ramsay somewhat more carefully than was his wont, if we are to judge by his usual field notebooks. He prepared about a dozen field sketches that were cut out and pasted into the journal. These were colored according to the supposed outcrops of the different kinds of rocks, and they served something of the purpose of geological sketch maps as well as memoranda.

Murchison and Ramsay could not themselves have walked over all the ground of the large areas depicted in these sketches, and much of their surveying was necessarily impressionistic. But this was Murchison's way of working. He would discover a few supposedly typical sections and make sweeping inductions on the basis of them. We need not suppose, however, that he was generally careless in his observations. Young Isabella was, we may hope, an isolated distraction!

In the next few days, Murchison (aged sixty-seven) and Ramsay (forty-five) toiled over a considerable amount of ground in the Assynt region, and ac-

---

8. At this spot, the thrusts curve round from their usual direction parallel to the Eriboll-Skye line, and run almost east-west, as may be seen in the lower-middle portion of plate 8 (the black lines on the outcrop of limestone, depicted in pale blue).

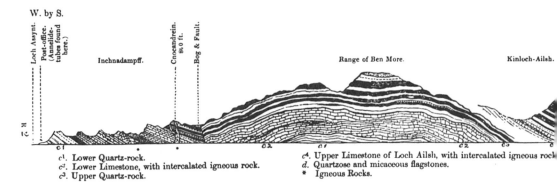

cᵈ. Lower Quartz-rock.
cᵈ. Lower Limestone, with intercalated igneous rock.
cᵈ. Upper Quartz-rock.

cᵈ. Upper Limestone of Loch Ailsh, with intercalated igneous rock
d. Quartzose and micaceous flagstones.
* Igneous Rocks.

Fig. 4.2a. Section of strata at Loch Assynt, according to Murchison (1860, 217).

Fig. 4.2b. Original in field notebook. Courtesy of the Library Committee of the Geological Society, London.

cording to Ramsay's field notes, "Sir R. walked gallantly" as they climbed to the bealloch (or pass) below Ben More Assynt.[9] Ramsay ascended to the summit, but the weather was bad and he only had ten minutes of view at the top for his pains. The point that Murchison particularly wished to establish, in his contest with Nicol, was that there were *two* quartzites or quartz rocks, not one, as Nicol now constantly asserted. This did not seem so very difficult to establish at Loch Assynt. A "lower" quartzite could be seen on the western shore near Inchnadamph, and to the north, running up the side of Quinag. Then as one climbed Ben More Assynt, or Ben an Fhurain, or walked over to Gillaroo Loch, one could easily find masses of quartz rock lying *above* the limestone, which seemed to provide ample empirical evidence for the existence of a second, upper quartz rock. Thus, wrote Murchison, "this excursion completely demolishes Nicol's notice of there being no upper quartz" (181). He claimed, moreover, concerning Cnoc an Droighinn, that he had made no mistake on his earlier visit to the region. There was "the clearest possible proof of a conformably overlying quartzose rock" (177). Indeed, this might be said to be the main finding of his Assynt work: the Cnoc an Droighinn hillside represented a conformable sequence of Silurian rocks, characteristic of the whole Northwest Highlands.

9. Imperial College Archives, KGA/RAMSAY/32, "1858[–1859] Highlands," f. 71r.

Even so, the two geologists must have had some doubts, or certainly plenty to think about. On August 4, the day that Nicol's ideas about the quartz rock were supposedly "completely demolished," they found a "great fold over" of the quartz rock under Ben an Fhurain (181). Below Ben More, the quartz rock was showing "much fracture and distortion" (182), and it was variable in its appearance at different exposures. Ramsay recorded the quartz rock below Ben More as "pitching all ways." [10] They also found some igneous rock in one or two places that might have seemed to be in accord with Nicol's idea of a "syenite," or some such rock, intruding along a line of fault where upthrow of the supposed "eastern" gneiss sequence occurred. Further, one of Murchison's sketches of Cnoc an Droighinn shows two major high-angle faults, reminiscent of Nicol's depiction of the general Northwest Highlands sequence. One of these sketches—a pasted-in figure—in fact shows a fault precisely where, according to modern theory, we expect to find the important Glencoul Thrust plane (176). But modern sections show a low angle for the dip of the thrust plane. [11]

Despite the problems that any geologist examining the Loch Assynt sites will find, owing to the complexities of the district, Murchison seemed fully confident in his interpretations. He supposed, broadly, that an ascending conformable sequence might be discerned, with two quartz rocks and two gneisses, rather than a doubling up due to folding or faulting. His section for the Cnoc an Droighinn hillside and further east is shown in figure 4.2, together with a copy of the original figure in his field notebook (176).

It is, however, difficult to know exactly where and what Murchison intended his section to represent. If one starts from Inchnadamph toward Cnoc an Droighinn, one should head northeastward. If one wants to go to Ben More, one should head approximately southeast. Thus, as Murchison acknowledged, he was offering a *generalized* section. But it is fair to maintain, on the basis of the Cnoc an Droighinn section, that there is the appearance of an "upper quartzite" as well as a "lower quartzite." [12] As to the igneous intrusion appearing in Murchison's section, it is impossible to say what or where it was. In his journal, Murchison stated that igneous rock was to be found at the top of the glebe (field) belonging to the Minister of Inchnadamph (177). This would place it somewhere on the Cnoc an Droighinn hillside. The rock was described as being "intrusive porphyritic and syenitic." No such material is marked on modern maps. [13] Be that as it may, inspection of Murchison's sec-

10. Ramsay, "1858[–1859] Highlands," f. 75r.
11. Cf. Institute of Geological Sciences (1965).
12. The "lower quartzite" does not outcrop on the eastern side of the lake at Inchnadamph. It may, however, be found at the southwest of Loch Assynt. The clearest appearance of a "lower" and an "upper" quartzite is perhaps to the north of Loch Assynt, to the west and to the east respectively of the road leading to Loch Glencoul. Two apparently distinct quartzites figure prominently in Murchison's colored frontispieces to the editions of *Siluria* of 1859 and 1867.
13. There are a number of outcrops of lamprophyre. But more probably Murchison was referring to gneiss, the identification of which caused considerable difficulty during the Highlands controversy.

tion shows that he had faults and igneous rock in it much the same as Nicol required. The question on which they agreed to differ, and which became central to their debate, was the issue of whether there was or was not an upper quartz rock.

Ramsay seemingly concurred with Murchison's findings. At least, there are no protests in his field notes. But these notes are meager, compared with Murchison's, and they are mostly concerned with observations of glacial phenomena, which was Ramsay's primary interest. So although he was willing to support Murchison at the Aberdeen meeting of the British Association that came a few weeks later, Ramsay never became specially interested in the Highlands controversy and published almost nothing on it. He concurred with Murchison's opinion, but he never embraced it with any particular enthusiasm.

Leaving Inchnadamph and Loch Assynt on August 8, the two geologists traveled north, crossing the ferry at Kylestrome, and reaching the village of Scourie in gneiss country on the west coast at 1.30 P.M., according to Ramsay's diary. There are some excellent exposures of quartzite, limestone, and so on, at Lochs Glendhu and Glencoul, which bifurcate at Kylestrome from Loch a Chairn Bhain. These rocks had previously been well mapped by Cunningham (see plate 2). Murchison's journal gives a fairly circumstantial description of the Glendhu-Glencoul area, but comparison with modern maps show that the account is somewhat garbled, and it is obvious from the time that the travelers spent in the district that they did nothing much more than drive along the road. Thus, I take it, Murchison's topographical and geological remarks were derived from sources such as Cunningham, or even Nicol. Nevertheless, he blandly remarked:

> The prevailing dip of the gneiss is N.N.W. Nicol's case of an eastern quartz being overlaid at Glen Coul by igneous rock was set aside by Ramsay's previous actual survey. It is the old gneiss, covered at a short distance only by Cambrian, which is there surmounted by limestone, and then followed by igneous rock. (191–92)

I am unable to reconcile this statement with modern maps. Moreover, it is unclear what Murchison meant when he mentioned "Ramsay's previous actual survey." Was this an impression that Ramsay formed with the aid of a spyglass from the summit of Quinag? Or did he nip out of the barouche from time to time as it descended toward Kylestrome? The point might seem unimportant, but it does cause one to question the *quality* of the evidence that Murchison was to bring with him to the forthcoming British Association meeting. It is unfortunate that the geologists did not spend more time at Loch Glencoul, for the structural relationships there are somewhat less complex than they are at Assynt, and more mappable. The Glencoul Thrust, as it is called today, may be seen very clearly, though in a somewhat inaccessible position. It is, of course, easy to interpret a thrust plane as a bedding plane. Indeed, there was no other way of regarding it before the concept of low-angle thrusting had been introduced to geological theory. So we can say nothing about what might have transpired if Murchison and Ramsay had given the Glencoul area more than a cursory examination. But a careful look at the sections at Glencoul could have been of value to Murchison.

From Scourie, Murchison and Ramsay went inland again to Lochs Stack and More. The latter had offered Nicol another important section, cutting through Murchison's "Silurian" sequence. Murchison gave a sketch of the section along the south shore of Loch More, again in accordance with his general ideas about the structure of the Northwest Highlands (193). The travelers stayed the night of August 9 at Stack Lodge and then returned to gneiss country on the west coast to travel via Rhiconich up to the very top of the Scottish mainland, reaching Durness the following day. On the road to Durness, Murchison noted quartzite on his left (to the west), which he knew was repeated in the region between Lochs Durness and Eriboll. Hence he correctly inferred the existence of a major fault, with downthrow to the northwest, roughly along the line of the road. This, in fact, is one of the faults by which the inlier of fossiliferous rocks of the Durness region came to be emplaced, and the structure of which Murchison apparently understood. It should not be confused with the main zone of thrusting to the east, running southwest from Eriboll.

At Durness, Murchison and Ramsay examined the rocks along the coast near Smoo Cave and the hamlet of Sangobeg, acknowledging "a scene of chaotic metamorphism" (195). On August 11 they looked round Far-out Head, and the following day they took a boat in calm seas to examine the awesome and inaccessible cliffs of Whiten Head. The geologists then moved on to Loch Eriboll, where they spent three nights at Eriboll House "with the excellent and hospitable Alexr. Clark of Eriboll, who gave us much instruction geologically" (202). This is a key area for understanding the geology of the northwest, and it was here more than twenty years later that Charles Lapworth carried out his meticulous map work that eventually led to the unraveling of the geological structure of the region.

Murchison's journal is at this point somewhat garbled, and it would appear that his amanuensis was unable to read his handwriting at some points, for there are a number of gaps in the text and the accounts of Whiten Head and Eriboll are partly conflated. Nevertheless, we have some interesting sketches. One of these shows a section of the hillside behind Eriboll House, on the eastern shore of Loch Eriboll (200). The general appearance of the area of contention is shown in figures 4.3 and 4.4.

This, like the section at Cnoc an Droighinn, was to become one of Murchison's favorite pieces of empirical evidence for his theory. His section shows a simple conformable ascending sequence from west to east: [lower quartz rock], limestone, upper quartz rock, upper limestone, "gneiss."[14] This last was

14. The lower quartz rock is exposed to the west of the loch and thus does not appear in Murchison's section at Eriboll House. I include it in the sequence here, since it formed part of his general theoretical structure.

Fig. 4.3. View looking southwest over Eriboll House and Loch Eriboll.

not labeled as such on the diagram, though it was depicted by a separate color. It was described in the notes as follows:

> The ascending order above the upper limestone is, first, decomposing ferruginous rotten beds; second, a little quartz rock, becoming more thin bedded; third, passage into talcose schists, here and there micaceous, and in parts becoming contorted and twisted on a small scale, then graduating into a sort of thin, flaglike gneiss, felspar layers intervening. Their transition and conformability are undeniable. (199, 202)

Clearly, the character of this rock was by no means certain, but Murchison was willing to see it as comporting with his belief in the existence of an "upper" or "eastern" gneiss. It may come as no surprise, though, to know that the rock is depicted on modern maps as "mylonised rocks and 'frilled schist,'" overlain by gneiss.[15] Murchison was walking over an exceedingly complex structure, and the repetitions of quartz rock and limestone (note that at Eriboll he now had to introduce an "upper limestone") have come to be construed in terms of overfolding and thrusting. Yet in a published section, shown in figure 4.5, he

---

15. The term "frilled schist" was coined by Charles Lapworth, as will be recounted in chapter 8. It is a metamorphic rock produced by the grinding or milling of rock at a fault surface.

Fig. 4.4. View of "Murchison's section" at Eriboll House.

showed a fundamentally simple arrangement for the strata.[16] The Eriboll
House section was cruelly deceptive.

The second section we should consider was even more problematical. It depicted the rocks at Ben Arnaboll, on the east side of Loch Eriboll, a few miles
to the north of Eriboll House. To the south of this mountain there is an unnamed valley, running east to west, which provides a useful opportunity to
view the mountain, and the region of thrusting, in section. Murchison depicted this section, showing his usual ascending sequence, but he also represented an intrusion of igneous rock rising right up through Ben Arnaboll,
thus giving the appearance of a great fracture through the mountain. The
sketch bears the note: "eastern dip not affected by the igneous rock" (201). Yet
his picture suggests very much the kind of theoretical structure that Nicol was
advocating: an igneous intrusion occurring all along his supposed line of
faulting in which the "western" gneiss was brought to the surface to the east.
What Murchison's "igneous" rock might have been in this particular spot, I am
unable to suggest with certainty on the basis of modern maps, though it was
in fact very likely Lewisian Gneiss. Murchison quite overlooked evidence of
earth movements and inversion of strata that may be seen quite easily at Ben

---

16. This section was in part based on information supplied by Harkness, subsequent to the Aberdeen meeting of the British Association (see page 91).

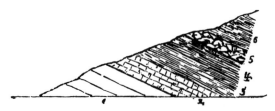

1. Quartz-rock.
2. Upper limestone.
3. Gneissose quartz-rock.
4. Gneissic strata.
5. Felstone (granitic).
6. Gneissic and micaceous schists.

Fig. 4.5. Section of strata behind Eriboll House, according to Murchison (1860, 231).

Arnaboll, such as is shown in figure 4.6, which was photographed on the western scarp of the hillside—that is, on the road side.

Years later, in reporting his version of the events of the Highlands controversy, Murchison stated that he and Ramsay traced the outcrop of quartz rock and associated strata from Loch Eriboll to Lochs Stack and More (Murchison 1867, 166). But this claim is not borne out by the journal, which says nothing at all about such a journey (although with three days at Eriboll the geologists might have looked into the valley to the south of the loch). More likely, I suggest, Murchison simply used the evidence of Cunningham's map. Again we have some reason to doubt the reliability of his observations.

Leaving Eriboll, the travelers took the usual weary route over the Moines to Tongue and on to Loch Loyal, where Ramsay climbed Ben Loyal, examining the granite and syenite of which that mountain is composed. Then on to Betty Hill, Melvich, Achintoul, and so on. We need not concern ourselves with the observations of the later part of the journey, apart from noting Murchison's insistence in his notes that the western and eastern gneisses were entirely distinct both in terms of lithology and of dip and strike. Thus he found Nicol's hypothesis wholly implausible; and if the appearances of the "two" gneisses were the only factors to be considered, his interpretation would be quite convincing. However, there remained the unsolved problem of how a metamorphic gneiss might come to be sitting on top of an unmetamorphosed limestone.

Murchison and Ramsay arrived in Aberdeen in time for the British Association meeting in September, with Murchison full of facts recently acquired or refurbished, ready to present a final demolition of Nicol's arguments. Nicol, meanwhile, as the leading geologist of the city where the meeting was being held, had been engaged in the numerous administrative details required for the successful organization of the geological section of the program.

There are several newspaper accounts of the proceedings of the geology section of the British Association meeting in 1859, but little by way of firsthand accounts of the events by the participants. Murchison himself gave the meeting a grand account of his views on the geology of northern Scotland, based in large measure on his recent investigations. But he said very little about it all in his journal. If fact, we only have the following few words:

> It seems that I was so busy at Aberdeen, concocting my great Lecture
> on the Geology of the North of Scotland, which I delivered before two

Fig. 4.6. Folded quartz rock on western side of Ben Arnaboll.

thousand [*sic*] persons in the New Hall, that I omitted to make any notes. (232)

The newspapers, however, did Murchison proud. He was reported in the national dailies such as the *Times,* and at great length in the local papers: the *Daily Scotsman,* the *Aberdeen Herald,* the *Aberdeen Free Press and Buchan News,* and the *Aberdeen Journal.* According to the historian of the Geological Society, H. B. Woodward (1908, 195), a "great contest" took place between Murchison (supported by Ramsay) and Nicol, and that contest was an unqualified triumph for Murchison. The newspapers reported his words largely verbatim. As the *Aberdeen Journal* commented:

> So plain and irresistible were the evidences presented in a number of transverse sections at Loch Broom, Loch Assynt, Loch More, Loch Eribol, &c. that the only wonder entertained [by Ramsay] was that any skepticism should have prevailed as to the order of succession.[17]

The geologists had taken considerable care to search for Nicol's igneous rocks, supposedly marking a great line of upheaval. And such material had indeed been found in a number of places. But, Murchison assured his listeners, "nowhere, including the flanks of Ben More Assynt, do such intrusions constitute a geological separation between two groups of rocks."[18] Thus geologists, and the good citizens of Aberdeen, could rest assured that the Survey officers

17. *Aberdeen Journal,* Sept. 21, 1859, 11.
18. Ibid.

knew what they were doing and had revealed an essentially simple structure underlying all the apparent complexities of the Highlands.

Murchison apparently illustrated his lecture with two diagrams—the first according to the old theory of Hugh Miller, with the eastern and western sandstones of Scotland regarded as belonging to the same formation and "framing" the rocks of the center; the other according to Murchison's views, which, he said, constituted "a 'Reform Bill' of the geological structure of the northern counties of Scotland" and "exhibited the structure that had been ascertained to prevail."[19] His address was acclaimed by the "very full audience," and a vote of thanks was presented by Sir Charles Lyell, seconded by the Oxford Professor of Geology, John Phillips, who spoke of the "high estimation in which Sir R[oderick] Murchison was held over half the globe as the master of the Silurian." The physicist Sir David Brewster then added his words of thanks and admiration for the address and presented Murchison with the Brisbane Medal of the Royal Society of Edinburgh, which was awarded particularly for Murchison's Highland work. Finally, the well-known amateur naturalist and publisher Robert Chambers read a eulogy, which included the remark that Murchison had "conferred a great favour on . . . [his] native country" for sorting out the stratigraphical relations of the Highland rocks.

There can be no doubt that all this must have given Murchison immense satisfaction. He was being honored in his homeland and seeing his claims for the extension of Siluria over vast tracts of northern Scotland ratified by some of the highest scientific authorities in the land.[20] But what of Nicol? As has been mentioned, he was one of the organizers of the geology section for the Aberdeen meeting, and he presumably had little opportunity to lobby on behalf of his views, even if he had had a mind to do so (which is unlikely, from what is known of his temperament). Yet given the fact that he was on his "home ground" in Aberdeen, one might have expected him to be accorded a fairly generous amount of publicity, and to have seen his views receive some public support; but Nicol's paper was scarcely reported in the press. He may, of course, have received some support in the discussions following his paper. But of such support, if any, we have no record.

The issue was, in fact, debated three days after Murchison's triumphant address, and the only extant record we have is the newspaper reports, such as the following from the *Aberdeen Journal:*

> Professor Nicol, F.R.S.E., gave an able and interesting notice on the
> relations of the gneiss, Red Sandstone, and quartzite in the North-West
> Highlands, illustrated by various sections. Professor Nicol had visited
> the Highlands, and had arrived at a different conclusion as to the suc-
> cession of certain crystalline and sub-crystalline rocks from that arrived

19. Ibid.
20. It is interesting to speculate what might have happened had Ramsay come to Aberdeen announcing his rejection of Murchison's observations and interpretations. But as we have seen, such a state of affairs was most unlikely. Ramsay had been effectively "locked in" to Murchison's doctrine.

at by Sir R. Murchison. He contended that the great series of rocks in question were of older date than that assigned to them by Sir R. Murchison, and endeavoured to prove, by a reference to the sections which he exhibited, that the order of super-position which he advocated was the correct one.

The President [Lyell] said, this question was a difficult one of interpretation, and that the burden of proof lay upon those who, like Sir R. Murchison, contended that the highly crystalline rocks were of newest date.

Sir R. Murchison, at considerable length, replied to Professor Nicol, referring to sections which he had prepared, and maintaining with great confidence that the order of super-position he had formerly contended for was the correct one. In company with Professor Ramsay, he had examined the country, and although they were aware of the difficulties of certain obscure sections here and there, he contended that, in no country he had ever examined, in any part of the world, had he ever seen a clearer order of super-position than that which he had endeavoured to point out—viz., the super-imposition of the quartz rock upon the limestone.

Professor Ramsay briefly confirmed the views of Sir R. Murchison, stating that he had noticed for miles the super-imposition of the quartz rock upon the limestone without any break, and felt not the slightest doubt upon the subject.

Professor Sedgwick, who had just arrived, rose amidst loud applause, and spoke in corroboration of the views of Sir R. Murchison and Professor Ramsay. Going hastily over the country, it certainly appeared to him that the order of super-imposition was that contended for by Sir Roderick, although it was perfectly possible that more extended observation might induce them to come to a different conclusion.

In reply to Professor Hopkins, Professor Ramsay said that he had observed the limestone pass under the gneiss for miles without a fault.

Professor Nicol briefly replied, holding that the result of his personal observations had been to convince him that the relations of the gneiss, red sandstone, and quartz were such as he had contended for in his paper, and referring to his sections in proof of the correctness of his theory.[21]

This report, though incomplete, tells us some interesting things about the manner in which geological knowledge came to be constituted in Britain in the mid-nineteenth century. Of the people present and recorded as participating in the discussion, only Murchison, Ramsay, and Nicol had fresh memories of the localities and only a handful of other people were in a position to have a view of the controversy on the basis of their own observations. Of

21. *Aberdeen Journal*, Sept. 21, 1859, 6.

these, Sedgwick, John ("Huge") Miller, and Peach were the most important. Peach had probably not been round all the sites, nor was he one of the inner circle of geologists whose opinions really counted; and in any case we do not know whether he spoke at the meeting. The same might be said of Miller. Sedgwick had gone over some of the ground, but rather hastily, and many years before, in 1827. His opinion probably proved decisive, however, since he was well known at that time to be locked in bitter debate with Murchison over the Welsh strata. But Sedgwick was not a wholly impartial witness in the Highlands. After all, he had written one of the first major papers on the area with Murchison. Perhaps more important, Murchison was offering him the western red-brown sandstones for the Cambrian system, the territory of which Sedgwick was presumably glad to see extended.

Of the other geologists present, Lyell was by far the most important. He seems to have expressed some reservation about Murchison's views, saying that the burden of proof lay with him. But this was not a sufficient objection to undermine Murchison's opinion in any way whatsoever. So Nicol had to battle alone. He had no one he could bring forward to corroborate his opinion. All he could do was reassert his observations and their interpretations. That wasn't enough. His social situation was inferior to that of Murchison, as was his reputation as a geologist. His position at Aberdeen was marginal both socially and geographically; and he was even indebted in part to Murchison for *that* position. Whereas Murchison was lionized in the press, the media gave Nicol negligible coverage.[22] The outcome was inevitable, then. So far as the world knew, the rocks of northwest Scotland were the way that Murchison said they were!

It is curious, however, that when the British Association *Report* of the 1859 meeting was published Murchison's paper was not included, whereas Nicol's was (Nicol 1860). I do not know the reason for this (though one is naturally inclined to suspect that Nicol found some way to exclude Murchison's paper). As for Nicol's paper, it makes some interesting points, though they were little appreciated at the time. As in his earlier paper, he contended that there was generally some kind of "igneous" rock to be found at the critical point where the eastern gneiss came in contact with the underlying rock, and he adduced further instances in support of this claim. But more interesting, he was obviously puzzled by the nature of the contact and discussed it in the following terms:

> The superposition of the red sandstone to the [western] gneiss can be
> observed over miles and miles of country; that of the quartzite to the
> red sandstone is no less distinct, being readily traced by the eye from
> mountain top to mountain top, from valley to valley; and again the
> limestone, though a small formation, everywhere clearly reposes on the

22. From the fact that Murchison's talk was reported in virtually identical terms in the different newspapers, I would conclude that he made his text available to reporters; or it was very carefully recorded by stenographers. However, Nicol also gave a paper on the geology of the area round Aberdeen, and this was fully reported.

quartzite. . . . But how is it with the next step in the supposed series? Nowhere is an overlap of more than a few feet or yards even said to be seen, though the supposed overlying rocks extend more than thirty miles to the east, the underlying fully as much to the west. The fact, too, that the eastern gneiss is brought into contact in one place with the limestone, in another with the quartzite, in a third with the red sandstone, according to the amount of denudation, and all within a few miles, prove that the junction is along a line of fault, and is wholly inexplicable on the supposition of conformable upward succession. (ibid., 120)

This is in fact a crucial argument, and one that Murchison never came anywhere near answering satisfactorily. The point is that the arrangement of strata in the field may on occasions be such that it is difficult to tell whether one is dealing with a thrust plane or a plane of unconformity (see chap. 2, note 40). A stratum, X, may lie at some places on A and at some other places on B. This may signal either a break in deposition and the presence of an unconformity below X. Or it may be that X has been thrust over A and B, so that it lies sometimes on A and sometimes on B. (In which case, X may be either older or younger than A and B.) So the mapped outcrop of a thrust plane can look rather like the mapped outcrop of an unconformity. (In the more typical kind of faulting, a stratum S may, by approximately vertical movement, be brought into contact with strata C and D, without producing the appearances of an unconformity on a map.)

Now in a sense, Nicol was trying to have it two ways, as indeed was Murchison. For Nicol, the way in which the "eastern" gneiss overlay a number of *different* rocks suggested faulting. But he didn't have the category of a low-angle (thrust) fault in his theoretical repertoire. So he settled for ordinary faulting—the eastern gneiss being the western gneiss brought somehow to the surface but with the junction obscured or confused by the presence of igneous intrusions. Unfortunately, this required him to turn a blind eye at places where the low-angle fault contact could actually be seen quite clearly—as, for example, at Loch Glencoul or at Knockan Cliff.

Murchison, on the other hand, insisted that there was a *conformable* relationship between the eastern gneiss and the Durness Limestone, and so on. This agreed with the observed nature of the (faulted) contact at Knockan Cliff. But, as Nicol emphasized, it left Murchison with no explanation of how the gneiss could lie sometimes on top of one rock and sometimes on top of another. Of course, if Murchison had not been so insistent on a conformable relationship, his problem would have been eased. On balance, Murchison's position was—regardless of his social advantages—probably somewhat stronger than that of Nicol in 1859. But it was by no means impregnable.

We find, then, that following the Aberdeen meeting Murchison urged another geologist of repute, Robert Harkness (1816–78), to make a quick visit to the northwest to endeavor to confirm Murchison's findings.[23] Harkness was a

23. On Harkness, see Goodchild (1882–83), Geikie (1878), Sorby (1879).

Lancashire man who had studied geology at Edinburgh under Jameson and Forbes, and who, during the course of his lifetime, obtained a sound reputation for his work on Palaeozoic rocks in the Lake District, Scotland, and elsewhere. As a young man, he made important collections of fossil graptolites from the rocks of the Southern Uplands and identified the strata as Silurian, according with Murchison's doctrines. We may assume, therefore, that he was in Murchison's "good books" in the early 1850s, and it comes as no surprise to find that in 1853 Murchison recommended Harkness to fill the chair at Queen's College, Cork, which had fallen vacant as a result of Nicol's recent move to Aberdeen. Murchison also nominated Harkness for a fellowship of the Royal Society in 1856 (Goodchild 1882–83, 152, 156).

It is evident, therefore, that Harkness might be expected to be favorably disposed toward Murchison's side of the argument in the dispute with Nicol, and a useful additional witness, being a competent but perhaps not wholly impartial observer.[24] On leaving Aberdeen, Harkness traveled north on his way back to Ireland, and a few weeks later he transmitted his interpretations of the sections at Eriboll House and Cnoc an Droighinn (Inchnadamph and Assynt) to Murchison. These were utilized by Murchison in the paper that he read to the Geological Society on November 16, 1859, in which he presented the results of his researches with Ramsay that summer in the northwest.

The sections, as construed by Harkness (or at least as described for him by Murchison), represented absolutely simple ascending conformable sequences, in perfect agreement with the Murchison view of the northwest structures (Murchison 1860, 221 and 231). Thus Murchison had successfully gathered in another adherent to his views.[25] But it is, I suppose, possible that the members of the Geological Society were well aware of Harkness's indebtedness to Murchison. And although Nicol was still his only active opponent, Murchison could not feel that his victory was complete. Or perhaps he was genuinely concerned that a geologist of Nicol's caliber should have serious reservations about his work. He may have been fully satisfied that the northwest was his but felt it time to move forward to conquer still further territory in the Grampians and elsewhere. To continue the metaphor, the ground had been claimed, but not fully subjugated. There was still work to be done.

On the whole, I am largely persuaded that the last of these interpretations is the most plausible. But the other possibilities certainly need not be excluded. Whatever the case, we find Murchison traveling north yet again in 1860, for his final attempt to "satisfy himself" of the correctness of his views concerning the rocks of the northwest. However, an account of this work requires us to bring the young Archibald Geikie into the story, and so it will be appropriate to turn to a new chapter.

24. Harkness gave a paper at Aberdeen, but it had nothing to do with the Highlands controversy.
25. Harkness (1861) also acquiesced in the correlation of the rocks of the Northwest Highlands and those further south (below the Caledonian Canal). In addition, he claimed correlations with Silurian strata in Donegal, Ireland.

5

The Murchison-Geikie Tour of 1860

One might have supposed that Murchison's triumph of 1859 was enough. The old geologist had received the plaudits of the British Association for his Highland work, and the Murchisonian picture of the structure of the Scottish Highlands had all been laid down solidly in the second edition of *Siluria*. He was still at the height of his fame, reputation, and influence, even though his physical powers were waning a little. But Nicol had by no means conceded defeat at the British Association meeting, even though he had attracted no recorded support. So in 1860, Murchison sallied forth into the wilds of Scotland to make his last major field trip, to try, yet again, to obtain the decisive evidence in favor of his theory, and to show Nicol, once and for all, the error of his ways. But this time Murchison adopted a different tactic. Instead of seeking to enlist a well-established and influential geologist to his views, he determined to take with him one of the most promising of the up-and-coming generation, Archibald Geikie (1835–1924), who was at that time just beginning to make his way as a young officer in the Scottish branch of the Geological Survey. Perhaps if Geikie could be enlisted as an advocate of Murchison's doctrine, the younger generation of geologists as a whole might be likewise influenced, and the Murchisonian view of Scottish geology would prevail. As we shall see, the idea proved to be a shrewd and successful move—though in certain respects it proved something of a disaster for British geology in the nineteenth century. Geikie, then, plays a role of the utmost importance in our account. It therefore behooves us to give some details of his life and character, especially since he has not been the subject of such detailed historical investigation as has Murchison.[1]

1. For literature on Geikie, in addition to his autobiography (Geikie 1924), see the Geological Magazine's Eminent Living Geologists series (1890), Houston (1925), Peach and Horne (1925), Strahan (1924, 1925, 1926), Horne (1926), Tyrrell (1926), Cutter (1974), Marsden (1979, 1980), Oldroyd (1980), and Browne (1981). For other sources on Geikie, there are considerable holdings of his papers at the Geological Society, London, the Archives of the Geological Survey at Keyworth, Nottingham, and the Library of the University of Edinburgh. Other material is distributed in smaller quantities elsewhere. Most interesting and important for the present inquiry are Geikie's field notebooks. Ten of these are held at the University of Edinburgh Library, one at the geology department of the University of Edinburgh, and sixteen at the Haslemere Educa-

Not long before he died, Geikie published his beautifully written autobiography (Geikie 1924). The book covers the full span of his career, and it was accomplished while he was still in full possession of his great literary abilities. The autobiography will always be the main source of information on Geikie's private life, for very little of his private correspondence with his family appears to remain. There is a considerable amount of other archival material, but this consists chiefly of his official government papers, and his correspondence with other scientists, mostly geologists. Unfortunately, even this correspondence is by no means complete. What we do have, however, is an excellent exchange of correspondence between Geikie and Murchison from 1860 until Murchison's death in 1871. We also have Murchison's journal for his expedition to the north of Scotland in 1860 and Geikie's field notes. So we are well placed to put together a detailed picture of the relationship that blossomed between the distinguished senior geologist and the young tyro as the two used each other, each trying to further his own interests.

Geikie was the son of an Edinburgh businessman, coming from a family with musical interests. He attended Edinburgh High School, and there he acquired a good grounding in and love for the classics, which he retained until his last years. He was fortunate to be taken into the Scottish countryside at an early age, and during family holidays in places like Arran he was quickly able to satisfy his taste for nature and begin some early untutored observations in geology.

A few years later Geikie was to write of his early geological rambles in a popular magazine, apparently suggesting to other young men what an excellent thing it would be for them also to improve their minds in a healthful manner by geological observations and field work (Geikie 1861a, 1861b, 1861c). There is also a manuscript of a "paper" he wrote based on one of his excursions, which strongly suggests that by the age of seventeen he already had ambitions to be a scientist and good prospects of achieving something in this line, for the document indicates a remarkable self-confidence, skill in observation, and reasoning power.[2] The observations were made in a fresh-water limestone quarry at Burdiehouse, about four miles southeast of Edinburgh, in the Carboniferous, from which Geikie extracted a number of fossil fish of reptilian appearance. It was no wonder that he was excited. This experience very

tional Museum at Haslemere in Surrey. It may be mentioned also that there is a biography of Geikie written in Chinese by Zhang Zi Ping and published by the Business House Press in Peking in 1935. It was one of a series of popular texts on science edited by Wang Yun Wu and Zhou Chang Shou, and one may suppose that it was intended to demonstrate to Chinese readers what the career of a successful scientist in the Western world might be like, and perhaps to inspire would-be Chinese scientists. The text appears to be based very largely on Geikie's autobiography (1924), and it has several illustrations reproduced from that work. (I am indebted to Yang Jing Yi for information on this point and for giving me a xerox copy of Zhang's book, together with some idea of its contents.)

2. A. Geikie, "On the Reptile Fishes of Burdiehouse, May 1853," National Museum of Scotland Archives, MS. 56 (41) 8.1 G, 29 pp.

likely determined him in his wish to join the ranks of the new profession of geologist.

The first visit to the quarry was undertaken while Geikie was still at school and led at an early stage to his being introduced to some men of science who assisted him in identifying some of the fossils (Geikie 1924, 17). He maintained a continuing interest in the topic, and following investigations in the quarry by Dr. Samuel Hibbert in 1825, Geikie began to prepare his own paper on the topic.[3] His essay shows that he tackled the investigation in a systematic manner, reading everything that had been written about the rocks of Burdiehouse and their fossil contents. By careful observations of the structures of his fossil fish, he endeavored to ascertain something of their life habits, and by examining the coprolites (fossil feces) of the organisms, the impressions thereon, and their shape, he gave some plausible suggestions as to a possible spiral structure to the animals' intestines. It was a remarkable exercise in reasoning on the basis of observational evidence for a self-tutored lad, though most of the ideas were probably second-hand.

Geikie had left school at fifteen in order to be trained in a lawyer's office, but he had little interest in the work. He associated with a number of Edinburgh intellectuals and made the acquaintance of the aging Hugh Miller, who was delighted to meet a young man who seemed likely to follow in his geological footsteps. Geikie also knew James Forbes, professor of natural philosophy at the university, and a number of other important scientists. After a couple of years, Geikie gave up the work in the lawyer's office. The obvious thing to do would have been to attend lectures at the university, but Robert Jameson, by then old and enfeebled, was still in charge of the mineralogical and geological work. Geikie felt no inducement to study science under Jameson, so he took private lessons in analytical chemistry under George Wilson and studied some other scientific subjects. Things seemed set for a change in 1854, when, following the death of Jameson, the former Survey palaeontologist Edward Forbes (1815–54) was appointed to the chair. This prompted Geikie to enroll at the university that autumn.

However, Edward Forbes died an early death at the end of 1854. At the same time the Geikie family fell into some financial difficulties, so that Archibald was unable to complete his degree and had to withdraw from the university, even though he had proved himself the outstanding student of the year. In the short time that he was at Edinburgh, Geikie chiefly studied classics; and with his incomplete degree he never obtained a full academic training in science. It mattered little, however, so far as his plans for a scientific career were concerned. He had already made field excursions to Skye and Arran. He had cultivated numerous scientific acquaintances, including Ramsay, who had come to Edinburgh in 1853 to make preliminary arrangements for the exten-

3. Hibbert, an Edinburgh physician and amateur naturalist, had been examining the rocks and fossils in the Burdiehouse quarry for many years. Geikie was certainly not beginning his observations in a wholly unknown locality.

sion of the work of the Survey to Scotland. Geikie showed Ramsay around
Arthur's Seat—the great extinct volcano on the eastern doorstep of Edinburgh—and obviously impressed him greatly, for even at that early date there
was talk between them of the possibility of an appointment with the Survey
should its activities be extended to Scotland. With the support of Hugh Miller,
Geikie also made a minor geological contribution by exhibiting and describing
some Liassic fossils from Skye and its adjacent island, Pabba (or Pabay), at a
meeting of the Royal Physical Society of Edinburgh (Geikie 1855). He was
making all the right moves for one who would wish to make a career for himself in geology.

When Murchison became director general of the Survey in 1855 and began
thinking of the extension of its activities to Scotland, he wrote to Miller to inquire whether there might be any suitable young man of his acquaintance in
Edinburgh interested in a career as a surveyor. This was exactly what Geikie
was looking for, and it must have been gratifying indeed to be informed by
Miller that he had written a warm recommendation of him to Murchison.
Geikie was interviewed by Ramsay and by Murchison, following the meeting
of the British Association at Glasgow in September 1855, which Geikie attended. (He therefore was present at the papers of Murchison and Nicol at
that meeting.) The following month he started work as an "apprentice" surveyor under Henry H. Howell (1834–1915) in the Haddington area, to the
southeast of Edinburgh (Geikie 1924, 40–43).[4]

In the early years of the Survey, it was the common practice to appoint
young persons with no formal training in geology, or even in science.[5] By and
large the system worked remarkably well. Young men of promise were engaged and put to work in company with an experienced geologist. In time,
they would gain enough experience to work alone in simple areas, gradually
moving on to more and more difficult terrain. The work was demanding, the
pay was poor, there were insufficient opportunities for promotion, and the
constant moves required were not conducive to married life. Nevertheless, for
some young men it offered a most attractive career, and the ones that didn't
like the work or were incompetent resigned or were eased out. For the most
ambitious and gifted, there was the possibility of obtaining a joint appointment as a Survey officer and as a university professor or lecturer, a combination that offered a quite adequate salary. So far as Geikie was concerned, a
position in the Survey was the ideal occupation, and in due course he succeeded in reaching the very top of the scientific establishment in Britain (president of the Royal Society). The esprit de corps of the "Hammerers" was
understandably high for most of the nineteenth century.

After a year's work, Geikie was no longer a probationer, and he was soon
doing independent work. He had a very considerable natural ability as an

4. On Howell, see further in chap. 10.
5. For histories of the Survey, see Flett (1937), Bailey (1952), R. B. Wilson (1977), and H. E.
Wilson (1985).

artist, which was of great value in map work.[6] He mapped the region round Edinburgh, did more work on the Mesozoic rocks of Skye, and extended his activities north of Edinburgh toward Stirling and through Fife. He soon found himself working among the maze of ancient volcanoes in that part of Scotland, which led to a lifelong interest in volcanic action, and eventually to his major achievements in geology. He was also gaining a firm grasp of the general geology of the Midland Valley of Scotland. He attended the 1859 meeting of the British Association at Aberdeen, and so he would have become fully aware of the ongoing debate between Murchison and Nicol. Also in 1859 and 1860 Geikie spent some of the winter months in the Survey headquarters in London, so he would have had the opportunity to become better acquainted with Murchison.

Right from the beginning of his career, Geikie was keen to make popular statements of his geological ideas, and to provide descriptions of the character of geological field work and the manner of reasoning of geologists. He published an interesting little book, *The Story of a Boulder* (Geikie 1858b), which showed the layman how a geologist might reason concerning a large glacial erratic. More particularly, it was an elementary essay in the philosophy and methodology of geology, in which Geikie proclaimed his adherence to the geological doctrine of uniformitarianism and, as he later put it, the principle that the "present is the key to the past" (Geikie [1905] 1962, 299). He was never a very philosophical thinker on these questions. The reasoning of geologists was always presented as essentially simple and straightforward, provided one was a careful observer. His general "philosophy of geology" never changed substantially and was stated clearly in his early tract of 1858.[7]

By 1860, then, young Geikie was already beginning to make a bit of a mark for himself in geological circles. From his autobiography we see that he had acquired a wide social circle of agreeable, interesting, and influential persons. He was reading voraciously, he was fit from all his outdoor work, and he was filled with zeal and ambition. His was to be the most successful scientific career of that period—if one measures success in terms of influence and social achievement.[8] This is not, of course, to deny the merits of the work he so suc-

6. Many of Geikie's beautiful watercolors have been preserved at the Haslemere Educational Museum. (Geikie retired to Haselemere and helped establish this interesting local museum.)

7. I have discussed elsewhere in some detail Geikie's philosophical views and their relation to his scientific work (Oldroyd 1980), and there is no need to go into these matters here (they will, however, be referred to in chapter 11). It may be mentioned that Geikie was a romantic with a great love for the natural world. His religious views were conventional, being typical of the natural theology of his day and age (though he was a Darwinian evolutionist). He saw science and religion as being in symbiotic relationship with one another, in that science was able to demonstrate what he took to be the divinely guided progress in the universe (Geikie 1869–70). Thus science for Geikie had a distinct moral, as well as intellectual, function.

8. Geikie's social success may be gauged from the following details. As early as 1865 he was elected a Fellow of the Royal Society. Two years later he was promoted to the position of local (regional) director of the Scottish branch of the Survey, and he also served that year as the president of the Geology Section of the British Association in Dundee. In 1871, he was appointed to the newly

cessfully carried out. Geikie was a prodigious worker throughout his life in his fieldwork, administration, and writing.[9] But the thing that was most certain about him was his ambition. Indeed, I shall argue, his work with Murchison in the Highlands in 1860 was a major factor in enabling him to fulfill that ambition.[10]

We may now look in some detail at the events of Geikie's tour with Murchison in 1860 and see how they contributed to the development of his career and stored up fuel for the second outburst of the Highlands controversy in the 1880s.

In his autobiography, Geikie wrote, concerning his tour with Murchison: "I was most unexpectedly called off to the Highlands in the autumn of the same year" (1924, 77). Part of Murchison's letter of July 18, inviting Geikie to ac-

---

established Murchison Chair in Geology at the University of Edinburgh, where he lectured successfully for a number of years and organized popular field excursions. In 1882, he obtained what was perhaps his chief goal by succeeding Ramsay as director general of the Geological Survey. In 1889, he was foreign secretary of the Royal Society, and in 1891 he was knighted and elected president of the British Association. He retired from the Survey in 1901 and two years later became secretary of the Royal Society. He was president of the Geological Society in 1906, and from 1908 to 1913 he was president of the Royal Society. Finally in 1913 he was awarded the coveted Order of Merit. Geikie retired from London to Haslemere the same year and lived there until his death in 1924 at the age of eighty-eight.

9. His bibliography (Cutter 1974) contains 246 entries, many of the items being major books, of which *The Scenery of Scotland* (1865), the *Text-book of Geology* (1882), and *The Ancient Volcanoes of Great Britain* (1897) were the most important. He also wrote very successful schoolbooks on geology and physical geography—indeed, he was a great supporter of science education—and also a major article on geology in the ninth edition of the *Encyclopaedia Britannica* (1879). In his capacity as director general of the Survey, he had to do a great deal of editorial work, and he also prepared a considerable number of book reviews and obituaries. He was a major early writer on the history of geology, with fine biographies of Edward Forbes, Murchison, Ramsay, and John Michell, a modest volume on Darwin as geologist, and editions of works by Hutton, Faujas de Saint Fond, and Seneca. His *Founders of Geology* (1897, 1905) has continued to be influential through its modern Dover edition (1962). Geikie (1882, 1904, 1905) published several interesting collections of his essays and also some purely literary pieces such as *The Love of Nature among the Romans* (1912) and *The Birds of Shakespeare* (1916) The range of his literary accomplishments was remarkable.

In addition, Geikie was a considerable traveler. He knew the terrain of Britain intimately, of course, but also much of Europe. He traveled as well in parts of North America and Russia. He was a regular attender at international scientific meetings, and had a worldwide network of correspondents, especially European and American geologists. He was a member of sixteen foreign academies and was awarded fifteen honorary degrees.

10. The man behind the ambition is less easy to grasp. To judge from his literary work and his autobiography, he was of an extraordinarily genial disposition. There are also many letters in his correspondence expressing gratitude for various acts of kindness, particularly the issue of testimonials. He had a very wide circle of acquaintances. But as we shall see in chapter 10, there is a lingering tradition in the Survey that he was an arrogant and by no means successful director. Sir Aubrey Strahan, director general of the Survey from 1914 to 1920, wrote of him that "a clear if somewhat cold judgment controlled his actions, but in his biographical work the coldness was masked by a studied kindliness of expression. Though he made many friends at home and

company him, is published in Geikie's *Life of Murchison,* in which the director's plans for the fieldwork are sketched (1875, 2: 229). The complete letter adds the interesting detail that all Geikie's expenses were to be "liquidated" by Murchison.[11] The preliminary correspondence also informs us that Murchison had learned from Peach that Nicol was again in the Northwest Highlands that summer.[12] But the contending geologists did not meet in the field. Geikie was instructed to familiarize himself with Macculloch's work on the Western Isles. Murchison stated in his July 18 letter that the main problem he had set himself was to try to use the sequence "established" in the Northwest Highlands to elucidate the structure of the southern parts of the Highlands; he wanted as well to do more work on the Fundamental Gneiss of the Outer Hebrides. But he also wrote that at Lochs Maree and Broom (where some of Nicol's best evidence was to be found): "I want to fasten a loose screw or two . . . where you will see the ascending order very clearly."[13] Whatever his other intended goals, Murchison was clearly desirous of persuading Geikie of the merits of his interpretation of the northwest.

The two geologists, aged sixty-eight and twenty-five, met at the port of Greenock, to take ship to the Hebrides on August 6. They went first to Islay, a southern island of the Inner Hebrides, where (as usual) Murchison took issue in his journal with some details of Nicol's map.[14] He also expressed disdain for Daniel Sharpe's ideas on foliation, saying: "Mr. Sharpe's foliation and lamination and his arches are all stuff when tested"![15] From Islay the travelers sailed to Oban, from where they made a number of side excursions. On the nineteenth, they were at Banavie near Fort William at the southwestern end of the

---

abroad, his sympathies with his fellow men were somewhat overshadowed by his love of nature and a passion for work. He did not seek collaboration, but preferred to work single handed, nor could he brook criticism" (Strahan 1926, 598).

Of Geikie's private life, we know surprisingly little, in spite of the fine autobiography. In 1871, he married Alice Pignatel, a lady of English-French parentage, and their married life was apparently tranquil and successful. But it is hard to judge on this point, since no Archibald Geikie family papers are to be found in public archives in Britain. We know next to nothing of the character of Mrs. Geikie and I have seen no likeness of her. All this is unfortunate, since it would have been of interest to have access to Geikie's personal thoughts on the Highlands controversy. Geikie's younger brother, James (1839–1915), also became a distinguished geologist and worked at the Scottish Survey, succeeding Archibald in the Murchison Chair at the University of Edinburgh. The younger Geikie was also a highly successful geological writer, specializing in studies of glacial phenomena. Numerous James Geikie family papers are preserved at the University of Edinburgh, but the archive contains little correspondence with Archibald. So information concerning the personal relations between the two brother geologists is very scant, and again we lack information about the elder Geikie's private thoughts.

11. Geological Society, London, Murchison Papers, E34, Murchison to Geikie, July 10, 1860.

12. Murchison to Geikie, July 24 (or 21?).

13. Ibid.

14. Geological Society Archives, Murchison Papers, M/J 25, "1861–62, France, Scotland, Bohemia, &c.," 17. This volume of Murchison's journal, which I use extensively in reconstructing the events of the Murchison-Geikie tour, is titled incorrectly. Within its covers, it has a section labeled "Scotland, 1860."

15. Murchison Journal, "France, Scotland, Bohemia, etc.," 19.

Great Glen, where again Nicol's map came in for criticism. Their further route lay through Glen Garry to Loch Quoich, from whence they crossed the watershed to the sea loch, Loch Hourn, which connects with the Sound of Sleat, separating Skye from the mainland. On one of the days, a boat picnic was enjoyed on Loch Hourn.

At Glen Quoich, the geologists stayed with a Mr. Edward Ellice, M.P., where they were able to enjoy the company of a large house party (Geikie 1924, 79). One of those present was the Frenchman, Prosper Mérimée, author of the libretto of *Carmen,* and an amateur artist. Mérimée produced a crayon sketch of Murchison and Geikie, which has been preserved in the archives of the National Library of Scotland, and which is shown in plate 4.[16] I think it shows something of the nature of the personal relationship that existed at that time between Geikie and Murchison, Geikie, a man of small physical stature, seeming to be ever eager to do the bidding of the distinguished elder statesman of geology. The sketch was not, of course, a great artistic accomplishment, and Geikie, who was fully entitled to make comment, being himself a fine artist, wrote in his autobiography:

> One of the company was Prosper Mérimée, whose chief occupation out of doors seemed to be sketching the scenery in body-color on coarse brown paper. He showed me some of his drawings, which filled me with surprise that so accomplished a man of letters should spend his time in such crude and inartistic efforts to represent the noble scenery around him. (1924, 79)

But perhaps Geikie objected as much to the manner in which he was depicted as to the artistic deficiencies of Mérimée!

From Loch Quoich, the geologists struck north over a steep pass into Glen Shiel and stayed with friends of Murchison at Inverinate on Loch Duich on the twenty-fourth; and at Balmacara, further down the loch, overlooking Skye, on the twenty-ninth. Murchison recorded numerous observations and made several suggestions as to how the disposition of the rocks might be interpreted, but there is no evidence in his field notes or journal that he did more than make a very cursory survey. Nothing in the way of mapping was attempted of course. He was in the land of his warlike ancestors and was as much intent on renewing old acquaintances as engaging in geological research.

Leaving Balmacara on the twenty-ninth, the road took the travelers north to Loch Carron, at the township of Strome Ferry (Jeantown). Thence they went to Loch Kishorn. They were at last approaching the southwest end of that interesting outcrop of rocks, running from Durness to Skye, mentioned in previous chapters, the interpretation of which formed the central problem

16. National Library of Scotland Archives, MS 15174/76. (The sheep are in the picture because certain glacial remains—which would have interested Murchison and Geikie—somewhat resemble sheep and are commonly called *roches moutonées*.)

for understanding the geology of the Northwest Highlands. We shall therefore examine what they did in more detail from this point on.

Loch Kishorn is a sea loch on the mainland pointing northeastward from
opposite Skye toward Loch Eriboll on the north coast, about ninety miles away
as the crow flies. To the northwest of the loch there is a large mass of red-
brown or "Cambrian" sandstone.[17] To the southeast there is a highly confusing mixture of this sandstone, and "eastern" gneiss. Within the valley of the
loch, there are extensive exposures of (Durness) limestone, overlying imbricated fucoid beds and Salterella grit, and these overlie quartz rock in the usual
manner. But there is also some quartz rock lying *above* the limestone, as
Murchison's theory—invoking two quartzites—required. According to modern maps (Institute of Geological Sciences 1975), an important thrust plane
(the Kishorn Thrust plane) runs northwest from the loch, and the area of
sandstone and gneiss to the southwest of the loch is a "nappe" consisting of
inverted Torridon Sandstone and Lewisian Gneiss.[18] It is a very difficult area to
construe.

What, then, did Murchison make of all this? He wrote in his journal:

> Nothing new to remark, except that Nicol's Map is inaccurate in not
> showing the lower quartz setting in above the head of Loch Kishorn,
> and the limestone about a mile long thinning out in the quartz rock.[19]
> The appearance at the head of Loch Kishorn of the sandstone (Cam-
> brian) overlying the limestone is entirely deceptive. It seems to me
> these two formations can be in contact through an upthrow of the
> older rock athwart the younger formations.[20]

I do not know exactly what Murchison observed, but he seems not to have
seen the "eastern" gneiss, which here lies *above* the "Cambrian" because of the
inversion. There is a contact near the road, but it is not clearly exposed today
and was probably obscured by drift in Murchison's day also. Chiefly, he seems
to have been reacting to Nicol's map rather than conducting his own survey.
He was aware that the Kishorn section, with "Cambrian" sandstone evidently
lying higher than the limestone, presented some difficulty for any straightforward interpretation. But beyond vaguely suggesting the existence of a fault
to account for this, Murchison had nothing to say about the matter. Geikie's
field notes made no mention of the Kishorn exposures. In fact, I suspect the
geologists simply drove along the road, with Murchison making a few notes as
he went.

At Shieldag, near Upper Loch Torridon, the travelers ran into magnificent

---

17. This is, of course, the Torridon Sandstone of modern geology. Loch Torridon, the type area
for this rock, is the next sea loch north of Kishorn.
18. A "nappe" is a recumbent fold or anticline—that is, one that has been pushed over onto its
side in such a way that the rocks of one limb of the anticline have become inverted.
19. For his earlier remarks about Kishorn, see page 52. See also page 59 for Nicol on Kishorn.
20. Murchison Journal, "France, Scotland, Bohemia, etc.," 44.

gneiss terrain; and then, following the road inland, and passing great thicknesses of "Cambrian" sandstone, they traveled to the little village of Kinlochewe at the head of Loch Maree, often considered the most beautiful of all the
Scottish lochs, with its many islands and the survival of a fair amount of original forest. Murchison and Geikie reached there on August 30, having walked
twelve miles over the watershed for want of transport, and arriving, as Murchison put it, "well soaked and heated and midged."[21] On the way Murchison
was somewhat puzzled about the relationships between the quartz rock and
the "Cambrian" near Loch Clare, where Nicol had claimed alternations of the
two kinds of rock, and Geikie was sent back a few days later to look into the
matter more closely.

Unfortunately, we do not have Geikie's notes for this part of his journey,
but the paper that the two geologists published on their excursion in 1861—
which was to become their major statement of their views on the geology of
the Northwest Highlands—gives us some information as to what he did
(Murchison and Geikie 1861a, 193–97). Even with an extensive published description of this part of the journey, however, there is considerable difficulty
in ascertaining just where Geikie went and what he saw. This raises the interesting question of the quality of geological description in the 1860s, and of its
effects on geological controversy. In the relevant part of the paper, Geikie
gave four sections, only one of them with a scale and none of them with compass indications. His description of the area is such that it is almost impossible
to gauge what he was referring to for much of the time. To replicate Geikie's
observations here, the geologist would have to wander round as in a maze,
hoping by chance to come upon the places that he was describing. In bad
weather it would be a hopeless task. In fact, to achieve a satisfactory replication, one would need to have Geikie act as guide to the excursion. Yet this
would scarcely yield an independent set of observations. Thus there was considerable need for improvement in geological communication. The members
of the Geological Society in London, or those attending British Association
meetings, listening to a paper describing a little-known region, would have
had to be guided by the reputations of the speakers, or by the eloquence and
confidence of their exposition, as much as anything else. It might be many
years before other geologists would go and check out these matters for themselves; and when they did they might find things rather different from what
they anticipated on the basis of the early reports, if in fact they could find the
relevant sites at all.

It is clear that Geikie climbed Sgurr Dubh (the "black sharp rock"), on the
south side of the road between Shieldag and Kinlochewe, at the watershed,
and he seems to have gone up the valley in which Loch Clare is situated,
southward toward the Dingwall–Loch Carron road. He definitely found good
evidence of unconformity between the quartz rock and the "Cambrian" at

21. Ibid., 49.

Sgurr Dubh, but this was not unexpected. He gave a section of the "Dun Tol-leah" Hills (not shown as such on modern maps) that indicates severe (high-angle) faulting and repetition of the quartz rock and limestone beds, but the direction of the section is not indicated. However, he did explain in the joint paper how this repetition might be represented and understood:

> The map of this region would therefore represent a series of irregular strips and patches of Silurian quartz-rock, inserted among Cambrian sandstones by the agency of longitudinal faults which seem to split off from or coalesce with each other. (Murchison and Geikie 1861a, 197)

Thus Geikie was not so far away from Nicol's "alternations" of strata.[22]

The trouble was that there was no detailed map, or none that was pub-lished. The mid-nineteenth-century geologists, on a reconnaissance survey, would swiftly proceed to a section—a highly theoretical thing—in giving a statement of the geology of an area. Of course, they had little choice if they were to say anything much about the supposed structure of a region before proper mapping was carried out. But it was not an ideal arrangement. For satisfactory analysis of the structure of a region, the map should really come first. Yet the participants in the debates were forever displaying their sections. Dozens of them of them were produced during the course of the Highlands controversy, and the main debate always revolved around sections. It is a pity, therefore, that they were often not executed with greater care. In fact, many of them were merely copies of what were sketched in the field in damp notebooks with cold hands! This can clearly be seen by comparing the published sections of Murchison and Geikie with those to be found in their field notebooks.

Let us return to Geikie and Murchison ensconced in the inn at Kinlochewe at the head of Loch Maree, noting that at Loch Clare Geikie had collected in-formation that was not uncongenial to Nicol's case. For much of August 30, the geologists were kept in by bad weather. But in the evening they sallied out to walk up Glen Logan to the northeast of the village.[23] They were now in what was one of the key areas for the whole Highland dispute (see fig. 5.1). Not far up Glen Logan, conveniently adjacent to the path leading up from the village of Kinlochewe, there is a small gorge in the river bed where one may find an extraordinary mixture of rocks, some of them obviously highly folded. The general appearance of the rocks at this point is shown in figure 5.2.

This area became a *locus classicus* for the Highlands controversy. According to modern maps, there are several thrust planes in the vicinity. One of them brings gneiss *over* limestone, fucoid rock, et cetera. Another brings (Moine)

---

22. Murchison was perhaps not too pleased with Geikie's independent work. In his autobiogra-phy, Geikie recorded that when shown his sections Murchison was "rather nonplussed, but thought they were merely local phenomena" (1924, 82).

23. This glen is not shown as such on modern maps. Instead, one finds the Gaelic name, Abhainn Bruachaig ("river of the bank"), for one of the two main rivers that run into the head of Loch Maree. The glen is also called Glen Cruchanie ("high-hill glen"?) by Murchison in his notes.

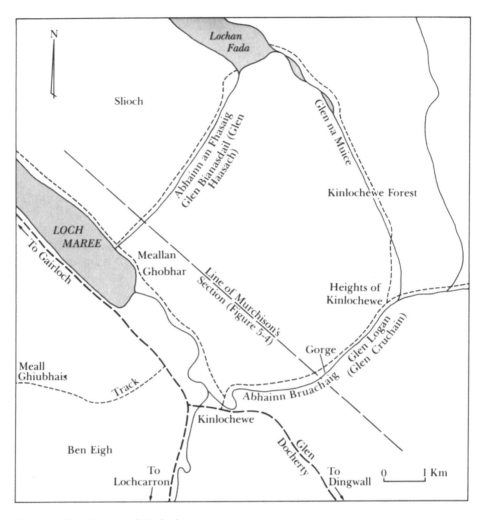

Fig. 5.1. Sketch map of Kinlochewe area.

schist *over* the gneiss from the east. But since schists and gneisses were not always clearly distinguished in the 1860s, there was ample opportunity for confusion. Indeed, the gneiss, quite altered in this area in the gorge, and lying *between* limestone and schist, might appear as some kind of igneous rock. Thus it was that Nicol could imagine that the gneiss in the river bed of Glen Logan was the igneous rock that he supposed he had found at all his junctions between the sediments (limestone, quartz rock, etc.) and superincumbent metamorphic rock ("eastern" gneiss or schists).

In Glen Logan, Murchison recorded nothing of special note in his journal during this, perhaps his first, visit to the place. He recorded limestone above

Fig. 5.2. Rocks in gorge at Glen Logan (Glen Cruchain).

"shale bands" (fucoid beds?) above quartz rock, all dipping southeast at about thirty degrees, which accorded well with his general views of the structure of the Highlands. He also reported annelid markings on both the upper and lower surfaces of the quartz rock, which might have led him to wonder whether some of the beds were inverted, but this idea did not seem to present itself.

Murchison visited the site alone a few days later on September 4. This time he recorded the presence of a rock of igneous appearance, which he described as "granitoid," to be found in the river bed and sweeping up the hill to the west.[24] In the published version of the field trip, there was also mention of an unusual kind of rock found in Glen Logan. "It is difficult to give this rock one specific name," said Murchison, "for . . . it varies greatly in mineral composition, even within a few yards. Near the limestone it is a serpentine; the green mineral then thins away, and quartz and felspar take its place, while to these is occasionally added hornblende. The proportions of the ingredients also vary to a large extent." (Murchison and Geikie 1861a, 192.) Murchison claimed to have described something similar in his paper recounting his northern journey with Ramsay (Murchison 1860, 238)—though from his description one would hardly know this. Whatever it was, it was held not to interfere with the general order of superposition of the rocks.

24. Murchison Journal, "France, Scotland, Bohemia, etc.," 59. (Today, this "granitoid" rock is mapped as gneiss.)

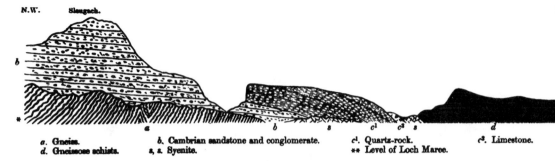

N.W.   Slougach.

b

a. Gneiss.   *b*. Cambrian sandstone and conglomerate.   *c¹*. Quartz-rock.   *c³*. Limestone.
d. Gneissose schists.   *s, s*. Syenite.   ** Level of Loch Maree.

Fig. 5.3. Section of strata at head of Loch Maree (north side), according to Murchison (Murchison and Geikie 1861a, 191).

What was this rock? Its character was ambiguous, the reference to serpentine being an indication of ignorance rather than knowledge. In the Logan valley, the geologists were in fact dealing with a region where some of the rocks had been considerably altered by thrust faulting. The mention of some kind of green material suggests the mineral epidote, which is characteristically found in the gneiss of Glen Logan.[25] The rock was presumably what later came to be called "Logan Rock" (until its nature as a gneiss was determined by studying it in thin section with a microscope). As mentioned previously, the interpretation of this rock was to cause considerable difficulties in the course of the Highlands controversy. But in 1860, Murchison and Geikie did not recognize that it represented a significant petrographic problem.

On September 2, while Geikie had gone back to examine Sgurr Dubh and Loch Clare, Murchison took himself along the side of Loch Maree to where a river runs in at right angles carrying water from a higher-level loch to the north (Lochan Fadda or the "long small loch"). The glen of this river, which runs under the fine "Cambrian" sandstone mountain of Slioch, is called Bianasdail (the "hide dwelling") on modern maps, and the river is called Abhainn an Fhasaigh. Murchison called it Glen Haasach ("desolate glen"). The results of his walk were figured in his notebook, and the sketch was cut out and incorporated in his journal.[26] It subsequently appeared in the published version of Murchison and Geikie's account (1861a, 191). (See fig. 5.3.)

It is instructive to compare the published section with the sketch in the notebook, which I have copied in part, and which is shown in figure 5.4. The published section shows Murchison's characteristic theoretical structure for the Highlands, with a regular sequence rising from quartz rock through limestone to "gneissose schists" ("eastern" gneiss), the thin but conspicuous fucoid beds being ignored. The gneiss in Glen Logan, which Murchison identified ambiguously in his notes as "granitoid," has been partly represented as "syenite," and for the rest has been conflated with the schist or

25. On epidote, see pages 180 and 194 and plate 6.
26. Murchison Journal, "France, Scotland, Bohemia, etc.," 52.

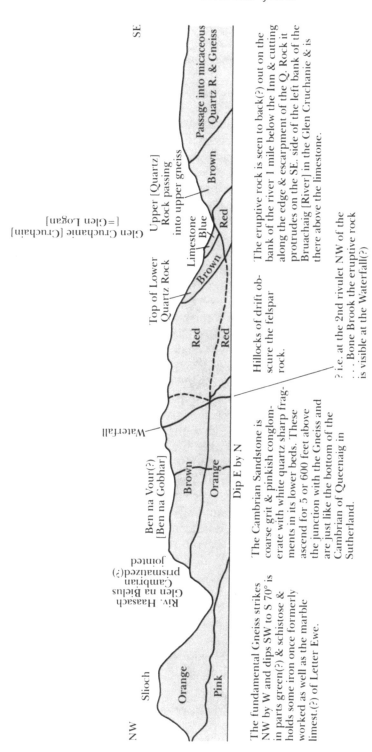

The analogy to Assynt & Sutherland—if not identity—is complete—Distance 4 to 5 miles.

SE

Passage into micaceous Quartz R. & Gneiss

Upper [Quartz] Rock passing into upper gneiss

Brown

Red

Blue

Limestone

Glen Cruchanie (Cruchain) [=Glen Logan]

Brown

Top of Lower Quartz Rock

Red

Red

Waterfall

Dip E by N

Hillocks of drift obscure the felspar rock.

? i.e. at the 2nd rivulet NW of the ... Bone Brook the eruptive rock is visible at the Waterfall(?)

The eruptive rock is seen to back(?) out on the bank of the river 1 mile below the Inn & cutting along the edge & escarpment of the Q. Rock it protrudes on the SE. side of the left bank of the Bruachaig [River] in the Glen Gruchanie & is there above the limestone.

Ben na Vour(?) [Ben na Gobhar]

Brown

Orange

The Cambrian Sandstone is coarse grit & pinkish conglomerate with white quartz sharp fragments in its lower beds. These ascend for 5 or 600 feet above the junction with the Gneiss and are just like the bottom of the Cambrian of Queenaig in Sutherland.

Riv. Haasach Glen na Bielus Cambrian prismatized(?) jointed

Slioch

Orange

Pink

NW

The fundamental Gneiss strikes NW by W and dips SW to S 70° is in parts green(?) & schistose & holds some iron once formerly worked as well as the marble limest.(?) of Letter Ewe.

Fig. 5.4. Section of strata at head of Loch Maree (north side), according to hand-colored sketch in Murchison's field notebook, 1860 (redrawn and with accompanying text reproduced). The colors were evidently intended by Murchison to represent particular strata as follows: Pink equals Fundamental Gneiss; Orange equals Red-brown "Cambrian" sandstone; Brown equals Quartz rock; Red equals Eruptive rock of uncertain nature; Blue equals Limestone. Courtesy of the Library Committee of the Geological Society, London.

misrepresented as quartz rock.[27] It should be noted that the "syenite" (*s*) is also shown outcropping at a low level between Kinlochewe and Glen Haasach, and the reader would be likely to assume a subterranean connection between that and the "syenite" in Glen Logan.

But the original sketch shows something substantially different. In the field section, Murchison gives the "syenite" as rising up to the top of the hill to the west of Glen Logan, breaking through both the "Cambrian" sandstone and the quartz rock, and clearly connecting underground with the "syenite" in Glen Logan. In his field notes he refers to an igneous rock to be seen by the side of the path along which he walked to Glen Haasach.

What was going on? First, on modern maps, the "syenite" on the path to Glen Haasach is marked as "gneiss," though it must be said that it is a good deal altered, and not very easily recognizable as such. But on Murchison's view, there was no reason for gneiss to be there. It should have been at a higher level and off to the east. So he construed it as "syenite". Second, modern maps have gneiss on the *top* of the hill, above the quartz rock, and curving down the hillside into the valley. It does this because of the inclination of the thrust plane. But Murchison, though he figured the whole of the hillside in his published section, clearly did not climb to the top of the hill to have a look to see what was there. He simply assumed that quartz rock ran to the top.

In fact, the situation was worse even than this. For according to modern maps (Institute of Geological Sciences 1962), on the top of this hill may be found Torridon Sandstone, lying atop the gneiss, also emplaced by a thrusting process. It had no reason to be there, of course, according to Murchison's theory. But we need not accuse him of faking his observations. It is perfectly clear that he simply did not climb to the top of the hill, and very likely it was obscured by cloud at the time that he walked by. But the point I made about the problem with sections for the presentation of geological results is particularly well supported by this example. There can be no doubt that Murchison figured in section strata that he had not properly examined in the field.

Finally, there is the question of the igneous rock, which appears so conspicuously in the field diagram, but which is largely edited out in the published section. Here it may be noted that if the igneous rock had been included, it would have been quite in keeping with Nicol's ideas. I believe that in making his walk and his observations Murchison may have been considerably influenced by Nicol's ideas, perhaps unconsciously. But subsequently, and on reflection, and after having discussed the matter with Geikie, he thought better of it, and the seeming massive intrusion, which he knew very well he had not traced to the top of the hill, was expunged. Of course the whole outcome might have been different if Murchison had had Geikie's younger pair of legs with him to climb the hillside. But whatever hypotheses or excuses one may make, the outcome

27. In his journal (page 60), Murchison said that the rock of granitoid appearance swept up the hill to the west of Glen Logan. But in his published section (see Fig. 5.3), it is quartz rock that does this.

was that a largely erroneous structure was published, which did much to add to the confusion attending the interpretation of the geology of northwest Scotland.

On September 4 Murchison made an excursion by boat down Loch Maree to Letterewe, a homestead on the northern side of the loch, inaccessible by road and site of the old iron industries in that part of Scotland.[28] Going westward, he was moving into the territory of Fundamental Gneiss, which, however, at Letterewe is remarkable for containing a substantial deposit of limestone. Nicol had confused this with the Durness Limestone, typically found in the northwest, and therefore his map was in error at that point. Murchison was out to demonstrate this error, which was not difficult to do, since the strike of the "Fundamental" limestone ran parallel to that of the "western" (Fundamental) gneiss, namely northwest to southeast, almost at right angles to the typical strike of the Durness Limestone. Nicol had indicated the Letterewe limestone as striking northeast to southwest. It was an agreeable little triumph for Murchison, as his journal demonstrated.[29] He also correctly observed that the line of Loch Maree must mark a fault parallel to the strike of the "Fundamental" limestone, since the exposures of "Silurian" rocks (quartz rock, etc.) did not match up on opposite banks.

On September 4 Geikie and Murchison parted company—Geikie to walk alone on a tremendous march of thirty-two miles, tracing the "Silurian" outcrops over wild and inhospitable country to Ullapool, and then (the next day) further north; Murchison to take the road west a few days later to the coast at Gairloch, whence he took ship to Stornoway in Lewis, in the Outer Hebrides. Here he was interested to examine more closely the Lewisian Fundamental Gneiss, but he also intended to enjoy the company of the local lairds and indulge in some deer-stalking. The two geologists rejoined in the middle of September at Balmacara on the mainland, after Geikie had done fieldwork in Skye. Since we need not be concerned in this study with either the geology of Lewis or deer-stalking, we may leave Murchison for the moment and follow Geikie's tracks.

Geikie was in such a hurry during his long march that he apparently took no notes at the time, only writing up his recollections the following day, and these were fairly brief.[30] We do not have his letters to Murchison while the two were separated, though we know that Geikie wrote to him, since Murchison mentions the fact in one of his letters to Geikie.[31] The best source of information about Geikie's journey is therefore the Murchison-Geikie paper of 1861,

28. See Dixon (1886, chap. 18): "The Historic Iron Works of Loch Maree."
29. Murchison Journal, "France, Scotland, Bohemia, etc.," 54 and 57. Murchison (page 57) found it "absolutely incredible" that any geologist would confound the two limestones.
30. A. Geikie, Field Notebooks, Haslemere Educational Museum, Notebook A, 5–6.
31. Geological Society Archives, Murchison Papers, E34, Murchison to Geikie, Sept. 9 1860.

though as we have seen, this is not always very clear in the matter of localities. A curious feature of the paper is that it presents all the observations exactly opposite to the way in which they were actually made, as evidenced by Murchison's journal and Geikie's field notes. Why this policy was adopted one can only surmise, but it may be that it was believed that the very best section in favor of Murchison's theory was at Knockan Cliff ("rock cliff"), a little south of Loch Assynt, Inchnadamph apparently being the furthest north that Geikie reached on this trip. So with Knockan Cliff described first, the whole argument might appear more convincing. Whatever the reason, there is no doubt that the journey was in fact carried out from south to north.

Inspection of a map suggests that Geikie's probable route was as follows. He would have left Kinlochewe and walked toward Loch Maree, turning up the track in Gleann Bianasdail to Lochan Fadda, with the "Cambrian" sandstone mountain of Slioch on his left and quartz rock topped by gneiss on the eastern side of the glen. Skirting the lochan, he would have headed toward a saddle (Bealach na Croise or "cross pass") leading to Loch Nid, there being no track here, and with dreadful boggy ground (gneiss). At Loch Nid he would have picked up a track leading due north past Dundonnell Forest to the valley (Strath Beag, or "small flat part of river valley") at the head of little Loch Broom. As he walked down Gleann Chaorachain to the Strath, he would have had the magnificent "Cambrian" mountain, An Teallach, on his left (west). He might have paused for tea at Dundonnell House in Strath Beag, before pushing north again to Loch Broom. He would have reached this by rounding the hill, Ben nam Ban, and arrived at dusk at the little hamlet of Allt na h-Airbhe on the loch opposite Ullapool, which he would have had to reach by means of a ferry. It was a long day's walk, over rough ground, and there can have been little time for much other than walking. It is not surprising that Geikie's field notebook is blank for this section of his tour.

If now we consult a geological map, it can be seen that for much of the time (after Lock Nid) Geikie would have been walking along the strike of the quartz rock, with "Cambrian" sandstone on his left (west) and (Moine) schist on his right. (Here for the most part the Moine Schist cuts out the fucoid beds and the Durness Limestone.) So he was walking right under the Moine Thrust plane for nearly twenty miles! He called the schist "gneissose schists" (Murchison and Geikie 1861a, 188, fig. 7). Thus a serious problem within the Murchison-Geikie theory was that no clear distinction was made between schists and gneisses. The "eastern" metamorphic rocks were all lumped together as one.

But Geikie seems to have been chiefly interested in the quartz rock and its relationship to the "Cambrian" sandstone. Some fine contacts can be seen in the valley of Loch Nid and on the slopes of An Teallach (see fig 2.1), and he depicted these in the paper, but again without scale and compass directions, so that it is only when one is actually on the spot with Geikie's sections in hand that they become meaningful and can be interpreted properly. Fortunately, in this case the localities are not particularly difficult to identify. The sections represent the "gneissose schists" as conformable with the quartz rock, even

though the limestone and fucoid beds were missing there.[32] It would have
been helpful if Geikie had examined the sections just before reaching Loch
Broom, for this was one of the main localities on which Nicol rested his case.
Unfortunately (though entirely understandably) he did not do so. Doubtless
very tired, and with evening drawing in, he took the ferry to Ullapool as
quickly as possible. Thus ended a great day in the field for Geikie. But to do
justice to that piece of terrain, he should not have attempted to do so much in
just one day. He was perhaps only the second geologist to cover the area near
Loch Nid. The outcrop of quartz rock there is shown on Cunningham's map.
Nicol may also have passed that way and possibly also Macculloch, though his
map is very inexact at that point. (See plate 1.)

The following day (September 5), Geikie explored with considerable care
the first couple of miles of the north shore of Loch Broom, inland from Ullapool. We have an exact account of his work from his field notes.[33] Loch
Broom was one of Nicol's key sections, where, he claimed, an igneous rock (a
"felspar-porphyry") could be found at the point of fracture where the eastern
gneiss had supposedly been forced to the surface. Naturally, Geikie was much
interested to examine this rock closely. He walked along the shore road a
short distance eastward from Ullapool to Corry Point, where a small burn,
Allt Corry (Corry stream), issues into the lake. The "porphyritic" rock was described thus:

> Where it overlies the limestone it is a greenish serpentine quartzose
> rock. Then the serpentinous matter thins out and pinkish felspathic
> matter takes its place with a plentiful admixture of quartz granules.
> Glancing over the exposed surfaces, of this rock I was surprised to ob-
> serve residual pebbles of red jasper. Scrutinizing [indeciph. word] I
> found the porphyritic-looking rock gradually assume all the aspect of a
> highly altered quartz and red jasper become very numerous, with the
> quartz granules come to form at least a half of the rock. The weathered
> surface showed how the felspathic matter had wasted away leaving a
> closely aggregated mass of smaller white & pinkish quartz granules all
> more or less rounded. In just a few instances they were arranged in
> rows, indicating lines of stratification.[34]

In the published version, Geikie described the rock as a "highly metamorphosed band of felspathic grit" (Murchison and Geikie 1861a, 186). On modern maps, the locality is represented as gneiss (Geological Survey of Great
Britain [Scotland] 1948).

So far as I have been able to ascertain, Geikie's description of the rocks in
the Allt Corry locality was quite satisfactory, in terms of the relative dispositions
of the different beds and the external appearance of the "felspar-porphyry" or

32. Presumably they may be present underneath the Moine Schists.
33. Geikie, Notebook A, 1–5.
34. Geikie, Notebook A, 1–2..

Fig. 5.5. Lewisian Gneiss ("Logan Rock") at Allt Cory, northern shore of Loch Broom, Ullapool.

"felspathic grit." In particular, he was perfectly justified in claiming that it exhibited an appearance of stratification, as figure 5.5 indicates.

Thus Geikie had good reason to deny Nicol's claim that the rock was igneous. As he put it: "We do not regard it as igneous at all, further than the gneiss is igneous" (Murchison and Geikie 1861a, 185). Nevertheless, it was really a rather peculiar rock. In hand specimen, some quite well-formed crystals of feldspar may be observed, and the seeming igneous appearance is more striking where the rock outcrops on the southern shore of the loch. *This* was, in fact, the locality on the basis of which Nicol had argued for the rock being igneous. But it will be remembered that Geikie could not have paused on the south shore at the end of his long march. So here we have an intriguing feature of the Highlands controversy. Geikie looked chiefly at the north shore of Loch Broom; Nicol's main evidence concerned the southern shore. They were indeed inspecting extensions of the same stratum. But the rocks they examined do not look exactly the same. Nicol could construe the rock as igneous. Geikie could see it as quasi-sedimentary, or perhaps some kind of altered gneiss. I suggest, therefore, that part of the difficulty attending the Highlands controversy was simply the fact that the participants to the debate did not always look at the same rocks, though they believed that they were doing so; and hence they imagined that their opponents were simply incompetent observers. The problem was exacerbated by the fact that the descriptions of the localities given in the published papers were often far from precise. The enigmatic rock of Loch Broom was, needless to say, just another exposure of the mysterious "Logan Rock."[35]

On the afternoon of September 5, Geikie went a little north of Ullapool into Glen Achall. At present, on the road to this glen, there is a huge quarry where limestone is extracted from the extensive deposits of the Durness Limestone that may be found there. However, Geikie gave attention to the quartz rock and its contact with the "Cambrian" sandstone. Moreover, he discovered once again the same "red quartzo-felspathic rock" that he had been observing by the lake in the morning. It looked similar, but it was not exactly the same. It appeared to him to have undergone a greater degree of metamorphism, which obscured its "original detrital character."[36] This might have alerted Geikie to the variable character of this rock, about which he and Nicol talked at cross-purposes. But such a consideration does not seem to have occurred to him.

The following day, Geikie pushed further north on the road leading to Loch Assynt, examining in some detail the area round Strath Kanaird, where he acknowledged that the "Cambrian" sandstone resembled the Old Red Sandstone, as Hugh Miller had earlier maintained. But also it sometimes seemed to resemble the "pseudo-porphyry" of Loch Broom! Also, the "upper quartzite" of Murchison's theory seemed to be represented here by a "gneissose quartzy schist."[37] Certainly there were considerable problems in unravelling the geology of this area. A great deal of the problems had to do with imprecise field interpretations of much altered rocks. But how could Geikie, single-handed, have carried all the needful specimens away for analysis? In any case, the Survey had not, at that stage, begun to use the technique of microscopic examination of thin sections for rock identification.

Geikie described some faulting in the Strath Kanaird area in his notes, but I have not been able to locate the place on the basis of his description. There was no suggestion that the faulting was in any way unusual. The faults were all interpreted as "dip faults."[38] Reacting to Nicol, Geikie asserted that there was no evidence of igneous disturbance.

Geikie's notes are undated for the next few days, but he appears to have made a very hasty push north to Craig an Knockan (Knockan Cliff) about five miles north of Strath Kanaird, where he figured in his notes and in the published paper a cliff section showing a conformable sequence:

> Gneissose schist
> [Upper] limestone
> [Upper] quartz rock

35. At present, there is a new road cutting at Corry Point, in which the observer may readily see evidences for thrust faulting in this area. Such evidence was not, of course, available to Geikie; and the section (as he would have seen it) does not "proclaim" itself as a thrust fault. It has to be construed thus with the help of theory.
36. Again this is mapped as gneiss on modern maps.
37. Geikie, Notebook A, 17.
38. Ibid., 20.

Limestone
Fucoid beds
[Lower] quartz rock

All this accorded with the Murchisonian theory. Modern publications concerning this classic site, however, only allow one limestone (Nature Conservancy Council 1986, 7). The junction between the "gneissose schist" and the limestone is today construed as the Moine Thrust, the upper rock being the Moine Schist.

Geikie seems to have reached Loch Assynt but to have spent almost no time there, since he made very few notes on this locality. There is just a mention of the important section of Cnoc an Droighinn (Drein) at Inchnadamph, but no figure or description. We may conclude, then, that Geikie never gained adequate first-hand experience of the rocks of the crucially important Assynt region, or for that matter the Durness region, in the early stages of the Highlands controversy. He took Murchison's word for it and was happy to press the Murchisonian case with vigor, as we shall see.

Returning to Ullapool, Geikie then spent a day or two exploring inland in the huge region of the Moine Schists. He traveled southeast to the head of Loch Broom, past the magnificent Corrieshalloch Gorge, over a bleak moor to Loch Glascarnoch and on down to Loch Garve. He was examining the general inclination of the schists, which seemed for the most part to have bedding-plane dips (not planes of schistosity) to the southeast, as Murchison's theory required. To be sure, things were not always regular. Sometimes the rock seemed more schistose, sometimes more gneissose. Sometimes considerable contortions and reversals of dip could be observed. But on the whole, the upper schist/gneiss was seemingly sedimentary in origin and such that it might reasonably be located above the Durness Limestone, lying in an essentially comfortable sequence. Lest it be thought that this is a wholly inappropriate interpretation, I offer a photograph (fig. 5.6) taken at the roadside near Corrieshalloch Gorge, which shows the eastern metamorphic strata displaying a rather sedimentary appearance. From Loch Garve, Geikie was able to take the road to the Kyle of Lochalsh, and the ferry to the Isle of Skye, where we meet him on September 11.

The geology of Skye can be roughly described as follows. The north of the island is made up of large areas of extruded rock, chiefly basalts of Tertiary age. There is a rugged central section made up of two main types of coarse-grained igneous plutonic rocks.[39] There are the Cuillin Hills to the west, made up of a basic intrusive rock called gabbro, which form the most rugged terrain in the whole of the United Kingdom and provide a mecca for rock climbers. Abutting these are the more rounded Red Hills, made of the quartz-rich

---

39. "Plutonic" describes deep-seated instrusions, which cooled slowly to form very coarse-grained rocks.

. 5.6. Moine Schists, near Corrieshalloch Gorge.

acidic rock, granite. Both the gabbro and the granite are also thought to be of Tertiary age. But it is a nice question which is the older. Did the gabbro come first, later to be intruded by the granite? Or vice versa? This question formed the topic of an acrimonious debate between J. W. Judd and Archibald Geikie, which had important implications for the Highlands controversy, and in a sense was part of it. We shall attend to that matter in a later chapter, but I need say no more about it here.

To the southeast of this massive region of plutonic rock, one finds a rim of much older rocks: the red-brown Torridon Sandstone and the Durness Limestone. But the area is much confused by earth movements, for in some places the sandstone is above the limestone and elsewhere it is below. Next, in the vicinity of the town of Broadford, we have quite a large area of fossiliferous Mesozoic rocks, chiefly Liassic (Lower Jurassic), but also some New Red Sandstone (Triassic).[40] Then in the southeastern portion of the island (on the Sleat Peninsula), we find that a whole series of nappes has brought along the ancient rocks again, chiefly the red-brown "Cambrian" sandstone (sometimes inverted), but also on the south*east* coast, an emplacement of gneiss, arising from the Moine Thrust. One locality on the western side of the peninsula is of special interest: the so-called Ord Window, near the tiny hamlet of Ord on the

40. These rocks occur at a number of other places on the island, but further details are not necessary here.

coast at Loch Eishort, looking over to the Cuillin hills. Here, as at Loch Assynt, the cover of overthrust sandstone has been eroded away (or perhaps it was very thin at that point) to reveal our old friends the quartz rock, the fucoid beds, the Serpulite grit and the Durness Limestone—but twisted and folded in such a way as to present a very complex problem to the stratigrapher.[41]

It is no doubt an act of historiographical anachronism to refer to these modern interpretations at this juncture. I mention them merely to give an indication of the huge problems that the field geologist encounters in Skye, and that the young Geikie sought to elucidate. It was not his first visit to the island. He had been there on holiday before joining the Survey, and again in 1856 during his summer vacation (Geikie 1924, 32, 42, 54), for so keen was he on geology that he customarily did additional fieldwork while on leave. In fact, he had already published a paper on the Mesozoic rocks of the southeast of the island (Geikie 1858a), which was a creditable performance, offering a geological map of the parish of Strath, running southwest from Broadford. However, in that paper Geikie presumed that the old Durness Limestone of Skye was a "marble"—a supposedly metamorhphosed piece of Liassic strata altered by the adjacent igneous intrusions.

It is evident from the Murchison-Geikie paper of 1861 that Geikie had grave doubts about his interpretation of the southern part of Skye, where one finds the termination of the long sequence of quartz rock, limestone, etc.—the "zone of complication"—running down from Durness. He mentioned "the limited time at our disposal, and the want of a [topographical] map having any approach to accuracy" so that "we did not attempt to work out the detailed structure of this region" (Murchison and Geikie 1861a, 199). Remembering that he worked alone on Skye, despite his reference to joint observations, it is instructive to see just what Geikie did.

His main focus of attention was the Sleat Peninsula on the southeast of Skye, where Nicol had previously visited. On September 11, Geikie walked northwest from Armadale Bay (on the southeast coast of the peninsula) along the main road to Ostaig, in gneiss terrain (of the Moine Nappe).[42] He carried on past Knock Bay to where the road branched off to the west to Ord. This road took him across the peninsula. He explored the area round Ord in some detail before turning south and walking over hills to Loch Dhughaill, where he met another road crossing the peninsula that led him back to Ostaig House, where he stayed the night. Thus he had gone round the middle of the peninsula in a large circle, anticlockwise. The next day he again took the main road leading northwest toward Broadford, but instead of branching off to Ord, he kept on through Isleornsay, thus making another traverse of the peninsula. Just before Broadford, he turned eastward toward Kyleakin, the main gateway to Skye from the mainland, where he stayed at Kyle House.[43]

41. For the geology of this region, see British Geological Survey (1976).
42. Geikie, Notebook A, 27.
43. It is perhaps worth noting that Murchison and Geikie virtually always managed to find accommodation at the grandest house in whatever village they passed through, or at the best inn. I do not believe they "roughed it," so far as accommodation was concerned. How the arrangements

On September 14, Geikie sensibly decided to have another look at Ord. He presumably went by carriage to Kilbride at the head of Loch Slapin (which separates the Sleat Peninsula from the rest of Skye), where according to his autobiography, he spent a day or two at the Manse with friends. From Kilbride, he was able to take a boat so that he could examine the strata near Ord from the sea, landing where necessary for closer inspection. His exact itinerary from there is not clear from the field notes, but he appears to have crossed the peninsula once again by the southern of the two traversing roads and returned to his accommodation at Ostaig House. On September 19, he was back on the mainland, at the harbor of Arisaig, well south of Skye, having joined up with Murchison at Balmacara (Geikie 1924, 83). His day or two of rest at Kilbride was surely well-earned.

Considerable interest attaches to Geikie's observations of the extremely intricate structures at Ord. It will be recalled that the general structure of the Highlands envisaged in Murchison's theory involved a fundamental (western) gneiss overlain unconformably by "Cambrian" sandstone, overlain unconformably by lower quartz rock, fucoid beds, Serpulite grit (often ignored), limestone and upper quartz rock, overlain by a gneiss or schist (eastern gneiss).[44] All the rocks above the "Cambrian" sandstone were claimed for the Silurian; all were thought to be more or less conformable; and all were thought to strike northeast to southwest and dip to the southeast. But the rocks at the Sleat Peninsula didn't preserve this sequence at all. If one starts at the east of the peninsula, one has (eastern) gneiss as required by Murchison. But traversing westward one encounters first "Cambrian" sandstone, not limestone. This sandstone *overlies* quartz rock to the west. Then comes the quartz rock again, followed by narrow bands of Serpulite grit and fucoid beds. The quartz rock forms a scarp overlooking Ord, where at a lower level by the sea there is quite a substantial deposit of Durness Limestone. Thus, apart from the eastern gneiss, the whole sequence is back-to-front, compared with the expectations of Murchison's theory. Worse still, walking a little to the south from Ord, "Cambrian" sandstone is encountered once again. But there is no western gneiss. One could hardly have a greater muddle, so far as Murchison's theory was concerned. In particular, it should be noted that the "Silurian" strata in Sleat generally seem to dip to the *west*, rather than the southeast, as is usual.

What did Geikie make of all this? It must be said first that he spent a little longer examining the ground (two days at least) than he had elsewhere during his tour with Murchison. His notebook shows a beautiful colored sketch map of the peninsula, which indicates that he had somewhat more information at his disposal than was usual.[45] However, this remained unpublished. During

---

were made for their housing the surviving document do not reveal. But Murchison had numerous friends throughout Scotland, and on presentation of his card he could hope to receive hospitality, even from those with whom he was not previously acquainted. However, when not traveling with the director, Geikie—and the other surveyors—certainly did not always have access to the finest accommodation available. See Geikie (1904, chap. 14).

44. There might also be two limestones at some sites.

45. Geikie, Notebook A, 41.

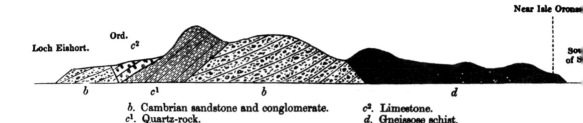

b. Cambrian sandstone and conglomerate. c^2. Limestone.
c^1. Quartz-rock.   d. Gneissose schist.

Fig. 5.7. Section of Sleat Peninsula, Skye, according to Geikie (Murchison and Geikie 1861a, 200).

this trip he also recognized the Durness Limestone for what it was, and no longer regarded it as some kind of marble.

Geikie's notebook shows as well a section across the peninsula from Ord to Isleornsay or Knock on the east coast. The published version (Murchison and Geikie 1861a, 200) is reproduced in figure 5.7.

It can be seen that this section is more or less in keeping with the disposition of the rocks as described above, though it should be noted that Geikie has quartz rock overlying the "Cambrian" sandstone, rather than the other way round as it appears in the field. It must be said, however, that the contact in the field (at the point where I have examined it at any rate) is obscured by drift, so that a mistake of that kind is quite understandable. The line of contact between quartz rock and sandstone in the figured section is drawn chiefly with the aid of Murchison's theory. If it were drawn sloping in the direction of Isleornsay, the whole diagram would look much more "modern" in appearance and might suggest thrust faulting—but Geikie did not have this concept to deploy.

The eastern boundaries of the Ord Window are marked by low-angle thrust faults, according to the modern interpretation. But to the southwest of the window there is a more usual high-angle fault. Geikie had no difficulty in recognizing this, and he figured it correctly in his unpublished sketch map. It can also be seen in figure 5.7, separating the limestone and the sandstone near Ord. Why was Geikie able to recognize high-angle, but not low-angle, faults? The reason was, I suggest, the way in which these manifested themselves in the field. For a high-angle fault, one might be tracing a bed, and then there might be a sudden switch to another bed along the line of the strike. Geikie found just this at Ord, where near the shore the limestone suddenly switched to sandstone. Or one might look at the two sides of a valley and see that the rocks on the opposite sides did not match up. (Murchison did this at Loch Maree, and he correctly inferred that the loch marked a line of fault.) Faults might be gauged also by the appearance of streams, rivers, or lakes, or even ridges on the ground, though these could often be eroded away. Sometimes the rocks at a fault surface exhibit a curious glassy, rubbed appearance, which can be a strong clue to the field geologist.

The reader will, on reflection, conceive that a vertical fault will traverse the countryside in an exactly straight line, independently of the topography. Conversely, the outcrop of a horizontal (thrust) fault will, like a horizontal bedding plane, follow a contour line exactly. So the outcrops of faults of intermediate angle will be almost straight for high-angle faults, and will be increasingly sinuous the less the inclination of the fault. In Geikie's day, geologists were well accustomed to recognizing high-angle faults, and Geikie did so without difficulty for the fault of that type at Ord. But he simply did not recognize the low-angle faults that are also to be found in the district. This is neither surprising nor reprehensible.

The structure that Geikie suggested for Sleat was that there had been an anticlinal arch, with the sandstone at the center, overlain by quartz rock, limestone, and gneiss (Murchison and Geikie 1861a, 199). Then, with much disturbance of the ground and faulting, the western side of the arch had supposedly fallen in to give the structure indicated in figure 5.7. It was not a very convincing argument, but after but two days of fieldwork in the area, with Murchison's theory like an albatross on his neck, and lacking the concept of low-angle thrust faults, it is not easy to see how Geikie could have done any better.

For the part of his fieldwork between Isleornsay and Broadford, Geikie was evidently wishing to test Nicol's claim that he had seen a rock of porphyritic appearance adjacent to the eastern gneiss. Geikie admitted that he found a rock on the roadside a little north of Isleornsay that might be thought to be a weathered porphyry. (He called it "a small pustule of porphyry soft with decomposing.")[46] But he denied it any significance and claimed that it in no way disturbed the general southeasterly dip of the gneiss. Modern maps show no such rock, and I am unable to suggest what it may have been. Geikie complained against Nicol that he seemed to have kept wholly to the road during his fieldwork in this district.[47] We may conclude, therefore, that Geikie himself was constantly deviating from his main route to make additional observations. Given his tremendous energy and enthusiasm, this is in no way surprising.

After Geikie and Murchison had joined forces with one another on the mainland, their route was as follows. From Arisaig on the coast they went inland to Banavie and Fort William, then south to Ballachulish, and through Glen Coe to Tyndrum. They then traveled northeast past Loch Tay to Pitlochry, before striking westward again along the River Tummel to Loch Rannoch. Turning north, they passed Loch Ericht and crossed a watershed to meet the main road from Perth to Inverness. Turning south again, they followed this road as far as Pitlochry, where they branched off to the east to Glen Shee and Blairgowrie. They then returned to the main road at Dunkeld and traveled easily to Perth and to Stirling, whence they took the train to Edinburgh, arriving there on October 1. In this latter part of the journey, Geikie instructed Murchison on the geology of the Carboniferous rocks of the Mid-

46. Geikie, Notebook A, 33.
47. Ibid.

land Valley, which he knew well from his activities in the Survey; and he
showed his prowess once again in Edinburgh by showing the director the geology of the city, which Geikie knew intimately.

It should be emphasized that the main objective of the tour so far as Murchison was concerned was to form a firm understanding of the "Silurian" strata
of the Northwest Highlands of Scotland, and to apply this well-established
knowledge to the even more difficult geology of the Grampian Mountains in
the middle of Scotland. Thus many of Geikie's notes, much of Murchison's
journal, and a good deal of their joint paper concerned this more southerly
part of Scotland. We have no need to go over all this in detail here. Suffice it to
say that Geikie seems to have found it almost impossible to make sense of it all
at the breakneck speed at which the two geologists worked. In Murchison's
journal, when they were near Pitlochry, there is reference to "doubts and difficulties of Geikie." But Murchison did not lose confidence: "I see nothing but
what I expected in approaching the Grampian jam."[48]

In Glen Shee, however, even Murchison began to lose heart somewhat: "I
am convinced that it is impracticable to trace the same order, which is so clear
in Sutherland, Ross, Islay, and Jura, and which we have been able to make out
across Argyllshire."[49] The rocks, he said; "could only be classified by receding
from the great region of disturbance, and by eschewing the instrusions of
Pluto and studying unaltered natural succession."[50] As for "poor Geikie," he
was "cracking his brains and exhausting his energies in trying to coax these
frightfully chaotic assemblages into the order of the north-west."[51]

Eventually Geikie acquiesced that some limestones in Glen Shee were analogues of those at Loch Assynt.[52] But even so he was not at all happy about the
extension of the Sutherland theory to the rest of Scotland. Thus in his biography of Murchison, he drew attention to the passages from Murchison's journal quoted above and informed his readers that Murchison "stuck to his
leading principle, from which no amount of contradictory detail would make
him swerve" (Geikie 1875, 2: 238). We may conclude, then, that the theory was
found wanting in its application to the Grampian region, even by Geikie. But
unlike Nicol, Geikie managed to retain his warm relationship with Murchison,
even in the face of some possible opposition—or at least "doubts." And Geikie
had no hesitation in firmly advocating the theory so far as its application to the
northwest of Scotland was concerned.

Thus ended one of the most important and influential pieces of geological
fieldwork in Britain in the nineteenth century. It laid its mark on the interpretation of the geology of Scotland for twenty years. Whether it was proper

48. Murchison Journal, "France, Scotland, Bohemia, etc.," 102.
49. Ibid., 105.
50. Ibid., 106
51. Ibid.
52. Ibid., 109.

that it should have done so, from what we now know of the matter in which the investigation was carried out, I rather doubt. But there is more to geological knowledge than simply the making, reporting, and interpreting of observations. It has much to do with the nature of the scientific community into which those observations and interpretations are launched. The nature of that community and its attitude toward different ideas require analysis if we are to understand the development of science. In the present case, the events that unfolded must be seen in the context of the particular relationships that developed between Geikie, Nicol, and Murchison in the last decade of Murchison's life and of the "politics" in general of the geological community in that period. We shall therefore turn to these matters in our next chapter.

6

Murchison and Geikie contra Nicol: The Establishment of the "Northwest Paradigm"

I should now like to examine in more detail the relationship that blossomed between Murchison and Geikie in the period after their joint fieldwork in Scotland, as well as their relationship with Nicol as it was affected by the Highlands controversy. I shall also discuss the implications of the Murchison-Geikie relationship for the history of the Geological Survey in the second half of the nineteenth century. My account is based chiefly on correspondence between Murchison and Geikie held in the archives of the Geological Society in London. But we must also explore the further development of Nicol's work.

A good deal of the Murchison-Geikie correspondence has to do with Geikie's obtaining a professorship at Edinburgh University. There is also much discussion of the geologists' joint work following the Highlands tour of 1860. Then there is material concerning the running of the Survey, and some discussion of fundamental theoretical issues in geology. Murchison's view of Nicol is a recurring theme.

The chair of geology at Edinburgh was eventually established with a considerable endowment from Murchison, and Geikie became the first incumbent in 1871. Yet as early as 1860, before their joint tour of the Highlands, Geikie was signaling to Murchison that he was interested in a professorship at Edinburgh. Thus in February that year, he wrote that he had heard a rumor from Ramsay that Murchison had some plans to endow a chair at Edinburgh, and that a man from Glasgow was a likely contender. Making plain his ambition, Geikie wrote to Murchison, in words intended to evoke a sympathetic response:

> Ever since the great truths of geology dawned on me, it has been the dream of my life to expound them in Edinburgh and to raise up again the school that was founded by Hutton, Hall and Playfair.[1]

And with extraordinary temerity, Geikie stated that he hoped the chair would not be created for several years. (Presumably, he wanted time for his career to mature so that he might obtain the position without difficulty.) He went on to suggest to Murchison that he feared his health would not stand up to

1. Geological Society Archives, M/G2/2, Geikie to Murchison, Feb. 17, 1860.

the strains of Survey life indefinitely and that he ought therefore to seek "some quieter post where such a constant tear and wear of . . . [the] physical frame . . . [was] not so constantly needed."[2] Given what we know of Geikie's exploits in the field only a few months later, this statement has to be treated with considerable reserve. Nevertheless, Murchison replied promptly and kindly, expressing sorrow that Geikie's health might lead him to look for a position with less fieldwork, and giving him some encouragement in aspiring to a chair.[3] Even at that very early stage, Murchison tested the opinion of fellow geologists such as Lyell concerning young Geikie's ambition and received some positive response. It is evident, therefore, that Murchison was well disposed toward Geikie almost from the beginning of their personal acquaintance. But things did not progress far at that stage, and a conversation on Geikie's behalf with a government official, Lord Greville, achieved nothing.[4]

Toward the end of 1860, we find Murchison and Geikie corresponding about their joint paper and also about plans to publish a geological map of Scotland with the Edinburgh publisher Keith Johnston, the intention being to bring out a Murchisonian alternative to the Nicol map. Preparation of the map (Murchison and Geikie 1861) involved discussion of the geology of the whole of Scotland, and the details need not concern us here.[5] But it is to be noted that Geikie was more than willing to express his opinions to Murchison with candor, and to disagree with him when he saw fit. He had such facility with his pen, and such tact, that he seems not to have caused Murchison any annoyance. We even find Geikie successfully toning down an attack by Murchison against Nicol in the introduction to the map, suggesting that there was no need to parade scientific controversies before the public: all that was needed was a statement that the new map offered a "reconstruction of his order."[6] It appears from the correspondence that Geikie did most of the drafting of the map, and that Murchison then reviewed his efforts. However, Murchison held himself responsible for the several parts of Scotland that he had visited but Geikie had not.

Meanwhile, Nicol was preparing another statement of his views. Indeed, it was to be his definitive statement. His paper was read at the Geological Society on December 5, 1860 (Nicol 1861a). An advance copy was made available to Murchison. At first he could not bring himself to read it and merely looked at the geological sections. But even these seem to have irritated him immensely, as we see from a letter to Geikie of November 8 containing triple exclamation marks. To understand what thus vexed Murchison, we must examine Nicol's paper in some detail.

As mentioned above, Nicol was engaged in fieldwork in the Highlands in 1860, at about the same time that Geikie and Murchison were also at work in

2. Ibid.
3. Geological Society Archives, Murchison Papers, E34, Murchison to Geikie, Feb. 21, 1860.
4. Murchison to Geikie, May 5, 1860.
5. For a reproduction of this map, see plate 5.
6. Geikie to Murchison, Nov. 21, 1860.

northern Scotland. But whereas we have copious documentary evidence as to the activities of Murchison and Geikie, virtually the only information about Nicol's work is to be found in his published paper. However, this is well furnished with details, and one can form a reasonable estimate of what he did in the field that year, though there is no certain way of telling the order in which he visited the several sites. My belief is that he started in the north at Durness and worked his way down to Skye, since this is the order of his published exposition, and he offers more detail in his exposition and argument for the northern exposures. Although Nicol's paper was presented to the Geological Society in December of 1860, it was not published until the following year. The joint paper of Geikie and Murchison was presented on February 6, 1861, and was published in the same volume of the society's *Quarterly Journal* as Nicol's paper.

An immediate difference between the two papers is that Nicol concentrated his attention wholly on the "Silurian" rocks from Durness to Skye, whereas Murchison and Geikie were much more ambitious and considered vast areas of the southern parts of the Highlands as well. It becomes evident, therefore, that Nicol had spent much more time covering a smaller region in greater detail. A second point is that Nicol's paper contains much more sustained and effective *argument* as to the merits of his observations and his theoretical interpretations, and it also has some suggestions of inadequacies in Murchison's theory.

By contrast, the joint paper of Murchison and Geikie was less polemical. This may come as a surprise. But there are two simple explanations. The first is that Geikie labored to restrain Murchison's polemics, for Geikie rarely engaged in public controversy during his career, and there is evidence in the correspondence to show that he endeavored to keep the arguments at a "scientific," rather than a "personal," level. The other thing is that Murchison and Geikie did not know what Nicol was actually doing in the field until they read the draft of his paper toward the end of 1860. Had they been aware of the arguments that Nicol was amassing against them, they might well have gone over some of the northwest sections with greater care, instead of trying to extend their work into the even more difficult terrain of the Grampian Mountains. But by the end of 1860, it was too late for them to revise their plans and go over a limited range of ground more carefully.

Nicol began his paper with a discussion of the strata of the Durness region. He described and figured a section running from Far-out (or Faraid) Head, north of Durness, southwest along the coast, and finishing just before Loch Eriboll. From one point of view, this was an unfortunate choice, since it involved Nicol in the very tricky problem of explaining how a fragment of the Moine Schist might be at Far-out Head, quite separated from the main body of schist to the east. On the other hand, the Durness area has an in-faulted patch, chiefly of limestone, the general structure of which was amenable to understanding with the help of the usual techniques available to field geologists in the 1860s. That is, most of the faults were of the high-angle type, and thus could be recognized without difficulty by Nicol. He could also see that the Far-out Head schist was similar to the standard Moine Schist ("mica slate") away to the east on the other side of Eriboll. Murchison was arguing that this schist lay conformably above the "Silurian" sediments and that it was Silurian also. Yet in the Durness region, though the schist and the limestone lay adjacent to one another at a faulted boundary of the normal type, it was evident that the schist did *not* lie on the upper surface of the limestone; and looking over all the exposures of Durness Limestone at Durness, Nicol could find no patches of schist lying on the limestone. Hence, he argued, the limestone was at the *top* of the sequence at Durness. Indeed, within the faulted inlier at Durness, the correct sequence of "Silurians" could be recognized: quartz rock, fucoid beds, [Serpulite grit,] and limestone.[7] Furthermore, he wished to claim, there was *no* upper quartz rock or "upper" limestone there—or anywhere else.

Nicol then sought to apply this argument to the Eriboll sections, a few miles to the east. There the faults are (we believe) mainly of a different character from those of the Durness inlier (emplaced by normal faulting), being low-angle thrust faults. But arguing by analogy with what appeared to be the case at Durness, Nicol wished to say that the eastern metamorphic rock was where it was by having been up-faulted from the depths (instead of, as we would hold, having been thrust there from the side). His reasoning was not completely sound. He didn't distinguish satisfactorily between schists and gneisses, and he was subsequently to argue that the "western" and "eastern" gneisses were essentially one and the same, the latter simply having been emplaced by upward faulting. In fact, however, the "western" schist at Far-out Head is a unique fragment, and it cannot serve as a general basis for an analogical argument for the identity of the western and eastern metamorphics. So we find that to the east, where both gneiss and schist may be found, Nicol did not distinguish these rocks as he might, and perhaps should, have done. Indeed, he did not always recognize the gneiss for what it was.

To put the matter another way, the argument for similarity between eastern and western metamorphics was based chiefly on a genuine similarity between eastern and western schist; but there is only one fragment of "western schist"—in the "klippe" of Far-out Head.[8] The "western" gneiss is really quite different from this. Thus Nicol's whole argument was by no means satisfactory. Nevertheless, he had arrived at an important truth on the basis of reasoning about the Durness-Eriboll sections—the correct "Silurian" succession was quartz rock, fucoid beds, [Serpulite Grit,] and limestone. There is no upper quartz rock. To support the idea of intense earth movements in the district, Nicol was able to point to undoubted evidence of disturbance to the limestone and other rocks along the coast east of Durness.

7. As before, I bracket Serpulite grit because it was not listed in the sequence.
8. The term "klippe" (not then used) refers to an isolated block of rocks, separated from the underlying rocks by a fault, usually of low-angle. An erosion fragment of a nappe will fulfill this condition.

Moving on to Eriboll, Nicol described a number of sections, west to east, on the eastern side of the loch near Eriboll House (one of Murchison's favorite sections), near the hamlet of Heilam, and through Mount Arnaboll—which runs parallel with the eastern side of the loch. This eastern side of the loch is perhaps the key area for the whole Highland problem. Certainly, as we shall see in chapter 8, it was the area where Charles Lapworth's mapping eventually began to unravel the whole mystery. But it is an area of great complexity and requires meticulous map work to understand the tangled structures. Nicol certainly did not grasp matters fully in 1860. Nevertheless, it is evident from his paper that he had gone over the ground with considerable care and was able to make some sense of his observations.

Taking a traverse from Camas Bay, on the eastern shore of Loch Eriboll toward Loch Hope (the next loch to the east—see fig. 7.11 below), Nicol depicted an ascending sequence of quartz rock, fucoid beds, and limestone, all dipping on the whole to the *west* (which is unusual). Then, going eastward, he found a region of faulting, with the rocks curving upward to form a scarp of the quartz rock. A little further north, a somewhat similar section was apparent, but the quartz rock could there be seen dipping to the *east*. This might not seem so very remarkable, but Nicol asserted—by the evidence of ripple marks and by the fact that the openings of the annelid tubes of the "pipe rock" were on the under surface of the quartz rock instead of the upper surface as usual—that *the quartz rock was here inverted.* Thus it might appear that the quartz rock had been folded upward and over on itself, and this might be the reason why there appeared to be an upper quartz rock. Murchison had been deceived because he had not noticed the evidence for the inversion of the strata.[9]

Near the top of the hills to the east of Loch Eriboll, and running northeastward toward Whiten Head, Nicol discovered a rock of curious appearance, which seemed to him to be igneous, but to which he found difficulty in assigning a name. He described it as follows:

> In general, it is a mixture of compact felspar and quartz, often with an imperfect laminar texture. With these, hornblende or talc or scales of bronzite become occasionally intermixed.[10] But in other places it passes into a distinct crystalline binary granite of orthoclase and quartz, or into felspar-porphyry or diorite, and where in contact with the limestone into a kind of serpentine. (Nicol 1861a, 89)

For want of a better name, Nicol called this peculiar rock a "granulite"—a term that has several meanings, but here apparently meant a rock made of a mixture of feldspar and quartz and various accessory minerals, with some foliation, having thereby a schistose or gneissose character. But at the time that Nicol was writing, it was an uncertain identification. In the nineteenth

9. I add this comment. It was not made here by Nicol.
10. Bronzite is a variant of enstatite, a ferromagnesian mineral belonging to the pyroxene family.

century, granulites were commonly thought to be a variety of granite and therefore might be deemed to be igneous, or the term was used in an indefinite way to refer to a gneissose rock.[11] Inspection of modern maps of the Eriboll region reveal that the rock that Nicol was calling a granulite is today regarded as altered gneiss, so that his identification and terminology were not unsatisfactory in terms of the theoretical concepts then available. But for Nicol the granulite was an instance of a supposed *igneous* rock, penetrating to the surface where the eastern metamorphics were supposedly faulted up to the surface. It was the equivalent at Eriboll of the "Logan Rock" we encountered at Loch Maree. The serpentinous matter mentioned as being found at the contact with the limestone suggests perhaps that Nicol had also encountered what was later to be termed "mylonite."[12]

Nicol's introduction of the term granulite into the discussions of the rocks of the Northwest Highlands was, I suggest, a point of considerable significance for the Highlands controversy. As will be discussed further in chapter 8, the type area for granulites was an area in Saxony, to the northwest of Chemnitz. Here one finds an extensive area of metamorphic rocks, which included ones that were called "granulites." Their metamorphic character was fully established by the researches of J. G. Lehmann (1884), using a microscope. But earlier the geologist C. F. Naumann (1797–1873) had supposed that these granulites could be regarded as igneous in origin (Naumann 1847, 1848, 1856).[13] It seems not unlikely that Nicol had picked up ideas such as those of Naumann concerning the nature of granulites when he was a student in Germany.[14] If this is so, it would help to explain how it came about that he imagined he could find igneous rock along the whole length of the "zone of complication" in the Northwest Highlands. It was thus always open to him to construe a specimen of "eastern" gneiss as igneous in character, and from this arose the whole confusion about the nature of the "eastern" gneiss and the "Logan Rock."

Unfortunately, however, the situation is even more complicated than this. The Saxon granulites are not at all similar in appearance to the Scottish rocks

11. In France in the nineteenth century, the word tended to be used for particular varieties of fine-grained granite. In Germany and Britain, the usual connotation was metamorphism.

12. In the Northwest Highlands, one finds that at the faulted contact between Moine Schist and Durness Limestone the latter often has a greenish appearance, perhaps suggestive of serpentine. For mylonite, see chapter 8 and figures 6.3 and 8.9.

13. Naumann's opinion did not receive universal acceptance. For example, the Austrian geologist Ferdinand von Hochstetter believed that the granulites of Bohemia were formed at the same time as the slates in which they occurred.

14. As mentioned previously, Nicol's diary for his student years in Berlin in 1840–41 shows that he attended lectures by the Prussian geologist H. von Dechen. But this offers us little assistance toward knowing what Nicol might have learned in Germany. Von Dechen (who had visited the Scottish Highlands in his earlier years, but not the northwest, and who is mentioned in Naumann's paper) is chiefly remembered for his work in mining geology. His lectures in Berlin might have made reference to the Saxon mining districts; but this is speculation. In any case, Naumann's paper of 1847 was translated into English the following year (Naumann 1848), and it is just as likely that this was Nicol's source.

to which Nicol was referring. They are the products of high-grade metamorphism and today are taken as type-specimens of rocks of the "granulite
facies" (corresponding to medium-to-high pressure and high temperature).
By using the word "granulite" in relation to Scottish gneisses, Nicol gave rise
to a British terminology that was different from that used by German authors
and that persisted in the writings of the Scottish Survey for many years,
though not always in the sense envisaged by Nicol. (The Scottish usage has
now largely dropped out of the literature.) One may wonder whether Nicol
ever saw any of the Saxon granulites, for if he had it would seem surprising
that he would have used the same term for the Scottish rocks. More likely, he
heard the term being used when he was in Germany, and thinking that it
referred to a gneisslike granite, which was perhaps igneous, the term seemed
to him to be appropriate for his Scottish rocks.[15]

However, it must be emphasized that this suggestion is something of a
speculation on my part. I can offer no documentary evidence to support the
conjecture that the confusions attending the "Logan Rock" or "eastern" gneiss
were actually the result of supposed analogies with Saxon rocks. On the other
hand, Nicol's line of thinking would be much easier to comprehend were he in
fact making use of some such analogy. As I say, it is not unlikely that he would
have made use of ideas that he came across as a student when he was later
carrying out his own researches. It may be mentioned also that some English-
language authors, notably Darwin, had similarly entertained the idea that foliated rocks might be igneous in origin.[16]

Be this as it may, Nicol contended strenuously that fragments of "mica
slate" (i.e., schist) could be found in the "granulite," and he supposed that
these fragments had got there when the "igneous" granulite had pushed the
eastern metamorphic rocks to the surface. On this view, then, the eastern
schists or gneisses could not be younger than the sedimentary "Silurians,"
contrary to what Murchison supposed. They could not be lying in sequence
over the sediments: if they appeared to do so this was because they had
been emplaced by faulting. There was, Nicol maintained, in contradiction of

15. I am grateful to Dr. Michael Katz for showing me some samples of Saxon granulites and to
Professor T. G. Vallance for information on the different terminologies used in Germany and
Scotland in the nineteenth century, and into the twentieth century. A modern definition of the
present usage of granulite is: "A high-grade metamorphic rock, usually light in color, medium- to
fine-grained and containing little or no mica. A characteristic strong layering is due to the occurrence of well orientated, flattened polycrystalline lenticles of quartz set in a fine-grained matrix of
quartz and feldspar, together with one or more of the following minerals: pyroxene, garnet,
kyanite, sillimanite. Spinel and rutile are possible accessory minerals" (Walton et al. 1983, 235).
The term was also used in France in the nineteenth century to refer to a fine-grained granite with
white mica.

16. In his geological investigations in South America, Darwin noted a number of observations of
manifestly igneous rocks showing foliations. He supposed (1846) that this might have arisen "by
the moving mass, just before its final consolidation, having been subjected (as in a glacier) to
planes of different tension". (See Darwin [1890], 441.) G. P. Scrope (see page 165) had also
offered similar ideas.

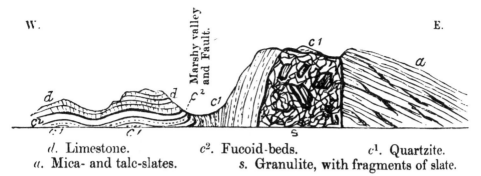

d. Limestone. c^2. Fucoid-beds. c^1. Quartzite.
a. Mica- and talc-slates.   *s.* Granulite, with fragments of slate.

Fig. 6.1. Section of strata to east of Camas Bay, Loch Eriboll, probably through Ben Arnaboll, according to Nicol (1861a, 88).

Murchison's theory, no upward conformable succession above the limestone. He pointed out, further, that north of the Loch Hope ferry, up toward Whiten Head, the fucoid beds and limestone were missing and either gneiss or mica slate (schist) could be found lying directly in contact with the quartz rock. This seemed unintelligible on Murchison's hypothesis of a conformable sequence above the limestone. The gap would therefore have to mark an unconformity or plane of fault.

Nicol's ideas about the relationship of the sediments, the "granulite," and the schists are shown in figure 6.1. Here one can see how Nicol imagined that the "igneous" granulite was *causally* connected with the emplacement of the eastern schists. Later theory regarded the granulite as a metamorphic rock, gneiss, that had become "wrapped up" with the sediments during the course of major lateral earth movements, while the schist was actually produced during the course of the earth movements. Thus it is not correct by any means to suggest that Nicol had solved the problem of the Northwest Highlands during the course of his travails of 1860, though some later commentators (for their own good reasons) represented his accomplishments in this light.

Nicol referred to a number of other sites near Loch Eriboll, all of which, however, need not be described in detail here. Only one should attract our particular attention. This is the traverse behind Eriboll House, which provided the key section in this region so far as Murchison was concerned. It supposedly displayed the existence of an upper quartz rock and a conformable sequence above the limestone up through this quartz rock into the gneiss and schist. But Nicol represented the section in the manner shown in figure 6.2. His section may be compared with that proposed by Murchison, reproduced in figure 4.5. (See also figs. 4.3 and 4.4 for an idea of the locality.)

Nicol's section for Eriboll House shows no "granulite" or any other "igneous" rock. But we can see how he imagined the eastern mica slate (schist) there was forced up from below, curving the "Silurians" to such an extent that they had become inverted. Note also that the quartz rock might be expected to

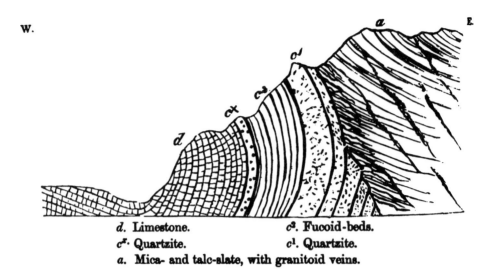

W.   *a*   E.

d. Limestone.   *c³.* Fucoid-beds.
cˣ. Quartzite.   *c¹.* Quartzite.
a. Mica- and talc-slate, with granitoid veins.

Fig. 6.2. Section of straia behind Eriboll House, according to Nicol (1861a, 92). (The quartzite, c^x, Nicol regarded as an atypical member of the sequence.)

outcrop again somewhere to the west—as indeed it does, to the west of Loch Eriboll. Finally, it may be noted that Nicol's figure is compatible with the idea later propounded by Lapworth that the eastern mica slate (schist) was pushed into position from the east (not from below, as Nicol supposed). The section is, of course, a highly theoretical statement, involving evidence that the quartz rock was overturned, as indicated by the annelid borings ("pipes").

Nicol next described sections at Loch More, which he used to illustrate further the conclusions he had reached at Durness and Eriboll, but it is not necessary that these sections be analyzed here. Further south he examined the terrain near Lochs Glendhu and Glencoul and the country south from there, over the watershed toward Loch Assynt. His section was taken from near Kylescu Ferry southeast toward the mountain of Glas Bheinn (Glasven). In fact, quartzite outcrops twice along this line of section, which Murchison would have ascribed to there being two quartz rocks. For Nicol, the explanation was that they were one and the same, but separated by a substantial intrusion of "syenite." For modern geology, the "syenite" is actually gneiss; and the repetition of the quartzite (as well as the presence of the gneiss) is due to a number of low-angle thrust faults from the east. It may be noted that according to modern maps (Institute of Geological Sciences 1965), Nicol's "syenite" (gneiss) extends southeast in a state of great confusion (for it contains many faulted-in slices of quartzite), into the range of mountains to the east of Loch Assynt, particularly Ben More. The "syenite" was, in fact, equivalent to the Logan Rock of Loch Maree, and perhaps also equivalent in part to the "granulite" of Loch Eriboll, though it was not precisely the same, for there is more mylonitization in the northern exposures than at Loch Assynt, and mylonite is chiefly found under the Moine Schists rather than in association with the gneiss. We find, then, that inexact rock identifications were the source of difficulty. But petrography was not, at that time, sufficiently developed to meet the needs of field geologists in a region as complicated as the Northwest Highlands.

For the Loch Assynt region, Nicol offered essentially the same kind of structure as that a little further north at Loch Glencoul. Again he denied the existence of any upper quartzite and had the limestone at the top of the sequence. He gave a description of the important section at the hillside, Cnoc an Drein (Cnoc an Droighinn), behind Inchnadamph, which is, however, not possible to reconcile perfectly with modern maps. Starting at the lake shore, one has quartzite,[17] a few slithers of fucoid beds and Serpulite grit, then a band of limestone, then quartzite at the top of the hill—which would certainly seem to accord with Murchison's hypothesis of there being an upper quartzite. However, Nicol contended that the quartzite was constantly increasing its easterly dip as the hill is ascended, until it was vertical at the top. He ascribed this to its having been forced up by "syenite" that he claimed to be present there. What would this "syenite" have been? It is hard to say. There is certainly no such rock in the vicinity. There is some gneiss a little to the north, which we have also been calling Logan Rock, and I think this must have been what Nicol was referring to. However, its disposition relative to the quartzite is not such that it might immediately be supposed to have caused the lifting up of the quartzite. So it might appear that Murchison's interpretation of the structure of the hillside was superior to that of Nicol.

It may be noted further that the hillside of Cnoc an Droighinn is actually made up of repeated "slices" of limestone and quartzite, according to modern maps, but these could only be discerned by the stratigrapher when subdivisions of the limestone and quartzite had been established. This subtlety was not evident to Nicol, so his proposed structure was at best very approximate. But at least he did not take the data completely at face value, and he realized that everything was not just as it seemed to be. He was able to mesh what he saw at Cnoc an Droighinn with his general theory of the Highlands. But this entailed regarding the "eastern" gneiss, or parts of it, as a syenite, which was an unhappy identification.

As to the general structure of the sedimentary rocks of the Loch Assynt region, Nicol regarded it as synclinal; and this, broadly speaking, was why the quartzite might often reappear to the east, sometimes seeming to overlie the limestone, but never in fact doing so. He offered the interesting argument that the limestone was porous to water but the quartzite was not. So some of the water running down into the limestone-filled valley above Inchnadamph would seep through until it reached the underlying quartzite and then emerge lower down as springs. Thus there was reason to believe that the quartzite went right under the limestone, forming a synclinal trough. As for Ben More,

17. The quartizite actually occurs to the west and south of the loch, and the contact with the overlying rocks in the series is obscured by alluvium. On the whole section, compare the map of Peach and Horne, shown in plate 8. The Cnoc an Droighinn section runs northeast from Inchnadamph.

Fig. 6.3. Thrust fault (Moine Thrust) at Knockan Cliff Nature Trail, showing Moine Schist thrust over altered Durness Limestone.

Nicol regarded it as "a great centre of elevation" of igneous rocks, with quartzite "thrown off" all round it.[18] But the adjacent rocks he called "granitic gneiss"; whereas by analogy with what he had seen a little to the north of Cnoc an Droighinn, they should, I suppose, have been called "syenite."[19] Or using an analogy with the exposures at Eriboll they might have been called "granulite." It can hardly be said that Nicol had a perfect understanding of the exceedingly complex Loch Assynt district. Nevertheless, he managed to make it appear as if it fitted into his general theoretical scheme. But this might equally be said of Murchison's work.

Nicol next explored the region round Loch Ailsh and Glen Oykell, south along the range from Ben More, but I shall omit consideration of this here. We may pick up the story at Knockan Cliff, near Elphin, on the road south from Assynt to Ullapool, where, it will be remembered, Geikie paid a very brief visit alone in the summer of 1860. Here the Moine Schists curve westward, so that they come to lie over the limestone, the fucoid beds, the quartzite, etc., according to how far the thrust plane extends westward at any given point. The sort of (faulted) contact between schist and underlying altered limestone that Nicol might have seen at Knockan Cliff is shown in figure 6.3. There is in fact *no* gneiss in this locality (no Logan Rock, no granu-

18. There are some intrusions of porphyry on Ben More, which offer a warrant for this suggestion.
19. Some twenty years later a visitor to the district, Wilfred Hudleston, was calling this same rock an everlasting expanse of Logan Rock. (See page 199.)

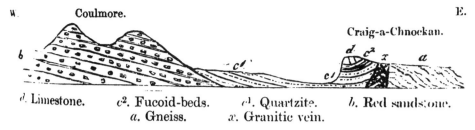

Fig. 6.4. Section at Knockan Cliff, according to Nicol (1861a, 101).

lite). Nevertheless, Nicol still supposed that he could see an igneous rock, in accordance with his general theory. He figured the cliff section as shown in figure 6.4. It will be seen that this time the igneous rock supposedly responsible for pushing up the quartzite has become a "granitic vein."[20] I am unable to suggest exactly what this might have been. As shown in figure 6.3, at Knockan Cliff proper (the site today of a nature trail), the Moine Schist lies directly on the limestone, with no intervening igneous rock. However, at the hamlet of Elphin, a little to the north, there is an exposure of marble, altered by an igneous intrusion; and a little to the south of Knockan Cliff there is a small vein of igneous rock running parallel with the strike that has not altered the limestone significantly. So although Nicol's section (fig. 6.4) is not exactly representative of the situation at Knockan Cliff, one can see how he might have offered the interpretation that he did. He also spoke of "fine-grained gneiss," by which I suppose he must have meant the Moine Schist. As I say, there is no gneiss at Knockan Cliff. Again we have the problem of inexact petrography.

We will pass over Nicol's work at Ullapool, referring the reader to his earlier work of 1856 (see chap. 3), as he did on this occasion. But we must look with some care at his observations at Loch Maree, where the section along the side of the valley, to the northwest of Kinlochewe, will again be the focus of our attention. Nicol's new interpretation of this district is reproduced in figure 6.5. (For Sleugach, read Slioch; and for Glen Laggan, read Glen Logan. The steep-sided valley to the east of Slioch is called Glen Bianasdail.[21] Cf. figure 5.1.)

Like Murchison, Nicol had no difficulty in depicting the relationships of the rocks at Slioch, and from there westward. He was also correct, according to modern maps, with respect to his representation of the eastern side of Glen Bianasdail. But further east, it has to be said that there were great discrepancies. For a start, the "igneous" rock in Glen Logan was now construed by Nicol as a diorite, that is, a medium- or coarse-grained igneous rock, characterized by the presence of the minerals hornblende and the type of feldspar called plagioclase. Of intermediate acid-base composition, diorites might possibly be thought somewhat similar to syenite, though they are usually darker in color.

20. Nicol (1861a,102) also called it a "granite or syenite." But this does not help much in understanding what he had been looking at.
21. Murchison had called it Glen Haasach.

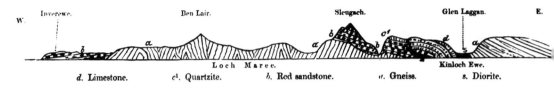

Fig. 6.5. Section of strata at head of Loch Maree (north side), according to Nicol (1861a, 104).

At least, they might conceivably be regarded as similar in hand specimens, though they differ in their feldspars. So Logan Rock had now become a diorite, having previously been called a feldspar-porphyry or a syenite; and it is mapped as gneiss on modern maps.

To the east of Glen Logan, Nicol had (eastern) gneiss dipping eastward. He called it "grey granitic gneiss." Today this is all mapped as Moine Schist. But as we have seen at Durness and Eriboll, Nicol was inclined to conflate gneiss and schist. There is some limestone in Glen Logan, which has been quarried at times. Nicol had it rising up the hillside to the west of the glen. Whether this is satisfactory depends on just where he intended his section to be drawn. There *are* patches on the hillside, but more of the fucoid beds and the Serpulite grit, which, however, were not depicted in Nicol's section. Further up the Logan valley, one also finds gneiss rising up the hillside. So again we have the problem arising from the depiction of the geology of an area primarily through sections rather than maps. Later geologists, visiting the region, would have difficulty in knowing whether Nicol's section was or was not a fair representation of the geological situation, for they would not have known precisely where the section was intended to be taken.

Beneath his limestone, Nicol gave quartzite, curving over toward Glen Bianasdail; but, supposedly because of a fault, by the time Glen Bianasdail was reached it formed only the cap of the hill between Bianasdail and Logan, the underlying rock being "Cambrian" red sandstone. Near the base of the hill (alongside the path from Kinlochewe to Loch Maree and Glen Bianasdail), there was supposedly an igneous rock (diorite?), as Murchison had also found. But this is gneiss, according to modern maps. It may be noted that Nicol sought to make sense of the whole structure of the hill between Bianasdail and Logan by the introduction of a fault. But it was of the high-angle variety, and there is no such fault there according to modern maps—though there are a number of thrust faults in this district. I do not think that Nicol climbed to the top of the hill between Bianasdail and Logan, any more than Murchison did. For if he had done so, it is hard to see how Nicol could have failed to notice the large patch of "Cambrian" red sandstone on the *top* of the hill—where it had no right to be according to his theory or that of Murchison. It is instructive to compare Nicol's section with that of Murchison, shown in figures 5.3 and 5.4.

Moving on, we find Nicol examining the terrain between Lochs Maree and Torridon. But instead of giving special attention to the area of Sgurr Dubh, near Loch Clare, as Geikie had done, Nicol seems to have examined the wild country between the mountains Ben Eighe, Sail Mohr, and Liathach, though his itinerary is by no means certain. Here he was interested in the relationship between the "Cambrian" red sandstone and the "Silurian" quartzite. In some areas the quartzite appeared to be below, rather than above, the sandstone. Nicol wondered whether this might have led earlier geologists to imagine that the western red-brown sandstones were actually younger than, and "framed," the rocks of the interior.[22] But for Nicol the section revealed massive earth movements:

> The red sandstone, always dipping to the east, and covered by its capping of quartzite, is again and again brought up by faults; . . . Had only a surface-view been exposed, as each fragment of quartzite seems to dip below the next one of red sandstone, it might have been supposed that the rocks alternated with each another [but] it is clear that the quartzite is mere fragments of the upper formation, brought down repeatedly by faults, and in some cases even forced in below the inferior red sandstone by enormous lateral pressure. This very clear natural section thus tells the structure of the N.W. Highlands, and the true nature of these apparently anomalous sections which have puzzled observers in other parts of the line of junction. (1861a, 105–6)

Nicol also visited the districts of Loch Kishorn and Skye, but we will not follow him thither on this occasion.

In concluding his paper, Nicol drew together all the possible arguments he could find to oppose Murchison's theory of the Highlands. He referred once again to the problem of overlap, which he had discussed at the Aberdeen meeting of the British Association in 1859 (see chap. 4), and argued strongly that the relationship of the eastern metamorphic rocks to the underlying "Silurian" sediments could only be explained in terms of faulting. He also objected that the notion of metamorphics lying conformably on unaltered sediments beneath was "contrary to all the established principles of geology." He even quoted Murchison against himself and in favor of Nicol's doctrine, for in his paper on the structure of the Alps, Murchison (1848) had supported a theory for the Alps quite similar to that which Nicol was now advocating for the Highlands—that is, overthrows or inversions of strata and "lateral extrusion."[23]

22. He was evidently thinking of Hugh Miller, and Sedgwick and Murchison, in their first Highland field trip.
23. This paper was, in a way, a very remarkable document, which makes certain aspects of the Highlands controversy difficult to understand, if taken out of their social context. For in his paper, Murchison explained how he had been accompanied by the Swiss geologist Arnold Escher von der Linth (see page 234), to a place called Martin Loch Pass in the Canton of Glarus, and had been fully convinced there of a "grand inversion" of strata, such that a limestone was overlain by a zone of talc and mica schist. He also rejected the hypothesis that the upper metamorphic rocks had been metamorphosed *in situ*. Thus, in effect, he had the main theoretical ingredients needed to explain the observations in the Northwest Highlands in terms of overfolding and faulting. In fact, he actually included in his paper reproductions of some figures by H. D. Rogers showing overfolding and reverse faulting, which were subsequently deployed by Charles Lapworth (see

Concerning a point to which Murchison attached great significance, namely, that the mineral composition of the "western" and "eastern" gneisses was substantially different, Nicol simply said that the western and eastern gneisses might indeed belong in part to two distinct geological periods. He had said this before, in earlier publications such as the explanation of his geological map of Scotland, and seemingly attached less importance to it than Murchison. It was scarcely compatible with Nicol's other claim that the eastern gneiss was in fact the western gneiss brought to the surface by faulting.

Another point that Murchison thought exceedingly important was the differences in strike between the western gneiss (northwest to southeast) and eastern gneiss (northeast to southwest), which had been emphasized long before by Daniel Sharpe. In answer, Nicol asserted that the two strikes were by no means always constant. He remarked, for example, on the "western" mica slate of Far-out Head, which strikes northeast to southwest, like a typical "eastern" metamorphic rock. This might well seem to have suited his case. But as we have seen, the "mica slate" of Far-out Head was in fact an unusually placed fragment of the "eastern" Moine Schists, existing as a "klippe" well to the west of the main body of the schists. In citing this special case, which seemed persuasive to Nicol, he was, in fact, making use of his mistaken but convenient conflation of gneiss and schist.[24] He used the same fragment at Far-out Head to argue also for the similarity in mineral composition between the western and eastern metamorphics!

In concluding his paper, Nicol asserted that no satisfactory evidence had been brought forward to show that the schists of the north coast of Scotland were one and the same as all the metamorphic rocks right down into Inverness-shire. These southerly rocks might just as well be equivalent to the ancient gneisses of the west coast of Sutherland, such as at Scourie or Loch Inver. Now here, perhaps, we come to the heart of the controversy, so far as Murchison was concerned. For if Nicol were right, Murchison would have no warrant for coloring in all those masses of metamorphic rocks in the Grampians, and so forth, as Silurian. They might very well be pre-Silurian, if not actually belonging to the Fundamental Gneiss of the west coast. Thus Murchison would be denied his great revolution in Scottish geology, and his territorial ambitions for the kingdom of Siluria would be severely curtailed. In their tour of 1860, Murchison and Geikie were indeed trying to find evidence that

---

chap. 8) as he worked his way toward a solution of the Highlands problem. It therefore cannot properly be said that Murchison could not have made better sense of the Highlands observations because certain important theoretical notions were unavailable to him. The difference between the Highlands and the Swiss case was, however, this: in Switzerland the inversions may almost be perceived by direct observation of the spectacularly folded and faulted strata. In Scotland, the earth movements have to be inferred from the analysis of careful traverses and detailed maps. Also, Murchison did not have in Scotland an Escher von der Linth to guide him, a companion who had already arrived at a theory of inversion of the strata by careful examination of a specific region during many years of fieldwork. Murchison's work was always too hurried.

24. But note that for Nicol some of the rocks today mapped as gneiss were igneous: syenite, diorite, porphyry, etc.

would justify their extension of the Sutherland succession into the southern Highlands. But as we have seen, they had little success in this, and Geikie was not fully convinced. Only Murchison had no doubts and reservations. In any case, at the time that Nicol was writing, Murchison and Geikie had not yet published their paper, and Nicol was quite justified in suggesting that a satis-factory correlation between the northern sections and those of the central and southern Highlands had not been accomplished. It was premature to be talk-ing about a "great revolution" or a "Reform Bill" in Scottish geology.

After Murchison was shown a draft of Nicol's paper in advance of its presen-tation to the Geological Society, he wrote a lengthy letter about it to Geikie on November 8, 1860. He was enormously angered by what he had seen. Nicol had "invented" faults. He had represented "ordinary flaggy mica slates" as part of the "Old Gneiss." He had not shown the faults at Loch Broom (Ul-lapool), perhaps because they were "too clear." What Murchison regarded as his "own tract of *Eriboll*," where he gave a simple conformable sequence at Eriboll House, was rent by a great "ravine of fault" by Nicol, which Murchison contended could not be seen at all. "I see," wrote Murchison, "that no terms are to be kept with him." Nicol's "last state is worse if possible than his first."[25]

Murchison wrote to Geikie again on December 1, a week before Nicol's paper was presented at the Geological Society. He had looked over it once more, and so too, it seems, had T. Rupert Jones (1819–1911), the society's assistant secretary, librarian, and curator and a personal friend of Murchison. Jones had told him that "the logic is clear & if the facts be so the inferences are . . . inevitable." Jones was a palaeontologist, but nonetheless a geologist whose opinion counted, and Murchison was evidently perturbed. He requested from Geikie a "brief outline of data respecting the sections(?) at Loch Maree & Loch Broom as will settle the point." He concluded his letter: "Pray send me as many crumbs of comfort as you can for it is evident that Nicol's hole and corner poking will lead many into the wrong path." In other words, Murchison was turning to young Geikie for ammunition in what he knew would be a tough battle at the Geological Society the following week.

It is interesting that for the most part in his letter of December 1, Murchison merely dismissed Nicol's arguments and evidence, accusing him of incompe-tent observation or extravagant speculation. But on one point, at least, Nicol did clearly worry Murchison, who wrote:

> The only point that bothers me is that [Nicol] says that the Annelid Tubes are upside down, i.e. trumpet-shaped upwards & that ? from the strata(?) of Quartz have been *overturned*. I really cannot remember the facts.[26]

25. That is, Nicol's argument leading to the conclusion that the Durness Limestone was Carbon-iferous.
26. Geological Society Archives, Murchison papers, E34, Murchison to Geikie, Dec. 1, 1860.

Fig. 6.6. "Quartz rock" showing "trumpet pipes."

The point being made here by Murchison requires elucidation. In the quartz rock of the Northwest Highlands, there are in many cases innumerable in-filled annelid borings, which can give the surface of the rock a curious but highly distinctive appearance. The borings are not all the same: mostly they are cylindrical, but some are branched; some only appear on the surface of the rock whereas others appear to penetrate through it; they vary in width; and some are flattened at the ends to give the appearance of small trumpets. Later geologists such as Peach and Horne used the various types of annelid boring to discriminate between different horizons of the quartz rock, and found them invaluable for the mapping of this unit. Figure 6.6 shows a typical surface of trumpet pipe rock.

Now in the matter of a trumpet, it is difficult to say what is "up" and what is "down." I would have thought that if the appearance were the same as in figure 6.6, one would describe the pipes as "trumpet-shape up." However, from a sketch in Murchison's letter, it seems that he meant the opposite. Anyway, Nicol had evidently found some the other way up from that shown in the photograph and hence argued for an inversion of the strata.

This worried Murchison, but in a postscript to his letter, he told Geikie that he had discussed the matter with the Survey's palaeontologist, Salter. He, apparently, had "laugh[ed] at the affair of the annelids"; and looking through his notes again, Murchison had reassured himself that the typical case for the trumpet pipes was like that shown in figure 6.6. Hence Nicol's very important piece of evidence for local inversions of the quartz rock was simply brushed aside. Where inverted trumpets occurred, they could simply be ascribed to minor and local faulting.

Nicol's paper was read to the Geological Society by Warington W. Smyth (1817–90), the Survey's first mining geologist who held a chair at the Royal School of Mines and rose to be chief mineral inspector to the Crown. Nicol was unable to be present, being at work in Aberdeen. On December 5, the evening after Smyth had presented Nicol's paper, Murchison wrote a lengthy account of the events to Geikie, giving some account of his response to Nicol, and he must have stayed up half the night to have done so. Murchison found his sections "cut to pieces" by his adversary—that is, great faults were introduced into them. He realized that Nicol's evidence of the annelid tubes at Eriboll was "sure to make some impression," but at the meeting Murchison had simply denied the evidence. He had also pointed quite rightly to the differences in strike and mineral composition between the western and eastern gneisses, and he had read out some information that Geikie had sent him concerning the large tract of land between Loch Maree and Loch Broom (which Geikie had covered in his thirty-two mile tramp). What this evidence might have been we do not know, since Geikie's letters to Murchison for this period are unfortunately missing. However, I doubt that evidence from that tract of country would have assisted Murchison greatly, especially when we remember how quickly Geikie had covered the ground. An interesting point raised by Nicol was that Cunningham had stated that some parts of the "eastern" gneiss in the interior of Sutherland had a strike similar to that of the "western" gneiss, contrary to the broad generalization of Daniel Sharpe. But this did not perturb Murchison. He simply answered that he would not be at all surprised if patches of old, fundamental, gneiss might be found to the east. It was, after all, supposed to be the basement rock for the whole stratigraphical series in Britain. A point that apparently impressed Rupert Jones was the one concerning the lack of overlap of the "eastern" gneiss on the rocks to the west. Murchison, it seems, could give no answer to this important argument.

We need not concern ourselves with the rest of Murchison's account of Nicol's work, since we have already been over the main details of his paper earlier in this chapter. Using his usual military idiom, Murchison wrote that "we have a fight in which our reputation & veracity are at stake & . . . we must 'buckle to' with due diligence for our Memoir."[27] But he was not sure that he had fully carried the day at the Geological Society, for "notwithstanding all that I did (& they told me I fought well and manfully) there is too much reason to fear that Nicol's campaign will produce a very adverse effect."[28] Thus he saw it as essential that he and Geikie should get out their paper as soon as possible, which they duly did at a meeting of the Geological Society on February 6, 1861; and the paper was published later that year. We have, of course, analyzed this in some detail in the preceding chapter, so there is no need to go

27. Murchison to Geikie, Dec. 6, 1860.
28. Ibid.

over the material again here. It may be noted, however, that in the published version of the paper Murchison added an appendix (also dated February 6), which specifically sought to answer Nicol's claims point by point, and thereby refute his theory (Murchison 1861).

At the section behind Eriboll House, Murchison asserted that he had seen no great line of fracture. But, in fact, Nicol's section was relying on what he had seen at Ben Arnaboll, a little to the north, where an unnamed valley, running east to west, enabled him to have a view of the strata in section (see fig. 7.11 below). As to Nicol's various "igneous" rocks, Murchison acknowledged that they did occur in a number of places, but, he maintained, they did not disturb the general sequence of the strata. On the question of the alleged "upper quartz rock" at Loch Assynt, Murchison reasserted its existence, and claimed the support of Ramsay, Harkness, and Geikie as witnesses. For the strata at Loch More, Murchison reasserted the difference between the western and eastern metamorphics. He maintained that the eastern rock was nothing more than a "flag-like schist" and he exhibited a specimen to prove his point. With respect to Nicol's section at Far-out Head, Murchison complained that its direction was drawn inaccurately and thus gave a misleading impression. With respect to a section by Nicol at Loch Ailsh, Murchison contended that it supported his case just as well as that of Nicol.

All this might sound quite convincing, but it should be noted that Murchison's position was rather like that of Nicol at the British Association meeting in Aberdeen in 1859. To a large degree, Murchison could merely reassert what he had seen and said on previous occasions. He had not revisited the sections in the meantime. However, he could still call on the support of three influential geologists who had visited the localities and were willing to lend their support to the Murchison doctrine. Nicol had no such advantage at Aberdeen.

By the middle of 1861, then, it might be said that Murchison and Geikie were about equally balanced with Nicol in arguing their cases before the Geological Society and the wider geological and scientific community. But it was the Murchison-Geikie theory that prevailed and became established in the textbooks. How did this come about, and how did Murchison accomplish the destruction of Nicol's theory? In fact, little new empirical evidence was brought forward by either side. So by what social process did Murchison and Geikie emerge as the "winners"?

An important first consideration was to bring the views of Murchison and Geikie to a wider public. It will be recalled that Nicol had published his geological map of Scotland in 1858, so that in a sense his ideas "held the field." It was necessary, therefore, that Murchison and Geikie bring out their version of the geology of Scotland in map form as soon as possible, and a good deal of their correspondence of 1860 and 1861 had to do with this matter. It must have been very difficult for them to accomplish the task, when Geikie was frequently in the field on Survey work, or in Edinburgh, and Murchison was in London. Nevertheless, their map appeared in the later part of 1861.

The map itself (Murchison and Geikie 1861), Murchison stated in his useful introductory "Explanation of the Map and Sections," was actually prepared by Geikie. In his introduction, Murchison gave a brief account of the history of geological exploration in Scotland and stated the names of the geologists on whom he and Geikie had drawn in their compilation. He adverted to his disagreements with Nicol and explained the differences in opinion that existed between them. But all this was done without perceptible rancor. Geikie had evidently succeeded in persuading his chief to moderate his language.[29] The map did not bring any new matter into the debate, but it gave a very public exposition of the Murchison-Geikie doctrine. It was a beautifully engraved and colored piece of work, and in many ways markedly superior to Nicol's production as a piece of geological cartography. It was also furnished with a number of clear sections, as may be seen in plate 5. Here, then, one may see the concrete expression of Murchison's views concerning the geology of the Highlands. Their remarkable simplicity should be noted, and also the vast extent of Silurian strata that the doctrine entailed. It was the greatest extension that the Silurian kingdom ever managed to achieve in Scotland.

But it was not enough for this map to be published. It was needful also that it be suitably reviewed, so that the interested public would be made aware of its presence and would be persuaded that its contents met with the approval of knowledgeable geologists. It would be unseemly to review one's own work in too obvious a manner. But with the nineteenth-century practice of unsigned reviews, it was not particularly difficult for Murchison and Geikie to obtain what they wanted. Thus we find that an issue of the *North British Review* in 1861 contained an interesting anonymous essay with the running head: "Recent discoveries in Scottish geology." This was ostensibly a review of all the geological maps of Scotland that had been issued, back to the time of Macculloch's work, and of other material relevant to the geology of Scotland, but in fact it was a grand "puff" for the recently published map of Murchison and Geikie and their developing theories. The "review" was actually written by none other than Geikie himself ([Geikie] 1861).[30]

As was his custom, Geikie did not attack his opponents, even under the cloak of anonymity. Rather, he took the opportunity afforded by his review to place his own recent investigations, and the new views about the structure of Scotland developed by Murchison, in the most favorable light. He emphasized the essentially simple structure of Scotland according to the Murchison doctrine, and the coherent theoretical picture that it offered:

> There can be few pleasures in a scientific life more intense than to
> mark, when once the clue to the geological structure of a difficult
> tract of country has been obtained, how district after district, like the

---

29. Murchison's letter to Geikie of Nov. 22, 1860, reveals that Geikie had been working to curb Murchison's invective against Nicol.

30. See Houghton et al. (1966, 686) and Geikie (1924, 88n).

detached portions of a puzzle-map, falls into its proper place, and how complete is the order, and how evident the arrangement, where before all order and arrangement seemed to be wholly absent. (ibid., 133)

It was acknowledged that there was difficulty in understanding how the *uppermost* rocks in the sequence in the Northwest Highlands might be metamorphosed. But "these and other obscurities [would] doubtless be cleared away at no very distant date" (ibid., 135).

Geikie also dilated on the fact that according to the new "theory" of Scotland, no cataclysms needed to be invoked. The structure seemed to indicate a long succession of events that had "proceeded according to definite laws"—which, he suggested, "in their beauty and symmetry, are still the mode in which the Creator regulates the economy of the world" (ibid., 156). This sentence is interesting, for it suggests an important reason why Geikie may have been antipathetic to Nicol's doctrine: it smacked of catastrophism and was not in accordance with Geikie's version of natural theology. In one sense it is curious that Murchison, who was not a uniformitarian, should not have been attracted to Nicol's doctrine. But as we know, Murchison had his own strong reasons for opposing Nicol. Thus Murchison and Geikie had rather different motives for their opposition to Nicol. Nevertheless, their overall interests coincided, and they united to produce a common force that easily vanquished the Aberdonian.

One of the most interesting features of Geikie's review is his reference to the merits of *regional* survey work: "The sporadic style of investigation which characterized much of the earlier research, has given place to a broader and more generalized method." No longer did one have papers on "the mineralogy of a single hill" or "the geology of a sequestered valley." Now there was a "more philosophical spirit and a more enlarged method of inquiry" (ibid., 155). Thus Geikie might seek to justify in the eye of the public the very rapid kind of reconnaissance work that he and Murchison had undertaken in the Highlands. But, I shall argue as this study develops, it was the great rapidity and superficiality of their Highland work that in large measure accounted for the problems and controversies that characterized the geological investigations in the far north.

We cannot suppose that the members of the geological community who interested themselves in Scottish geology could not perceive Geikie's penmanship behind the essay in the *North British Review*. And from the point of view of such "insiders," the essay would not have made much difference to their opinions. But clearly the intended audience for such an exercise was something rather different. Geikie was aiming at the general educated public who might be only marginally interested in an obscure controversy over the geology of the Northwest Highlands. With the publication of such a review, the Murchison-Geikie theory might be made to seem the uncontroversial orthodoxy—which every educated person with scientific tastes might assume as a commonplace. In short, Geikie was seeking to legitimate his ideas in the public mind. The publication of his essay was a highly significant maneuvre in the Highland campaign.

Nicol's opportunities for reply were limited if a prestigious Scottish journal such as the *North British Review* had opened its pages to Murchison and Geikie, and I know of no effort of a similar kind from Nicol. We may wonder, then, what efforts Nicol did make on his own behalf. Information on this is scarce, but we must return to the curious fact that despite the great success of Murchison's paper when it was presented at the meeting of the British Association at Aberdeen in 1859, and its wide dissemination in the press at that time, the paper did not appear in the published report for that year. One might think that Murchison had himself withdrawn it from publication for some reason. But a letter he wrote to Geikie in July 1861 indicates that this was not the case.

> In truth, I am somewhat amazed that what I recorded before the British Association is entirely omitted in the volume of 1859, whilst all Nicol's trash is printed! [31]

One is forced to suspect, therefore, that Nicol, who organized the geology section of the Aberdeen meeting, had somehow found a means to ensure that Murchison's paper did not get published.

Supposing this was the case, it is possible that Murchison then managed to get his own back, though I have no concrete evidence to support this suggestion. On June 18, 1862, a paper of Nicol's on the geology of the Grampian Mountains was read at the Geological Society (Nicol 1863). This dealt with the difficult problem of the attempts to extrapolate the Northwest Highlands stratigraphy to the complex region of the Grampians, which as we have seen is what Murchison and Geikie had been trying to do in their explorations in Scotland in the summer of 1860. There was a delay in the publication of Nicol's paper, and only a short note concerning it was issued in 1862 (Nicol 1862). It appears that Nicol wished to append a note giving a reply to Murchison's criticisms of him (Murchison 1861), but this was disallowed by the society's council on the not unreasonable grounds that Nicol's paper was about the Grampians, not the Northwest Highlands (Woodward 1908, 196). Even so, one may suspect that Murchison exerted what influence he could behind the scenes to ensure that Nicol's reply remained unpublished. He had, of course, attacked Nicol's sections in the appendix to his joint paper with Geikie of 1861.

Yet even if some of the things that Nicol wanted to say were disallowed for publication, he still managed to say quite a lot. His Grampian paper was chiefly concerned with the description of sections in the main body of the Highlands—an immensely difficult task, since there were no fossils at all to guide him, and the differences in lithologies between the different kinds of metamorphic rocks ("clay slates," "mica slates," etc.) were by no means clear-

---

31. Geological Society Archives, Murchison Papers, E34, July 4, 1861.

cut. However, he approached the problems in somewhat the same manner as he had done in the Northwest Highlands, finding evidences for inversions and repetitions of strata, large-scale faulting, and intrusion of igneous rocks along fault lines. There were also some scarcely veiled criticisms of the work of Murchison and Geikie.

For example, Murchison and Geikie, in a paper that they published in 1861 (Murchison and Giekie 1861b), had argued against the ideas of Daniel Sharpe (which, it will be recalled, Murchison regarded as "all stuff"). The problem was that theory had to accommodate at least three types of layering: ordinary "bedding," "cleavage" (as in slates) arising from pressure, and "foliation" (as in schists and gneisses). The foliation was of doubtful origin. In the gneisses of Scotland, one frequently sees bands of material of different mineral composition, texture, and color. These bands might be construed either as arising from the effects of pressure (as Sharpe supposed) or as vestiges of the original sedimentary layerings (as Murchison and Geikie believed). Thus Murchison and Geikie maintained that when they were examining the metamorphic rocks of the Highlands they could in fact make out the original bedding— whereas very often what they were looking at was lamination arising from some process that had occurred long after the rocks were originally deposited. All this had much to do with their general theory of the Highlands, for their assumption about foliation allowed them to suppose that the gneiss or schist seen lying on top of the Silurian sequence (of quartz rock, limestone, etc.) formed one regular ascending sequence—"truly one great series" (ibid., 236)—for they were assuming that foliation was one and the same as bedding, or at least intimately related to it.[32] Hence Nicol's arguments about faulting could be discounted.

In his Grampian paper, Nicol took the opportunity to refer to the "great merit" of Sharpe's ideas, saying that "it has too often been assumed, without proof, that all the marked division-lines in the metamorphic strata of the Scottish Highlands are lines of deposition, and much error and confusion has thus been produced" (Nicol 1863, 193). This was an obvious reference to the ideas of Murchison and Geikie.

Toward the close of his paper, Nicol made some rather indefinite remarks about the relationship between the metamorphic rocks of the main body of the Highlands and those of the Northwest Highlands. He added the thought that since the schists and gneisses of the more southerly rocks were evidently quite distinct, the same might hold good for those of the northwest. In fact, this marked a fairly substantial shift of opinion on the part of Nicol, since, as we have seen, much of his argument for the Durness-Eriboll region depended on the conflation of schist and gneiss. In a footnote, Nicol indicated that he did not necessarily think that the western gneiss and certain (eastern?) "mica- or chlorite-slates" were identical. Indeed, they might well be of wholly different ages. He claimed that he had held this view in his earlier publications. Yet

32. See figure 5.6. This shows that on occasions the assumption was by no means implausible.

Murchison could well be forgiven for having understood Nicol the way he did, and it is not surprising that in a letter to Geikie he should have written:

> I have not seen [Nicol's] Memoir, but if I understand Jones's brief Abstract he seems to be backing out by revising(?) his old specimens of metamorphism & by saying that though Gneiss is Gneiss he does not mean to assert that there are not gneisses of different age in Scotia![33]

Nicol (1863, 208–9) concluded his paper with some rather oblique remarks concerning the antiquity of the "eastern" metamorphic rocks of both the Northwest Highlands and the more southerly districts. He did not come out and challenge Murchison directly on this point, but a challenge was implicit in the whole of his theory. For if the "eastern" metamorphics were indeed upcast representatives of the "western" metamorphics, Murchison's case for the Silurian age of the Grampians and elsewhere would be grievously damaged, and all the brown coloration of the Highlands of the Murchison-Geikie map (see plate 5) would have to be replaced by something else.

Nicol made no further attempt to press his views before the Geological Society, but it appears that he continued to put forward his case to the best of his ability in his home territory. Thus in a letter to Geikie dated January 16, 1865, Murchison described a report he had heard of a lecture by Nicol "which is a violent attack upon our map of Scotland." Murchison was by this time utterly incensed:

> Will you not condescend to give him a good back-hander or two. He gives you a famous opportunity by his glaring mistakes.
>
> Besides, if he really shewed in the first place that the Applecross red rocks [i.e., the western red-brown, or Cambrian, sandstones] dipped under all the mica schist & gneiss of the east how can he make these rocks to be the same as the Laurentian?
>
> He is a bitter biter—and I rue the day that I had him with me & pushed(?) him etc(?) through the Highlands.
>
> He was not only what the French call "Mauvais fois" by the way; but he slyly went by himself the next year & then tried to turn me over & get all the credit himself.
>
> I wish you would hammer him.[34]

What Geikie was supposed to do about all this is not clear. It was hardly in his temperament, as I read it, to engage in direct public conflict with other scientists. Perhaps Murchison had in mind that Geikie would challenge Nicol at any lecture he might give in Scotland that Murchison was unable to attend. However, I have no evidence that Geikie did any such thing. Yet something similar seems to have cropped up the following year, for we find Geikie writing to Murchison about a "reply to Nicol" that he had prepared at Murchison's

---

33. Murchison to Geikie, June 29, 1862. Nicol's Grampian paper (1863) was read on June 18, 1862.
34. Murchison to Geikie, Jan. 16, 1865.

request, for publication in the *Scotsman*.[35] Apparently the newspaper's editor did not accept Geikie's response to Nicol, but Geikie was unperturbed. "The more I think of it I am the more convinced that the best way of answering Nicol is to take no notice of him." Perhaps surprisingly, Murchison did not disagree: it was best to let Nicol "simmer away."[36]

We can infer, then, that Nicol did not withdraw from the battle after his Grampian paper in the Geological Society journal. But he certainly retired from the London scene to Scotland, where he may have been able to gain a more sympathetic hearing. His last statement on the Highlands controversy was given in an appendix to a small book entitled *The Geology and Scenery of the North of Scotland* (Nicol 1866). This contained the texts of two lectures Nicol gave to the Philosophical Institution in Edinburgh, and it was very likely one of these lectures that raised Murchison's extreme ire, and to which Geikie was expected to furnish a response. The appendix, Nicol stated, discussed the points he had wished to include in his addendum to his Grampian paper, which had been disallowed by the Geological Society.

In the appendix, Nicol took his readers round each of the contested sites and gave his interpretation of what he saw there, explaining his differences with Murchison. Not much new material was introduced, so we need not cover all the ground again here. But a few points are worth making. At Assynt, Nicol argued persuasively against the existence of an "upper quartz rock" by explaining how he had been able, starting from a location on the quartz rock on the slopes of Quinag, to walk continuously on this rock either to Murchison's "lower" or to his "upper" quartz rock in the vicinity of Ben More (Nicol 1866, 90). If this were true, it would have been a wholly convincing argument. Unfortunately for Nicol, however, inspection of a modern map shows that there is a narrow outcrop of limestone, running north from Assynt, that must be crossed if one wishes to reach the upper quartz rock. (And there is almost certainly more hidden under alluvium.) Thus his argument at this point was based upon an inference from a faulty observation. The error remained unnoticed and unchallenged.

For Loch Maree, Nicol published a partly revised section (ibid., 41), which is shown in figure 6.7. It should be compared with figure 6.5 above. It may be noted that the outcrop of the limestone (*l*) has been reduced—which brings it into better agreement with modern maps and with what may be seen in the field. There is now a substantial fault represented in Glen Bianasdail (Glen Haasach), which Nicol here calls "Avon Inassie."[37] The outcrops of the sandstone (*f*) and quartzite (*g*) on the eastern side of this glen are improved. But it is apparent that Nicol still had not climbed to the top of the hill between Bianasdail and Logan, for the Torridon Sandstone to be found on the summit

35. Geikie to Murchison, Jan. 21, 1865.
36. Murchison to Geikie, Jan. 20, 1866. (It may be noted from the dates that the correspondence here is incomplete. Murchison refers to letters of Geikie's that are not in the archive.)
37. The Gaelic name for the burn in the glen in Abhainn an Fhasaigh.

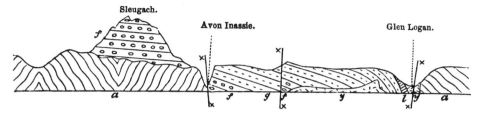

Fig. 6.7. Section of strata at head of Loch Maree (north side), according to Nicol (1866, 41).

is not depicted.[38] It is also to be noted that Nicol has now introduced a major fault into Glen Logan itself, which should really have been there before to be in keeping with his general theory. We may note particularly that what had been called a diorite is now described as being a syenite. (In his text, Nicol also called the rock a "prismatic porphyry" [ibid., 42].) On the southern shore of Loch Maree, round Ben Eighe, Nicol referred to the huge degree of faulting that may be inferred there, which Murchison had overlooked, regarding the whole area as evidence of a regular ascending sequence. Nicol was quite right. There is another klippe there, formed in association with the Kishorn Nappe (Anderson 1983, 36). However, as we know, neither Murchison nor Geikie examined this particular locality to the south of Loch Maree.

Finally, Nicol gave a statement of his general theory of the geology of the Highlands in the form of a hypothetical history of the area. This is worth stating almost in its entirety, since hitherto his doctrine could only be inferred from various fragmentary statements:

> We have caught dim glimpses of a dark ocean where the debris of unknown lands were laying the foundations consolidated by internal heat,—the mud and sand changed into gnarled gneiss. Then, this gneiss planed down by the slow waste of ages to a great platform, on which, as the bottom of another sea, thousands of feet of new strata are accumulated. And now this vast mass of strata is raised up over an axis running for full an hundred miles from Whitten Head, on Loch Eriboll, to the Sound of Sleat, in Skye. Along all this line the rocks are broken by faults,—some longitudinal, some transverse—into huge fragments, which have slipped down into the yawning chasms that opened in the vast arch, rent asunder as it rose. Veins and lumps of igneous [*sic*] matter—granite, granulite [*sic*], serpentine, and porphyry—have been violently forced in among these fragments. But there is no reason to regard this axis as the result of a mere sudden or solitary convulsion. Rather, as nature's analogy would teach, the arch was often forced up and fractured,—often gave way, sunk down and became welded anew by intense heat. For long ages the Titanic conflict may

38. Nicol was now using the term "Torridon Sandstone" (*f*).

have gone on. Is it then wonderful that along this line the strata should be crushed, contorted, thrown into apparently discordant positions? That masses of one age should be brought into contact with masses of another, even widely different age? That wedges of the higher lime-stone or quartzite should lie alongside, or even lower down than the once deeper-seated Torridon Sandstone or the gneiss? (Nicol 1866, 46–47)

This was Nicol's last public statement on the Highlands controversy. It may be seen that his argument was one that modern geology would have no diffi-culty in accepting in broad outline, for he certainly had a general understand-ing of the great earth movements that have subsequently been called the Caledonian Orogeny.[39] Moreover, modern geology would agree whole-heart-edly with Nicol that it was improbable that a gneiss or a schist could have somehow been "deposited" on a limestone in a conformable arrangement, as Murchison's doctrine required.

It is known that Nicol continued to make field trips in the Highlands into his old age, for in his obituary to the Geological Society, Henry Clifton Sorby (1880, 35) wrote that Nicol "never ceased to wander . . . year after year, among these Highland rocks in search of fresh evidence in support of his view." And though "he never published the results of these later investiga-tions, he remained fully convinced of the correctness of his own view[s] . . . till the day of his death."[40] But living up in distant Aberdeen, remote from the main centers of "geological power" in Britain (London and Edinburgh), Nicol found himself unable to influence events any further. He seems to have found the forces ranged against him too strong, and he quietly withdrew from the contest. He was apparently unsuccessful in gaining any adherents to his views that carried weight, except for the professor of natural history at Aberdeen, Henry Alleyne Nicholson, and J. W. Judd, some of whose work we shall look at in chapter 10. In summarizing the Highlands controversy in 1885, Judd recalled that he had visited Assynt and Loch Broom in 1877 with a predisposi-tion to accept Murchison's theory, but he became convinced that the doctrine was untenable. On meeting Nicol up there, and after hearing his ideas ex-pounded, he became a convert to Nicol's doctrines (Judd 1886). So it is pos-

39. That is, the great earth movements occurring in cyclic fashion between the late pre-Cambrian and the Devonian periods, which culminated during the time of Murchison's "kingdom."
40. It may be noted that Sorby—a petrologist who took no part in the Highlands controversy—acknowledged the assistance of H. Bauermann, C. Lapworth, H. G. Seeley, H. B. Woodward, and W. S. Dallas in preparing the obituary. According to Bailey (1952, 109), the obituary was largely written by Lapworth. This would make sense, given that Nicol was eventually allowed a generous obituary by an organization from which he had largely disassociated himself in his later years. Lapworth, of course, was to be perhaps the principal actor in the eventual overthrow of the Murchison theory of the Highlands. Evidence of Nicol's late fieldwork in the Highlands is to be found in the archives of the Elgin Museum, Morayshire, in the papers of the Reverend G. Gordon, an important local amateur geologist: Nicol to Gordon, May 7, 1863; Aug. 11, 1864; Aug. 22, 1864; Aug. 4, 1866; Aug. 25, 1875.

Fig. 6.8. Portrait of Nicol
in his later years, from
Anderson (1899, plate facing
page 83). Courtesy of
Aberdeen University
Library.

sible that the tide might have begun to turn for Nicol, but it was by then too
late. He died in April 1879 without gaining the recognition that his single-
handed intellectual and physical efforts entitled him. I reproduce in figure 6.8 a portrait of Nicol from the archives of Aberdeen University, which
shows, I think, the strain under which he must have lived during the latter
part of his life. To what extent this may be attributed to his breach with Murchison and Geikie, I cannot say. But I am inclined to think that it may have
been a significant factor.[41]

We may leave Nicol now and return to a consideration of the activities of
Geikie and Murchison. Murchison's goal was to see his ideas about the geology
of Scotland enshrined in a new edition of *Siluria,* and to have the newly expanded Silurian domains accepted and depicted in the standard textbooks.
Geikie's goal was to get a chair somewhere, preferably at the University of Edinburgh. It is of the highest interest to see how these goals were made to re-
inforce one another, and how both aims were achieved.

In 1863, we find Murchison writing to Geikie that when the appropriate
time came he would "delight" in recommending Geikie to a chair in the Royal
School of Mines.[42] In the same letter, he encouraged his protégé with the

41. The portrait, a copy of which is to be found in the Aberdeen University Archives, has previously been published in A[nderson] 1899 plate opposite page 83.
42. Murchison to Geikie, July 6, 1863.

thought that should Ramsay retire (which was thought quite possible because of his ill health), he would have Geikie in view for Ramsay's position. Neither of these plans came to anything.

But the following year there was a brief period of dissension between Murchison and Geikie. Geikie was preparing a geological map of the British Isles, with an accompanying memoir (Geikie 1864), and looking over the draft of this Murchison expressed his displeasure that Geikie had failed to mention the fact that Murchison had been the first to introduce the term "Fundamental Gneiss." This, said Murchison, was the thing of which he was "most proud."[43] We do not have Geikie's reply, but it must have been unrepentant and perhaps couched in uncharacteristically undiplomatic language. But after some further exchange he must have climbed down, for we find Murchison writing in a manner that is worth transcribing at some length. His letter reveals so clearly the manner in which the relationship was now developing between the two men:

> Your letter written on St. Andrew's day has quite rejoiced me & *has entirely* removed from my mind the vexation of spirit which your previous letter had occasioned.
>
> I can assure you that my brief note to you when I first glanced at your Map & description, was only caused by surprise that the omissions should have occurred, & attributing them to inadvertence, I thought verily that you would at once have set all to rights by a candid explanation.
>
> *Loving you as I did* [emphasis added] & as I again do now, you may conceive my grief when I got your angry & irritable reply; for I never thought of accusing you of an ungenerous & selfish deed!
>
> I am glad now that you have calmly looked over your introduction & map, that you see how the general reader might construe it in the manner that was far from your mind & ? it is most satisfactory to know that you will make the amends in your power.
>
> Nothing can be easier than to tell Keith Johnston to engrave under the little type section across the North Highlands "by Sir R.I.M."
>
> Rely upon it that the inadvertence on your part has not in the slightest degree altered my regard & esteem for you now that you have written me a full & [candid?] explanation of the matter.
>
> I am not one of those who readily forget my friends or the excl. services to Science they have purposed & I am truly happy to find, that you are the same Archibald Geikie as when I formed the highest opinion of your ability & expressed my gratitude for your kind co-operation & support.
>
> Yours sincerely
> Rod I. Murchison[44]

43. Murchison to Geikie, Nov. 16, 1864.
44. Murchison to Geikie, Dec. 4, 1864.

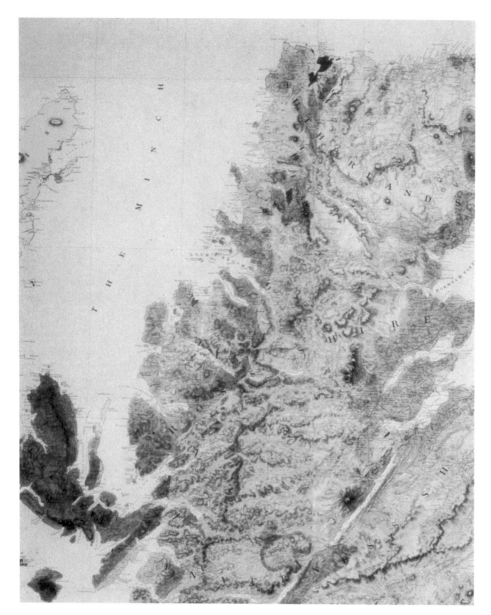

Plate 1. Part of Macculloch's *Geological Map of Scotland* (1843 issue). Courtesy of the British Library.

Plate 2. Portions of Cunningham's hand-colored [*Geological*] *Map of the County of Sutherland* (1838). Above: part of northwest Sutherland, showing region round Loch Durness and Loch Eriboll; below: part of southwest Sutherland, showing Loch Glencoul and Loch Assynt. Courtesy of the National Museums of Scotland.

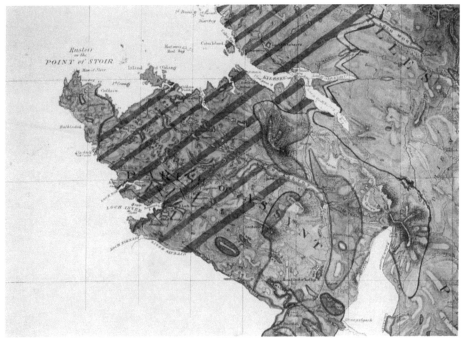

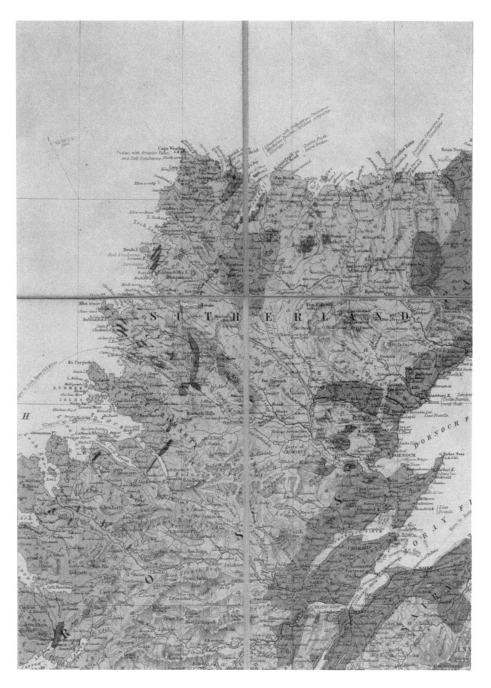

Plate 3. Part of Nicol's *Geological Map of Scotland* (1858). Courtesy of the Library Committee of the Geological Society, London.

Plate 4. Cartoon of Murchison and Geikie by Prosper Mérimée (1860). Courtesy of the Trustees of the National Library of Scotland.

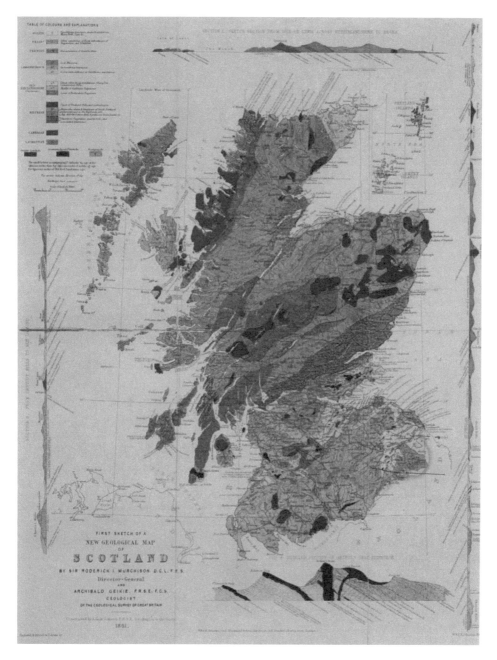

Plate 5. Murchison and Geikie's *First Sketch of a New Geological Map of Scotland* (1861), showing (section 1) the structure of the Northwest Highlands according to the Murchisonian theory. Courtesy of the Library Committee of the Geological Society, London.

Plate 6.a. Specimen
of "Logan Rock"
(Lewisian Gneiss).
Photograph by Belinda
Allen.

b. Photomicrograph
of thin section of 6a,
viewed in ordinary
light. Photograph by
F. I. Roberts.

c. Photomicrograph of
thin section of 6a,
viewed with polarized
light. Photograph by
F. I. Roberts.

Plate 7. Part of Heddle's *Geological and Mineralogical Map of Sutherland* (1881). Courtesy of the Library Committee of the Geological Society, London.

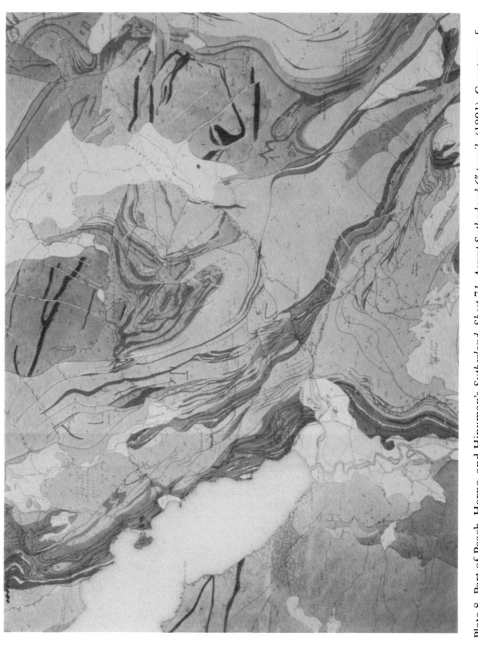

Plate 8. Part of Peach, Horne, and Hinxman's *Sutherland, Sheet 71. Assynt Sutherland 6″ to mile* (1891). Courtesy of the British Library.

It must be acknowledged that this was an extraordinary letter for a man like Murchison to have written to one of his junior staff members. I know of no person, other than Lady Murchison, whom Sir Roderick would have said he loved—certainly no other of his geological colleagues. Interestingly, one can see in Murchison's letter that it cost him some pains to write the word "loving." It is almost scrawled out in his seeming inhibition and embarrassment at writing the term of endearment. Nevertheless, the word is certainly legible, and on this occasion at least he opened his heart to young Geikie.

It is most unfortunate for the student of the Highlands controversy that we do not have the complete correspondence between Geikie and Murchison; but it is evident that Geikie never made the mistake of crossing his chief a second time, though, as we shall see, he was no mere sycophant and was perfectly prepared to express differences of opinion with Murchison. What appeared to rile the old man was any failure to grant him his full share of recognition for any geological achievement. But thereafter we do not find him at any time weakening in his resolve that his mantle should pass to Geikie, and never again did Geikie fail to give credit where credit was claimed to be due.

In 1865, Geikie published his first major geological text, on the geology and scenery of Scotland (Geikie 1865). This book—a semipopular work though an important contribution to the literature on Scottish geology—was a characteristic piece of tapestry of Geikie's philosophy of science (Oldroyd 1980). It linked the scenery of a region to its physical structure, and ultimately to its geological history. In subsequent publications, Geikie (1905) would endeavor to link all this to human history too. The book contained a synoptic view of the structure and geological history of the Highlands (Geikie 1865, 91–113), and Geikie made it quite clear that Murchison was responsible for the notion of "Fundamental Gneiss" and its equivalence to the "Laurentian" of the North Americas. Moreover, he obliquely rejected Nicol's doctrines in a manner that would have been congenial to Murchison:

> If we were still vaguely surmising that the gneiss and the schist had been thrown down upon the floor of a primaeval thermal ocean, and had been broken up when that hardened floor was upheaved into the first dry land, there might be some excuse for a belief that the ancient convulsions had been the means of throwing up our northern mountains and tearing open their glens. But when these rocks are discovered to be only a modification of ordinary sedimentary deposits, and to reveal their geological age by their enclosed organic remains, all such vague conjectures must cease, and the rocks must be examined and determined by the ascertained laws which govern the arrangement of masses of stratified rocks. (ibid., 95)

Geikie's book was a fine work, and ran through three editions, though for the second edition of 1887 he would have to rewrite the material on the Northwest Highlands completely in the light of the recent discoveries, and would finally have to repudiate the old Murchisonian doctrine.

In one of Geikie's field notebooks, held at the Haslemere Educational Museum, there are three drafts that he scribbled with tentative dedications of his new book to Murchison.[45] They are worth reproducing here since they provide further evidence of the personal relationship between Geikie and his chief.

1. My dear Sir R. To Some who read the following little work and who know that its whole aim and purpose are directly opposed to views which you have strenuously upheld throughout your long and active scientific life might be surprised that it should be inscribed with your name. It is this very antagonism of opinion however which has prompted the present dedication for I would fain have it known that in the great national undertaking over which you preside perfect liberty and independence of thought do not interfere either with the harmonious carrying out of the objects for which we are combined . . .

2. My dear Sir R. To no one can a volume on Scottish geology be more fitly inscribed than to you to whom Scottish geology owes so much. And yet to some who read the following chapters and who see that their aim and purpose are at variance with views which you have strenuously upheld throughout your long and active scientific life it may be a matter of some surprise that this little work should have your name on its front. But in the national undertaking over which you preside we have found that the utmost freedom and independence of opinion interfere neither with the harmonious carrying out of the common objects for which we are combined nor with a pleasant and friendly intimacy.

3. To no one can a new volume on Scottish geology be more fitly inscribed than to you whom Scottish geology owes so much. There are grounds of a personal kind also on which I gladly avail myself of the opportunity of placing your name on this page. Many points of the scenery here discussed were first visited by me in your company, and the problems then suggested regarding the history of the surface of the country have been a fruitful subject of reflection during the years which have since slipped away. That the views to which I have been led are sometimes directly at variance with your own would only serve, if need were, as an additional motive for dedicating this volume to you.

The dedication as eventually published was

4. My dear Sir Roderick,
To no one can a new volume on Scottish Geology be more fitly inscribed than to you, to whom Scottish Geology owes so much; nor perhaps can such an inscription come with greater appropriateness than

45. A. Geikie, Notebook F, 1865. The drafts are on unnumbered leaves at the back of the notebook.

from one who has worked with you, hammer in hand, over many a league of Highland ground, and who is further bound to you by many acts of courtesy and kindness. You are of course in no way responsible for the opinions expressed in the following chapters. That they may sometimes be at variance with your well-known views of the same subjects would only serve, if need were, as an additional motive for offering you this expression of respect and esteem.

The second and third editions of Geikie's book, published after Murchison's death, and after the demise of his Highland theory and its abandonment by Geikie, carried no dedication at all.

These draft dedications, and the final version, could well be a fruitful object of inquiry for a psychoanalyst. But we need not undertake such a task here. Suffice it to say that Geikie's first version, and presumably the one that most expressed his spontaneous feelings, gives indication of significant differences of theoretical opinion between Geikie and his patron. The question is, however, whether these differences had to do with the Highlands controversy or with other topics of theoretical geology. Geikie had views very different from Murchison on questions of glacial theory and on uniformitarianism, for at bottom Murchison still sided with the older school of geology of William Smith, Georges Cuvier, and Adam Sedgwick, who imagined that certain geological events of the past were fundamentally different in kind from those occurring in the present, with convulsive or paroxysmal forces such as had no equal today being the major determinant of topography and scenery. By contrast, Geikie was of the school of James Hutton, Charles Lyell, and Charles Darwin. He was a "gradualist": nature moved with dignity. The linchpin of Geikie's whole methodological and theoretical paradigm was that—in his own words—"the present is the key to the past." So I do not think that the Highlands controversy was at the front of Geikie's mind when he wrote the first curiously grudging drafts of his dedication.[46] In any case, his tactical sense soon got the better of his theoretical reservations toward Murchisonian geology, and the jolly "hammer in hand" spirit took over. Note that Geikie said that he was "bound" to Murchison. This was certainly true until the coveted chair was attained, and indeed until Murchison's death not long after. On April 6, 1865, Murchison wrote to Geikie informing him that he had been elected that day to a fellowship of the Royal Society, Murchison having actively supported the candidature.

On December 26, 1865, we find Geikie writing to Murchison, congratulating him on the announcement of his elevation to a baronetcy, and saying that he had heard from Ramsay that Murchison was making efforts toward

46. I am informed by John Thackray (personal communication, Dec. 16, 1988), who is in possession of Murchison's annotated copy of Geikie's book, that the marginalia show that Murchison was in fact by no means pleased with what Geikie had written. There was indeed a profound difference between Geikie's "slow Huttonianism" and Murchison's catastrophist explanations in terms of faulting and drifting icebergs. It is in a sense a paradox that Nicol's theory, invoking massive faulting, had so little appeal to Murchison.

the establishment of a chair at Edinburgh. Murchison replied encouragingly on the thirtieth, saying: "I hope to succeed in the business of the Scottish Geological *Chair* & to live to see you in it." In June the following year there was news of a possible professorship in natural history at Glasgow University, and Geikie wrote to Murchison seeking his advice on whether to apply.[47] Murchison wrote recommending that Geikie not consider it. It would involve the teaching of botany and zoology, and although Murchison had no doubt that his protégé could "get up" these subjects (which with Geikie's abilities was doubtless true) he advised against it, pointing out the distraction there would be for his proper subject, geology.[48] By June 10, Murchison had heard that the Glasgow vacancy had been filled, so the matter was quickly dropped.

In August, Murchison informed Geikie that he was planning a new edition of *Siluria*, and he requested the young man's assistance in the considerable task of bringing the volume up to date.[49] Though the book appeared under Murchison's name, in fact he made use of a number of collaborators, including T. Rupert Jones, who effectively saw the book through the press (Thackray 1981). Geikie expressed his willingness to cooperate and to provide information on Scottish rocks.[50] In a further letter, discussing the updating of *Siluria*, we can see Murchison revealing his schemes for the establishment of an Edinburgh chair. Obviously, the two issues were connected in his mind, and in Geikie's:

> Be assured that I will push the D[uke] of Buccleuch on.[51] At my request, he accepts the Chair of the British Association in 1867 & it will be a fine card to play if he can assist in the establishment of a Geology Chair in Auld Reekie [Edinburgh].[52]

By October, Geikie was hard at work on his contribution to the revisions of *Siluria*, but candidly expressed to Murchison the fact that he found difficulty in concentrating because of his anxieties respecting the Edinburgh Chair. Then expressing a nice understanding of scientific politics, he urged Murchison to push things along with the duke before Parliament became wholly engrossed in the forthcoming debates concerning the Reform Bill.[53] At the

47. Geikie to Murchison, June 1, 1866.
48. Murchison to Geikie, June 2, 1866.
49. Murchison to Geikie, Aug. 10, 1866.
50. Geikie to Murchison, Aug. 12, 1866.
51. Walter Francis Montagu Douglas-Scott (1806–84), duke of Buccleuch, duke of Queensberry, and earl of Doncaster, was an influential politician and patron of science at that period. He was Lord Privy Seal from 1842 to 1846 in Pitt's ministry and president of Council in 1846. Among the other positions he held were president of the Society of Antiquaries (1862–73); president of the Highland and Agricultural Society (1831–35); chancellor of the University of Glasgow (1878–84); and as mentioned here by Murchison, president of the British Association for the Advancement of Science (1867). Clearly he was an eminently suitable person for Murchison to attempt to lobby on Geikie's behalf.
52. Murchison to Geikie, Aug. 16, 1866.
53. Geikie to Murchison, Oct. 23, 1866.

end of the month, Murchison was writing again and sending his draft for the section of *Siluria* dealing with the Northwest Highlands, for which Geikie had provided a geological section.[54] Apparently, the draft contained a polemical history of the researches in the region. Geikie thought ill of it and did not hesitate to say so: "It is too long, too detailed, too personal, and you do not show how it is the metamorphosed rocks of the NW that are repeated fold after fold across the whole breadth of Scotland."[55] He recommended an account that was "short, clear, and as firm as a diamond," for in his opinion "the world ha[d] had enough of the personality of the warfare." He then offered to write "what occurs to me in the proper way of finishing off your share in the controversy."

Murchison seems to have taken Geikie's strictures to heart, for on November 11 he wrote: "I dare say I am too difficult(?). I had certainly no intention of being egotistical." He accepted Geikie's offer of help. We may presume, therefore, that Geikie had a considerable hand in revising that portion of *Siluria* that referred to the Northwest Highlands. The new edition appeared the following year.

At about this time negotiations for the establishment of a chair at Edinburgh were overshadowed by Murchison's efforts toward the enlargement of the Survey. There was some dissatisfaction both within and without the organization as to its rate of progress, and clearly, from the geologists' point of view, the best solution to the problem was an increase of staff. Murchison worked mightily toward this goal, and with very considerable success. To attain his ends, he had to negotiate with the civil servants of the Department of Science and Art, under whose jurisdiction the Survey lay, and "lobby" various members of the government and other influential persons with whom he was acquainted. Murchison was extremely adept at this.

At Geikie's suggestion, Murchison recommended to his board that the salaries of four senior geologists should be raised from £350 to £500.[56] On reflection, he added Geikie's name to the list, on the grounds that the Scottish maps would be under his surveillance—even though Geikie had not at that time reached the top of his salary scale of £350.[57] This was good news from Geikie's point of view, and was no doubt what he had in mind when making his suggestion to Murchison, but his further aim was to see a separate branch of the Survey established in Scotland, as there had been in Ireland for some time. Accordingly, he wrote a long statement to Murchison setting out the (strong) case for the establishment of a Scottish branch office. He felt that it was necessary for the person in charge to be resident there: at present Ramsay (whose duty it was to oversee the Scottish activities) merely rushed through and didn't really know what was going on. "His 'inspection,'" wrote

54. Murchison to Geikie, Oct. 31, 1866.
55. Geikie to Murchison, Nov. 5, 1866.
56. See Geikie to Murchison, Nov. 11, 1866. The four senior geologists were William Aveline, Henry Bristow, Henry Howell, and Edward Hull.
57. Murchison to Geikie, Nov. 18, 1866.

Geikie, "is a simple farce."[58] Obviously, he was hinting his own suitability to Murchison.

Needless to say, Ramsay was loath to have any of his responsibilities taken away from him. Murchison's plan, then, was that he himself should be director general of the whole organization; Ramsay should be director for England and Wales; and Geikie director for Scotland.[59] In lobbying the duke of Buckingham (Murchison informed Geikie), he did not "demand" the establishment of a separate Scottish Survey, but said that if such an organization were established Geikie would undoubtedly be the man to take charge of it.[60] Within a few days, it was all worked out even better than might have been hoped. There was to be a director general; a senior director for England and Wales on an increased salary of £700; a local director for Ireland as before; a new local director for Scotland on a salary of £400, rising to £600; two extra senior geologists, or district surveyors, for England and Wales, one for Ireland, and one for Scotland. Ramsay was to be director for England and Wales (with increased salary but reduced work-load), the new district surveyors for England and Wales were to be Aveline and Bristow, and Howell was intended for Scotland. Jukes carried on in Ireland. Geikie, of course, was to be the new director for Scotland.[61] (Murchison moved swiftly to remove the word "local" from the title.) There were, besides, to be increases in the number of the more junior staff.[62] Hull was to be left out of the promotions, but in fact he became a district surveyor in Scotland and he gained the Irish directorship a couple of years later. It was all a great "coup d'état," as Murchison put it. He finished his letter with an urgent request from Geikie for information about the Scottish coalfields—which may remind us that the success of Murchison's plan may have had much to do with the fact that the government had at that time a royal commission sitting to inquire into the British coal industry, and the economic value of the Survey was becoming increasingly evident to the authorities.[63] The changes to the Survey had to be ratified by the chancellor of the exchequer, but this was accomplished without difficulty, and the great expansion began the following year.

Not surprisingly, Geikie was euphoric over the news of his impending promotion: "Your letter has completely taken my breath away!!! I can hardly believe that this coup d'état is anything but a dream of my own. Let me thank you again and again for so thoughtfully remembering me in the allocation of

58. Geikie to Murchison, Nov. 21, 1866.
59. There were problems for Murchison in Ireland at that time. The local director, Joseph B. Jukes, was becoming increasingly mentally disturbed. He died in 1869 and was replaced by Edward Hull. See Herries Davies (1983, 186–91).
60. Murchison to Geikie, Nov. 23, 1866.
61. Murchison to Geikie, Nov. 26, 1866.
62. In the years immediately following the great expansion of the Survey, thirty-three new assistant surveyors were appointed, nineteen of them in 1867. The remarkable geologists Benjamin Peach and John Horne, whose names were to figure so prominently in the Highlands controversy, joined the Scottish branch of the Survey in 1867 (Flett 1937, 75).
63. Ibid., 76.

the new posts." He then promptly suggested the need for the appointment of
a palaeontologist in Edinburgh, and announced that he was off to Ayrshire to
examine Silurian rocks.[64] A few days later, he was already giving thought as to
who might make suitable new appointments in the Edinburgh office.[65] He was
actually appointed to the post of Scottish director on April 1, 1867.

In the winter and spring of 1866–67 there was a fairly continuous exchange of correspondence between Geikie and Murchison concerning the affairs of the Survey and the publication of *Siluria*, the full details of which need
not be entered into here. Exactly what role Geikie played in the revisions of
the book is unclear from the correspondence, but it is evident that he wrote a
significant portion of the revisions for the Scottish rocks, particularly for the
"Silurians," and provided Murchison with some figures for sections; in addition he assisted with the correction of the proofs. There was also some discussion in the correspondence of general geological theory, particularly on the
question of the possible intensity of geological change in the past. Murchison
wrote that he was glad to see that Lyell was coming round to the idea of "*moderate convulsions*" and Geikie replied that he was "not a rigid uniformitarian,"
that he had "long felt that Lyell and his school ha[d] pushed the doctrine of
uniformitarianism to the extreme," even though he himself did not "believe in
cataclysms."[66] Evidently, the two men were trying to reach some common
ground in the matter of geological theory.

In another act calculated to advance Geikie's career, Murchison proposed
him for the chair of the geology section of the British Association for its meeting in Dundee in 1867, and the recommendation was duly accepted.[67]

It is clear, then, that by 1867 Geikie was very much in Murchison's debt,
though he had, to be sure, rendered him considerable assistance in getting out
the revised edition of *Siluria*. It might be thought that the publication of a
third edition of a book was not a matter of especial significance.[68] But one
should remember the example set by Lyell's *Principles* and his *Elements*, and
Darwin's *Origin*, which, coming out in repeatedly updated editions, set the
terms of debate on their subjects, and dominated whole areas of science both
theoretically and empirically for many years. Murchison, I suggest, had similar ambitions for his *Siluria*. He wanted it to be the unquestioned authority
concerning Palaeozoic rocks. It was, one might say, the instrument by which
he sought to rule his Silurian kingdom. Consequently it was essential that it be
kept up to date, so that authority might be maintained. That, I suggest, was
why Murchison invested so much time and energy in the task during the closing years of his reign, even though he had to invoke the assistance of others to
attain his ends.

64. Geikie to Murchison, Nov. 27, 1866.
65. Geikie to Murchison, Dec. 4, 1866.
66. Murchison to Geikie, Dec. 14, 1866; Geikie to Murchison, Dec. 16,1866.
67. Murchison to Geikie, March 18, 1867. See also: Geikie (1974, 117).
68. It was called the fourth, but whether it was third or fourth depended on whether one counted
Murchison's *Silurian System* (1839) as the first edition. This was really a different book.

In point of fact, the last edition of *Siluria* was not a publishing success. It lost money and it took nearly twenty years to clear the stocks (Thackray 1981, 39). But this was in no measure due to lack of effort on Murchison's part to ensure its well-being. There were two major points that he wanted to fix in the public mind: first, that he had established the "Fundamental Gneiss" (Lewisian or Laurentian) as the basement rocks for the whole stratigraphical system in Britain; second, that the seqeunce in the Northwest Highlands was an ascending conformable pattern, which provided the key to understanding the structure of the Highlands and a whole. This, of course, entailed the huge territorial claims for the Silurian kingdom in Scotland that we have seen in the Murchison-Geikie map of 1861 (see plate 5). That Murchison attached very great significance to the evidence from the Highlands may be gauged by the fact that he gave a view across Loch Assynt to Quinag as frontispiece to both the second (1859) and third (1867) editions. But the second edition (appearing in January) was issued before Murchison's field trip with Ramsay that year, and of course before his trip with Geikie in 1860. Thus it hardly gave expression to Murchison's "Great Reform" in Scottish geology; and in consequence he attached much importance to the third edition of 1867.

To assist his claims, Murchison chose to dedicate the new edition to Sir William Logan of the Canadian Geological Survey—Logan had suggested the idea of the Laurentian system and had accepted Murchison's correlation of the Canadian and British basement rocks. By contrast, the first edition (1854) was dedicated to Sir Henry De la Beche, and the second to the European geologists Edouard de Verneuil, Alexander von Keyserling, and Joachim Barrande.[69] All these were judicious choices. Although De la Beche was Murchison's old foe, he was getting near the end of his directorate in 1854 and his mantle fell on Murchison the following year.[70] Keyserling and Verneuil were old colleagues of Murchison and co-authors of his *Geology of Russia in Europe* (1854).[71] Barrande's palaeontological work in Bohemia was of fundamental importance for the whole of Murchison's Silurian work, particularly with respect to the exposures in the Southern Uplands of Scotland, as will be discussed in chapter 8. In fact, the choice of dedicatees could on occasion be a matter of critical importance in securing or maintaining patronage, or gathering adherents to one's views.

69. It is interesting that Murchison eventually made up his quarrel with De la Beche but never reconciled his differences with Sedgwick and Nicol. The simple explanation is probably that De la Beche's surveyors were constantly generating new information that could be utilized by Murchison. By contrast, the two busy university geologists had nothing of equivalent value to offer him.

70. It must be said, however, that the initiative for Murchison to be appointed director came in the first instance from within the Survey, and specifically from Ramsay, who prompted the elder statesman of geology of that time, William Fitton (1780–1861), to petition Lyon Playfair and the Board of Trade to appoint Murchison. (See draft of document to be signed by Ramsay and other Survey officers, dated April 18, 1855: Ramsay Archives, Imperial College, London, KGA/RAMSAY/8/413.)

71. De Verneuil extended Murchison's "kingdom" in France, Spain, and America, and Keyserling did likewise in Russia.

Murchison was not content simply to attract adherents to his views by means of his "inscriptions." He also took active steps to ensure that he gained sympathetic reviews for his treatise. For the second edition he had sought the assistance of Professor James Forbes (1809–68), by then principal of St. Andrews University. Perhaps, Murchison inquired, Forbes could organize a review in the pages of the *Quarterly Review* or the *Edinburgh Review?* But Forbes courteously declined the suggestion.[72] Now, for the third edition, Murchison tried a similar tactic, with Geikie as his potential reviewer. Here he met no resistance, and so we find that Geikie composed no less than three complimentary reviews of the third edition of *Siluria.* The manner in which this was fixed up is of considerable interest, and is illustrative of the art of action within the "agonistic field" of science, a topic about which I shall make some more general remarks in my concluding chapter. Unfortunately, however, at this point we only have Murchison's half of his correspondence with Geikie.

From a letter from Murchison to Geikie of November 14, 1867, we may infer that Geikie had already written something for Murchison, and that some discussions had taken place about putting it into the *Edinburgh Review.* We also learn that Geikie was putting something together for the *Pall Mall Gazette.* Then from a letter of December 10, a day after the appearance of the *Pall Mall* review, it appears that Murchison had been conversing with the editor of the *Times* about yet another review from Geikie.

The *Pall Mall* essay ([Geikie] 1867) was a gem from Geikie's ever-fluent pen. It indicated in a charming manner something of Murchison's life and character, stated that the "fourth" edition of *Siluria* was virtually a new book, and emphasized its general significance. And just as Murchison would have wished, Geikie adverted to the importance of the work in the Highlands:

> Among the important additions to the present edition comes the great discovery, made by Sir William Logan and his associates in the Geological Survey of Canada, of organic remains in the lower portion of a vast series of crystalline rocks, older than even the most ancient of the fossiliferous deposits of Britain. The position of a similar group of crystalline rocks in Scotland, and the arrangement of the metamorphosed silurian strata by which that older group is overlaid, are now clearly set forth.

Yet at the end of his review, Geikie was not afraid to express—with extraordinary tact—his disagreement with Murchison about the possible intensity of geological activity in the past. For all their mutual interests, it remains a fact that Geikie was a uniformitarian (or "quietist") while Murchison was a catastrophist (or "convulsionist").

Murchison was very happy with the *Pall Mall* review, but he detected some omissions, and in letters of December 12 and January 7 he mentioned some

72. St. Andrews University Archives, Forbes Papers, 1858/68, Murchison to Forbes, Feb. 8, 1859. See also Forbes to Murchison, March 7, 1859 (Letter Book 5, 534).

matters that he thought would be "satisfactory to me to have mentioned."[73] Some of these had to do with points of Murchison's early career, and the fact that he was still active as a geologist. He also desired that Geikie should mention his work on early fossil fish, which appeared to come suddenly into the stratigraphical column. This (he thought) was a convincing refutation of Darwin's doctrines. In his *Times* review, Geikie did add some points about his chief's early career, but said nothing about the Silurian fishes. On the other hand, he said more about the "Fundamental Gneiss," about which he knew Murchison was constantly agitated:

> To Sir Roderick himself we owe the discovery of rocks of this age
> [Laurentian] in Britain; a discovery which, with its accompanying
> deductions, whereby the geology of more than half of Scotland was
> revolutionized, is certainly his greatest achievement in science—the
> Silurian system always excepted. ([Geikie] 1868a)

Once more Geikie referred (with consummate tact) to the ongoing debate between catastrophists and uniformitarians. He peppered his text with military metaphors, to which he seems to have been just as much addicted as was Murchison. He also referred to the excellence of the figures in *Siluria*, some of which, of course, he had prepared himself.

Murchison was well pleased with Geikie's effort, describing it as "apposite and well written" and as treating geological theory in the "*fairest* and *ablest* manner."[74] Extraordinarily (I would have thought), it was Murchison who sent Geikie his fee of £10 from the *Times*.[75] But yet more could be done. In a sense, Geikie's two reviews so far were mere journalism. What was needed was a commendatory piece in a substantial journal such as the *Quarterly Review*. Murchison got to work again. In a letter to Geikie of March 1, he mentioned that the editor of this publication had received a review by some other author but had decided to "put the article . . . into the fire" because of its dullness.[76] Whether Murchison then (or possibly even before) twisted the editor's arm, I have no means of knowing, but the upshot was that Geikie was commissioned to write yet again, on Murchison's recommendation.

The next development in this little saga was really the most remarkable of all. It appears that in preparing his piece for the *Quarterly Review* Geikie had written to Murchison inquiring about the old meaning of the (Wernerian) term "Transition series." In his reply, Murchison enlarged upon details of his career for Geikie's edification, and enclosed three foolscap pages of suggestions as to what might usefully go into Geikie's essay.[77] In effect he was seeking to write his own review, as the beginning of this extraordinary document reveals:

73. Murchison to Geikie, Dec. 12, 1867.
74. Murchison to Geikie, Jan. 17, 1867; Murchison to Geikie, Jan. 27, 1867.
75. Murchison to Geikie, Feb. 6, 1867.
76. Murchison to Geikie, March 1, 1867.
77. Murchison to Geikie, April 10, 1868.

Although the Silurian Classification as worked out by Sir Roderick Murchison after seven years of labour in Wales & the adjacent counties of England beginning in 1831 & finished in 1838 was reviewed by us [the *Quarterly Review*] on some(?) former occasion(?) this new Edition of his work entitled "Siluria" has been so brought up to the present state of geological knowledge respecting the Palaeozoic formations that we are bound to call public attention to it.[78]

Murchison carried on in this style for some time, and then concluded with seven specific points that he wished Geikie to mention in "his" review. It should be said that none of these had anything directly to do with the Highlands controversy as I have described it in previous pages (though there was an issue relating to the Laurentian rocks that will be mentioned in chapter 7). The main points that were troubling Murchison's mind were to do with the quietist/cataclysmic issue and the claimed empirical refutation of Darwin's theory with the Silurian fish beds. Nicol was definitely not mentioned; but he was, I suppose, a vanquished foe in Murchison's mind by this time, or at least one who had withdrawn from the scene of battle.

The reader will naturally be interested to discover what Geikie did in fact put into his review for the *Quarterly* ([Geikie] 1868b). It was another fine essay, discussing in a general way a number of the more important theoretical issues that were exercising the minds of geologists at that time. I am glad to say that Geikie did not demean himself by utilizing directly the suggestions that Murchison had so improperly put to him. Nonetheless, what Geikie wrote was beautifully crafted to the Murchisonian doctrines and Murchison's goals. In a manner that was characteristic of Geikie's whole approach to science, he set out a valuable synoptic history of Palaeozoic researches—which necessarily involved a considerable exposition of Murchison's views, for, as Geikie put it, "'Siluria' forms in itself a sort of cyclopædia of palaeozoic geology" (ibid., 191).

While disagreeing with many aspects of Murchison's "convulsionism," and his antievolutionism, Geikie undoubtedly gave a sympathetic exposition of his chief's views. Here we must be primarily interested in what he had to say about the history of researches in northwest Scotland and about James Nicol. As might be expected, Geikie set out the history presenting almost exclusively Murchison's side of the story. Rather amusingly, he stated that Murchison, "in order to satisfy himself more surely, and to augment the data for his generalizations, . . . revisited the Highlands, in company with geological friends of long experience" (ibid., 196). One of these friends was, of course, young Geikie, in 1860. But having cautioned Nicol against invoking "dislocations" wherever required to explain recalcitrant observations ("Nothing is so easy, and nothing unfortunately is more common, when facts are wanting to support a favorite theory" [ibid.]), Geikie indicated that Murchison was guilty of doing just this in the more southerly parts of the Highlands. We know that

78. Document accompanying Geikie to Murchison, April 10, 1868.

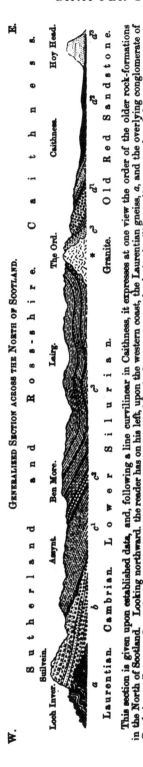

Fig. 6.9. Structure of North Scotland, as represented by Murchison (1867, 169).

Geikie was not at all happy with Murchison's attempted extrapolation of his northwest findings into the Grampians. Part of the problem for Geikie was that Murchison—in keeping with his "convulsionism"—was inclined to invoke some kind of fracture for every Highland glen, rather than applying the Huttonian doctrine that valleys were, generally speaking, excavated by processes of erosion. As the *Quarterly* review shows, Geikie was not afraid to speak his mind against Murchison. Indeed, when we consider the grave theoretical differences that separated the younger Geikie from his aging chief, it is remarkable that they worked in partnership as well as they did. But it must be reemphasized that they had interests that mutually enhanced each other, and clearly they got on together extremely well at the personal level.

Murchison thanked Geikie for his efforts in two letters written on the same day.[79] He thought the review was written with "great spirit" and he was grateful to have so "lively a reviewer." He was, it appears, a little put out that Geikie had spent so much time on the "chaos of ice basins and denudations." Nevertheless, he was glad to have "a good fine criticism." Murchison did not advert to his list of suggestions for things that he would have wished Geikie to put into the review. Seemingly all parties were content with the outcome.

Having said so much about the reviews of Murchison's book, something should be said about its contents, though not much will be required since for the most part the discussion of the Northwest Highlands was a history of the researches there, which I have already gone over in some detail. Fortunately, with Geikie's help, the account was virtually free of polemics against Nicol, and there was honest acknowledgment of what Murchison conceded as Nicol's contributions, such as the recognition of the unconformity at the base of the quartz rock. The text contained a hypothetical section across the north of Scotland from the west coast to the east, which is shown in figure 6.9 (Murchison 1867, 169). This became the paradigm for understanding the geology of the region for the next twelve years or more.

One issue that is worth mentioning here is the fact that the text gives some hint of how Murchison's ideas concerning the relationships between the rocks of the Northwest Highlands, the Central Highlands, and the Southern Uplands of Scotland were developed. Early on, Murchison had attempted to show that certain schists of the southern zone of the Highlands were metamorphosed equivalents of the Silurian greywackes of the Southern Uplands (Murchison 1851). That is, they were regarded as being essentially of the same sequence, having dipped under the younger sedimentary rocks of the Midland Valley, forming the industrial region of Scotland. These views had been supported by Nicol (1852). We may see, then, how Murchison (1867, 172) sought to demonstrate a unity of the Scottish Silurians all the way from the Northwest Highlands right through to the Southern Uplands, by means of an idealized section (fig. 6.10). This had actually been drawn by Geikie.

It will be noted that the sequence of "Silurian" rocks identified in the northwest (quartz rock, limestone, gneiss/schist, etc., all labeled *b*) was supposedly

79. Murchison to Geikie, July 14, 1868.

GENERALIZED SECTION ACROSS SCOTLAND, FROM THE COAST OF ROSS-SHIRE ON THE NORTH-WEST TO THE CHEVIOT HILLS ON THE SOUTH-EAST.

N.W.

S.E

Applecross.

Great Glen.

Ben Nevis.

Ben Lawers.

Loch Tay.

Ochil Hills.

Fife.

Edinburgh.
FirthofForth

Midlothian
coal-field.

Southern Uplands.

d^2. Coal-measures, Millstone-grit, and Carboniferous Limestone. d^1. Calciferous Sandstone series of the Carboniferous Formation. c^2. Upper Old Red Sandstone and Conglomerate. c^1. Lower Old Red Sandstone. conglomerate, and interstratified volcanic rocks. b. Lower Silurian shales and greywacké, with occasional limestone-bands, of the Southern Uplands. b'^2. Lower Silurian schists and gneissose rocks of the Highlands. b'^1. Lower or quartzose series with bands of limestone and quartz-rock, forming base of b'^2. a. Red and chocolate-coloured sandstones and conglomerates of Cambrian age. a. Fundamental or Laurentian gneiss. T. Igneous rocks, contemporaneous and intrusive. G. Granite and porphyry. f. Fault.

Fig. 6.10. Correlation of strata in Highlands and Southern Uplands of Scotland according to Murchison and Geikie (Murchison 1867, 172).

repeated several times, all the way down to the Midland Valley (where Old Red Sandstone and Carboniferous rocks appeared), and then turned up again in the Southern Highlands. This was what Murchison's Silurian kingdom supposedly looked like in Scotland when seen in its essentials. Note that it was in the Northwest Highlands that the broad stratigraphical sequence could seemingly be discerned. Then this could supposedly be applied as a key to the understanding of the whole.[80] It is interesting that Geikie is stated to have furnished this section, even though, as we know, he had considerable doubts about the extrapolation of the northwest sequence to the main body of the Highlands.

Leaving *Siluria,* we may inquire how Geikie's ambitions toward acquiring a chair at Edinburgh were getting along. Geikie seems to have given Murchison every satisfaction in his running of the Scottish branch of the Survey, and he quickly gathered around him a group of very able young geologists (see fig. 6.11), several of whom figure extensively in the chapters that follow. In March 1869, we find Murchison sending Geikie some autobiographical notes, which I take it indicate that already plans were being made for Geikie to write a biography of Murchison.[81] That year also Edward Hull transferred from the Scottish branch of the Survey to take charge in Ireland, and shortly thereafter he applied for a chair in the Royal College of Science in Dublin. The government authorities inquired of Murchison whether it was feasible to hold the two positions simultaneously. Murchison supported Hull, and he was successful in gaining the position he sought. (Apparently, Hull didn't want to see a separate professor in Dublin, possibly teaching the students ideas at variance with those propounded by the Survey.) In recounting these matters to Geikie, Murchison told him, "I did all this with a view to *your own probable future state.*"[82] That is, he wished there to be a precedent if and when Geikie sought to occupy a chair in Edinburgh while continuing to serve as director of the Scottish Survey.

In 1869, the aged geologist George Poulett Scrope (1797–1876), a contemporary of Lyell, one of the leading uniformitarians, and an expert on volcanoes, being then too old to undertake active fieldwork, invited Geikie to travel to Italy the following year on his behalf, to make field observations in the Lipari Islands and test certain theories that he had been developing.[83] Scrope, who was a wealthy banker, undertook to meet Geikie's expenses. The

80. By today's understanding, this was almost entirely wrong. In Scotland, Silurian rocks are only to be found in the Southern Uplands—and most of these hills are now classified as Ordovician. However, in 1867 this unit had not been proposed, and it was at that time quite reasonable to depict the whole of the Southern Uplands as Silurian. It was also reasonable to refer the quartz rock, limestone, etc., of the northwest to the Silurian on the basis of the Durness fossils. But all the rest was insecure generalization, based neither on the evidence of fossils nor on a sound understanding of geological structures.
81. Murchison to Geikie, March 3, 1869.
82. Murchison to Geikie, Aug. 30, 1869.
83. Geological Society Archives, LDGSL 790, Scrope to Geikie, Dec. 23, 1869.

Fig. 6.11. Scottish Survey staff in 1868. Standing (left to right): J. Geikie, J. Horne, J. Croll, C. R. Campbell, B. N. Peach, D. R. Irvine, T. M. Skae, R. L. Jack. Seated: E. Hull (left), A. Geikie (right). Reproduced from R. B. Wilson (1977, 6). By permission of the Director, British Geological Survey: British Crown copyright reserved.

idea was that the journey would be undertaken during Geikie's period of summer leave, but an extension of this was required. Murchison readily acceded to Geikie's request for this.[84]

The trip proved to be a disaster for Geikie. He was taken seriously ill, perhaps because of some kind of malarial infection, and apparently almost died. It was only with great difficulty that he was able to get back to Britain, and even then he only made a slow recovery (Geikie 1924, 137–40). But even before the illness things were not, it seems, going exactly as might have been hoped, for Scrope expressed (in the mildest terms) some disappointment with the results that Geikie was obtaining.[85] Following Geikie's illness, both Scrope

84. Murchison to Geikie, Feb. 3 and Feb. 12, 1870.
85. Scrope to Geikie, April 19, 1870.

and Murchison were exceedingly solicitous for Geikie's welfare. Fortunately, by August he was back in the saddle again in Edinburgh, though not at full strength, and the disease kept recurring for the next few years. Eventually, whatever it was, he threw it off completely. He was never again seriously ill until his final illness at the age of eighty-nine.

Nearing the end of his life, Murchison had begun to think about the disposition of his property. It appears that his first intention had been to leave a sum of money for a chair at Edinburgh in his will. But that was scarcely satisfactory to Geikie: Murchison could very well live for many a long year more. We find, then, Geikie dropping various hints and suggestions that it might be better to act sooner rather than later. In particular, the resignation of Professor G. J. Allman from the chair of natural history prompted Geikie to write urging action, for if the vacancy were filled immediately it might inhibit subsequent moves to create a geology chair. He already knew that Murchison was intending to nominate him as the first incumbent.

> When a chair of geology in Edinr. is talked of I always hear your name mentioned in connexion with it. My own earnest desire is that it should be *The Murchison Chair*—a fitting and noble memorial of you and your labours, in the chief university of your own native country, and if anything could add to the gratification of that desire it would be that the chair should be founded in your lifetime that you might yourself see the beginning of a movement which more than anything else you [could do] would advance our favourite science in Scotland, and that you might receive the congratulations of the community at large for this crowning benefit to Scottish geology. The probable coming of the British Association to Edinburgh next year offers perhaps an additional argument.[86]

Fortunately for Geikie, Murchison took no exception to his subordinate's presumption. But by that time Geikie knew pretty well how to pander to his chief's vanity, and as he said in his autobiography, Murchison treated him more or less as a son (Geikie 1924, 149). We find that Murchison acted almost immediately, for two days later he wrote to Geikie with the news he had had some negotiations with the influential Lyon Playfair. Murchison had long written in his will that a sum of £6,000 was to be left to the University of Edinburgh for a chair of geology.[87] He now reported to Geikie that he had determined to make a gift of the money immediately, on the understanding that he should be allowed to nominate the first incumbent. He would find it "*very very* gratifying to me" if the chair were named after him.[88] Thus the arrangements were set in train. It may be noted that a sum £6,000 invested at 4 percent would yield an income of £240 per annum, which was not really sufficient for a chair. Hence it was necessary to get the government to provide

86. Geikie to Murchison, Aug. 26, 1870.
87. Copy of Murchison's will, British Geological Survey Archives, Keyworth, GSM 1/133.
88. Edinburgh University Archives, Gen. 1425/387, Murchison to Geikie, Aug. 28, 1870.

some funds also—which was why particular negotiations with Playfair as intermediary were required.

According to Geikie's autobiography (ibid., 142), it took some time to settle the arrangements, for there were objections from the Home Office to Murchison nominating the first incumbent. Indeed, the government was not disposed to see the foundation of chairs in the manner that Murchison was pursuing. There were also objections from the Department of Science and Art about Geikie holding two positions simultaneously: one at the university and one with the Survey. (No doubt the precedent of Hull in Ireland was then vigorously invoked by Murchison.) Moreover, toward the end of 1870, Murchison had the first phase of the illness that ultimately led to his death the following year. But fortunately for Geikie, Murchison recovered sufficiently to maintain his efforts on his protégé's behalf and his contacts were sufficiently powerful to ensure the success of his endeavor. So it came about that on February 15, 1871, Geikie eventually heard from Playfair that he had been appointed to the chair, and he took up his new position on March 25 (ibid., 143, 145). Thus his great ambition was eventually achieved. Had Murchison not acted during his lifetime, at Geikie's urging, it seems that the outcome could well have been different, for Geikie could hardly have carried out a campaign on his own behalf directly, the way Murchison did.

As we have seen, Murchison was already sending Geikie material for a biography, to be completed after his death.[89] A codicil had been added to Murchison's will on March 10, 1869, altering matters in such a way that on his decease all Murchison's papers were to be passed to Geikie, who would be responsible for writing Murchison's memoir of his life. For this task he was to receive a sum of £1,000 from Murchison's estate.[90] Geikie surely earned this sum, with compound interest, for his biography of his old chief was a notable literary achievement and one of the finest specimens of scientific biography produced in the nineteenth century. How he managed to find time to undertake such a major task, among all his other responsibilities, including the occupancy of a chair and the running of the branch of the Survey responsible for elucidating the most difficult geology in the United Kingdom, it is hard to conceive. It should be remembered also that Geikie was still not fully fit at that time, and he was just beginning the responsibilities of married life. But Murchison's biography was successfully completed in 1875 in two volumes, 780 pages in toto. Geikie himself referred to it as his pièce de résistance (Geikie 1924, 160).[91]

In commissioning Geikie to write his biography, Murchison clearly had it in mind that he would be gaining the services of one who would present his achievements in the most favorable light and who would set the seal on his

89. Geological Society Archives, Murchison to Geikie, Spring 1871 (undated), letter 187.
90. R. I. Murchison, Will. Before the addition of the codicil, Murchison had named the publisher John Murray and the curator of the Geological Survey, Trenham Reeks, as his literary executors.
91. It was republished in 1972 by Gregg International Publishers Ltd., Farnborough.

numerous accomplishments. In particular, I suggest, the hope or intention was that the major points of theory about which Murchison had battled so strenuously and successfully during his lifetime—especially concerning the Devonian system, the Silurian system, the disposition of gold-bearing deposits in different parts of the world, and the structure of Scotland—would receive full posthumous acceptance as a result of Geikie's efforts. Murchison's views would become paradigmatic; they would become received and unquestioned *truth*. In addition, his character would be presented in the most favorable light. Geikie differed from his former mentor in the matters of quietist versus paroxysmic geological progress, biological evolution, and glacial theory. But he could be relied on to differ with the utmost tact.

In all this, Murchison's aims were more than fulfilled by Geikie's literary accomplishment. For many years the world has chiefly looked back at Murchison through the lens of Geikie's narrative. Only in recent times, through the work of Rudwick, Secord, and Stafford has a very different picture begun to emerge—of Murchison as a skillful manipulator, an unforgiving foe, and one who would take immense pains to ensure that his views always prevailed. In particular, we now know much more about how the patronage system operated in Victorian science; and how Murchison, perhaps above all other scientists of his day, was adept at using it to his advantage and for the advancement of his views. Nevertheless, despite the scrutiny that has been given to Murchison's work in recent years, the picture of him as a man of distinction, full of reckless energy, and enjoying life to the utmost, as Geikie portrayed him in his biography, has rightly survived. Figure 6.12 shows an engraving of the geologist in his old age, which was prepared to accompany an obituary in the *Illustrated London News* (1871).

Following Murchison's death, the question immediately arose as to who was to be his successor as director general of the Survey. Geikie might have sought the position, I suppose, had he not just embarked on a new life as professor at Edinburgh. But to the best of my knowledge he made no attempt to obtain Murchison's position. Geikie was then only thirty-six. So Murchison's mantle passed to Ramsay, then fifty-seven. In terms of achievement in geological science, this was entirely fitting. Ramsay was then much the most experienced and eminent officer in the Survey, and there were no university geologists in Britain at that time of equal standing. However, Ramsay had been ill, on and off, for a number of years, and although he was only about the same age as Murchison had been when he took over the reins of the Survey, he seems not to have had Murchison's energy in the latter part of his career. Nevertheless, the Survey's activities were pushed along actively, which was feasible with the considerably increased staff that had been made available as a result of Murchison's coup d'état in 1867. Ramsay's period in office was marked particularly by the large amount of fieldwork accomplished and the publication of geological maps, though the memoirs that should have accompanied them were often not issued (Flett 1937, 91). It is reported, however, that towards

Fig. 6.12. Portrait of Sir
Roderick Murchison in his
later years (Illustrated
London News 1871).

the end of his directorship the Survey became "sadly disorganised" (Glasgow
Herald 1924). When Ramsay retired in 1881, Geikie was his natural successor, and so he eventually achieved his ambition of succeeding to the director
generalship at the very suitable age of forty-six. As usual, Geikie's career ran
like a well-oiled machine.

Meanwhile, Geikie was proving a great success at the University of Edinburgh. One cannot tell with certainty, but if his literary style is anything to go
by, I imagine he would have been an outstanding lecturer. He instituted fieldwork exercises for the students, which were something of an innovation and
proved exceedingly popular. He put on separate classes for ladies and even
took them on field excursions away from Edinburgh, but separately from the
men and under the chaperonage of Mrs. Geikie. One exercise recalled by
Geikie (1924, 165) that must have been exceedingly interesting and instructive for students of physical geology was to trace a river from its source
to the sea. On another occasion he and his students traced and mapped
the Highland Border Fault (which separates the Highlands from the Midland Valley of Scotland) from coast to coast.[92] In addition, he introduced the
latest petrographic methods with the microscope into the laboratory classes
(ibid., 153).

92. National Museum of Scotland Archives, bound collection of papers of Archibald Geikie, 55
(081)/Geikie/G: sections and notes inserted into back of volume.

In all this work, Geikie was teaching the view of the geological structure of Scotland that he and Murchison had developed in their fieldwork and in their publications. Thus a set of lecture notes from his first year of teaching at Edinburgh, preserved in the Edinburgh University Archives, gives the Murchison-Geikie theory exactly, and as if it were wholly uncontroversial. Ralph Richardson, the student taking the notes, recorded that "the structure of the Scotch Highlands is not, as was once supposed, a mass of chaos, but a simple continuation of anticlinal and synclinal folds."[93] A section was recorded also, from the Minch, through Quinag and on to Brora, exactly in keeping with the Murchison doctrine, and evidently based on the Murchison-Geikie map of 1861.[94] The students were further informed that the Highland schists were the equivalent of the Llandeilo series in Wales, which Murchison held to be Lower Silurian.[95]

The theory was also received without demur into works such as Ramsay's *Physical Geology and Geography of Great Britain* (1878, 285–88) and Lyell's *Elements of Geology* (1866, 762–63), and into semipopular geological texts written in the 1870s, such as W. S. Symonds's *Record of the Rocks* (1872, 23–28 and 42–45). For Lyell, the theory was particularly acceptable, for in thinking that there was a relatively recent epoch when the metamorphism of the "Silurian" strata of north Scotland occurred, and that such changes were not restricted to the special conditions of "primeval" or "primitive" times, the doctrine seemed to accord with his notion of the uniformity of the processes of geological change. Amateur geologists, visiting the Northwest Highlands, reckoned that the rocks looked the way Murchison said they were (Crosskey 1865).[96] The theory was found (naturally enough) in Geikie's influential article in the *Encyclopaedia Britannica* (Geikie 1879), which formed the basis of his great textbook of 1882.[97] The theory also underpinned his *Geological Map of Scotland* (1876). Most important, it was assumed to be correct by the official Survey. To a large extent as a result of Geikie's literary and cartographic work, it became the received wisdom in the geological community, being taught by him in the leading university in Scotland. It was favored by Ramsay as late as 1880 in his presidential address to the geological section of the British Association at its meeting in Swansea (Ramsay 1881), in effect providing a public statement of the Survey's continued adherence to Murchison's doctrine.

In fact, Nicol having retired from the battle, there was apparently no

93. R. R., "Lectures on Geology delivered by Archibald Geikie LL. D., FGS., FRS L. & E. &c. during the session 1871–72," 5 vols., manuscripts bound as one volume (Edinburgh University Archives, Geikie Papers, Gen. 694), vol. 4, 72.
94. Ibid., 74.
95. Ibid., 94. Today, the Llandeilo Series is classified as Ordovician, but his unit had not been defined in 1871-72. See Secord (1986a, 287).
96. On Crosskey as amateur geologist, see Lapworth (1895).
97. Uncautiously, Geikie wrote in the Britannica article that Murchison's generalization that the upper schists and gneisses were really metamorphosed Lower Silurian rocks "had been completely confirmed by subsequent investigations" (1879, 333).

significant questioning of the doctrine for most of the 1870s, and the geological community at large for a time lost interest in the matter. Apart from
efforts toward fossil collecting, there are no records of any significant geological research in the remote northwest region until 1878, when the whole
question was reopened by the geologist Henry Hicks.[98] For the moment, the
whole issue died down, and geologists turned their attention to other matters.
There were perhaps private doubts, but Murchison's efforts to impose his
views on the geological community, aided by Geikie, seemed to have proved
eminently successful. A paradigm had been established for interpreting the
strata of northern Scotland. Murchison himself, of course, held to his views
unswervingly to the end.[99] But, as T. G. Bonney put it in his presidential address to the Geological Society (Bonney 1885, 56), the Murchison hypothesis
lay like an incubus on the geological investigators in the Highlands, impeding
progress; and it was invoked like a fetish by its proponents.

98. William Jolly organized efforts to find further fossils in the Northwest Highlands, and he
reported his results to the British Association in 1872, 1873, 1874, and 1879. He engaged local
teachers, clergymen, and so on, to search the potentially fossiliferous sites with care. A considerable number of specimens were found, but no major advances were made as a result of these
efforts.
    As previously stated, Nicol continued his fieldwork in the northwest until the end of his ca-
reer. But his late observations remained unpublished.
99. Murchison's last fieldtrip to the north of Scotland was made in the summer of 1869, when he
stayed with friends near Ullapool. His observations were made chiefly from a boat on Loch
Broom, from which vantage point the rocks seemed to comport with his theory most satisfactorily. See Murchison (1870).

7

The Battle Rejoined and the Collapse of the Murchisonian Paradigm

I come now to what is in many ways the most intricate but also the most interesting part of the Highlands controversy. The Northwest Highlands were visited by geologists during the early 1870s, but as we have seen most of them seemed satisfied with Murchison's interpretation of what might be seen there. Geikie and Ramsay did not revisit the area duing this period, and the officers of the Survey had not yet begun to move into this remote and difficult terrain to begin the huge task of mapping it in detail. The geological maps, such as they were, were very general. Sections remained the main visual medium for the transmission of ideas about geological structures for that part of Britain. (Maps were sometimes exhibited at geological meetings, but they did not get published.)

It is not perhaps a coincidence that the person who reopened the debate concerning the Highlands rocks that had simmered for more than a decade was the able amateur geologist, Henry Hicks (1837–99), for he became one of the leading critics of the alleged insensitive attitude of the Survey toward outsiders.[1] The complaint was that the Survey regarded itself as the ultimate and sole source of authority and the arbiter on all controverted questions in British geology. Indeed, through the later years of the nineteenth century, there were somewhat strained relations between the Survey and the Geological Society of London, which represented the older amateur tradition (Woodward 1908, 220).

Hicks came from a medical family from the small city of St. David's, at the southwestern extremity of Wales. He followed his father's profession, studied medicine at Guy's Hospital in London, and took his M.D. at St. Andrews University. In the 1860s he practiced medicine at St. David's and then moved to London in 1871. A keen and accomplished amateur geologist, he became specially interested in Lower Palaeozoic and pre-Cambrian rocks. Some of the latter he discovered in his home territory near St. David's, though their pre-Cambrian character was contested by Ramsay, who had worked in that region in his early days at the Survey. In the 1880s, Hicks again came into dispute

---

1. On Hicks, see the Geological Magazine series on eminent geologists (1899b), Woodward (1899–1900, 1908), Whitaker (1900), Bonney (1905).

with officers of the Survey concerning the interpretation of the rocks in the St. David's area. On that occasion his chief opponent was Geikie (1883).

While working in St. David's, Hicks undertook the care of John Salter (1820–69), the Survey's former palaeontologist, who, it will be recalled, provided Murchison with identifications of the Durness fossils. Salter's career at the Survey had been going down-hill rapidly, partly owing to a form of mental illness; and being forced to resign from the service, he made a number of visits to South Wales to try to recuperate (Secord 1985). Whatever the rights and wrongs of the case, Salter certainly felt himself a victim of the government bureaucracy in the matter of his pension, and we may reasonably suppose that he conveyed to Hicks his dislike for the powers-that-be in London. Be that as it may, during the last few years of his life, Salter formed quite a close relationship with Hicks, and they did some fruitful joint work in Pembrokeshire. In fact, it was Salter who really introduced Hicks to geology and showed him how to do palaeontological work in the ancient rocks. Some of their results were communicated to the British Association at meetings in 1864 and 1866 (Salter 1865a, 1865b; Hicks and Salter 1867).

After Salter's death, Hicks continued with his researches in the ancient rocks, and after his move to London he turned his attention to North Wales, where he worked during his vacations. He was responsible for identifying pre-Cambrian rocks in several previously unsuspected places in Britain, and he published a number of influential papers on the ancient rocks (Hicks and Harkness 1871; Hicks 1872, 1873–86). Hicks became a respected member of the Geological Society, serving as a member of council for three periods and as president in 1896–98. He was elected Fellow of the Royal Society in 1885. By 1878, when he entered the Highlands controversy, Hicks was already a force to be reckoned with in the geological community in Britain, for he was a council member of the Geological Society. But he was not directly linked with the Survey and had crossed swords with Ramsay (Woodward 1908, 220). Remembering that Salter had been forced out of the Survey in 1863 by Murchison, on the insistence of Thomas Henry Huxley (Secord 1986a, 288), we need have no difficulty in understanding why Hicks was the first person since Nicol seriously to question publicly the "official" view of the structure of the Northwest Highlands.

Hicks's paper, which reopened the whole question of the geology of the Highlands, was delivered to the Geological Society on May 22, 1878 (Hicks 1878). He did not state when his fieldwork was carried out, but it seems reasonable to suppose that it was done the previous summer. We know that Judd met Nicol in the Highlands in 1877 (Judd 1886, 999) and was at least a partial convert to the old geologist's views there and then. It is possible that Hicks also met Nicol there the same year, or Judd may have passed on his incipient change of opinion to Hicks. However, Hicks is recorded years later, in 1890, as claiming that he had come to think that there was something in Nicol's views as a result of finding a "silvery mica schist" on the west coast that seemed to resemble the Moine Schist of the interior.[2] This rock was apparently found on the shore near Gairloch ("between the Strath Hotel and Gairloch Kirk"). Gairloch is a fishing village on the coast directly to the west of Loch Maree. Examining the rocks on the shore between the church and the hotel, I have myself found Torridon Sandstone—in agreement with what is indicated on modern maps for the area. However, the observer is not far from (western) gneiss at that point, and I was able to locate some bands of schistose material adjacent to the sandstone; also some rocks very like Geikie's "sedimentary porphyry" at Ullapool. Pebbles of dark micaceous schist may also be found on the beach, which could easily have been derived from the hornblende schist referred to on modern maps as outcropping a little to the north and to the east and having nothing to do with the eastern (or Moine) schists.

It would be foolish for me to pursue further any attempt to ascertain what Hicks actually did see.[3] The point is that he *thought* he had seen a rock in the territory of "western" gneiss that resembled the eastern metamorphics, namely, the Moine Schists; and thus he began to think that there might be something in Nicol's claims that required his investigation. Accordingly, we find Hicks traveling inland to look at the disputed exposures near Kinlochewe at the head of Loch Maree. The results of his inquiries were embodied in two sections, shown in figure 7.1. These should be compared with the corresponding sections published by Murchison and Nicol. (See figs. 5.3, 6.5, and 6.7.)

Comparison of the sections of Murchison, Nicol, and Hicks, shows that the newcomer to the debate was introducing a very different way of looking at the tangle of rocks in Glen Logan (which Hicks called Glyn Laggan). Whereas Murchison had the metamorphic rocks on the eastern side of the glen as different in kind from the gneisses and other rocks to the west, and Nicol had them the same (thus according with their respective theories), Hicks had it both ways, so to speak. That is, he had "blue flags, sandstones, &c." along the lower part of the eastern side of the glen, and "argillaceous, quartziferous, and micaceous flaggy beds" above and extending well to the east.

In fact, this is not out of order, in the sense that there are two main thrust faults to be reckoned with in Glen Logan according to modern maps (Institute of Geological Sciences 1962). Lewisian Gneiss has been thrust in so as to form the lower part of the eastern side of the valley of Glen Logan, and Moine Schist has been thrust over this. But while Hicks extended the "western" gneiss to the east, so as to form "Ben Fyn," he seemingly had difficulty in offering a coherent interpretation of the situation east from Glen Logan.[4] Thus we see in figure 7.1 that a fault is introduced rather tentatively, and the relationship between the two sets of eastern metamorphics is sketched in very hesitantly

2. British Geological Survey, Edinburgh Office, Archives, SOA 1/491, J. J. H. Teall to C. T. Clough, Feb. 28, 1890. (On Teall, see page 252. On Clough, see page 306.)
3. Clearly it puzzled Teall too, for he wrote: "What is this rock? To the best of my recollection you have not sent up any rock which could be described as a silvery mica schist."
4. "Ben Fyn" is marked as Fionn Bheinn on modern maps.

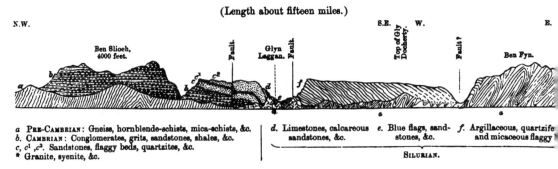

a PRE-CAMBRIAN: Gneiss, hornblende-schists, mica-schists, &c.
b. CAMBRIAN: Conglomerates, grits, sandstones, shales, &c.
c, c¹ ,c². Sandstones, flaggy beds, quartzites, &c.
* Granite, syenite, &c.

d. Limestones, calcareous   e. Blue flags, sand-   f. Argillaceous, quartzite
      sandstones, &c.   stones, &c.   and micaceous flaggy

SILURIAN.

Fig. 7.1a. Section of strata at head of Loch Maree (north side), according to Hicks (1878, 812).

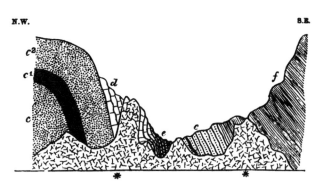

Fig. 7.1b. Enlargement of same, showing detail of geological structure in Glen Logan, according to Hicks (1878, 814). For explanation of symbols, see fig. 7.1a.

near the location marked "Top of Gly[n] Docherty." We see that a major assumption—very likely acquired by picking up the piece of schistlike rock on the beach at Gairloch (Strath)—has found its way into Hicks's thinking in such a way as to dominate his whole section. That is, he held the eastern metamorphic rock and the western metamorphic rock to be one and the same, as Nicol had long maintained. It is to be noted that Hicks claimed that the strike of the metamorphic rocks he observed on Ben Fyn ran northwest to southeast, in contradiction of the view long held by Murchison, a view central to his argument that the western and eastern metamorphic rocks belonged to two quite different series.

   The great trouble with Hicks's interpretation, however, was this. His Gairloch rock was a schist and thus, we might suppose, should have been related to the upper rock on the eastern side of Glen Logan, for that is where the Moine Schist resides. But Hicks related it to the metamorphic rocks at Ben Fyn. These we regard as belonging to the Moine Schist; but Hicks, it can be seen, did not regard the upper rock on the eastern side of Glen Logan (f) as a schist. It was not a gneiss either. Now Nicol's theory had to do with the supposed identity of a western and an eastern *gneiss,* not a schist.[5] Actually, the way Hicks envisaged the matter, the *upper* rock to the east of Glen Logan (f) was Silurian and was scarcely metamorphic. It supposedly rested on an underlying metamorphic rock (a), which he thought of as a (pre-Cambrian) schist—but which we call a gneiss. Thus Hicks's interpretation hardly satisfied the petrographic characters of the several rocks in question, in that there was equivocation as to what was and what was not metamorphic, and what was a schist and what was a gneiss. But at least it got rid of Murchison's problem: a metamorphic rock (Murchison's eastern gneiss) lying atop unaltered sediments in a seemingly conformable sequence.

As for the situation at the floor of Glen Logan, near the stream bed, Hicks now developed the idea of there being a substantial igneous instrusion there. (See fig. 7.1b and compare with fig. 5.2.) This was once again the "Logan Rock." As can be seen from Hicks's section, he represented it as "Granite, syenite, &c." It will be recalled that this rock is today mapped as gneiss, thrust in by earth movements. So some of the Logan Rock (gneiss) was represented by Hicks as granite/syenite, and some of it was probably in what he was calling "flags and sandstones," or perhaps even the "micaceous flaggy beds." Certainly his lithological categories and boundaries did not coincide with those that are adopted today, and it is difficult to mesh what may be seen in the field with what Hicks described. Moreover, the suggestion that the whole curious structure at Glen Logan might somehow be attributed to the postulated igneous intrusion of "syenite" was mechanically most unlikely. Not surprisingly, one of his audience, the Survey officer Charles de Rance, is recorded as having made precisely this point in the discussion that followed the presentation of Hicks's paper. Ramsay also took issue with some of the arguments made by Hicks, and it would appear, therefore, that the Survey was seeking to defend itself against questioning of the accepted Murchisonian doctrine.

Looking at Hicks's first section again, it may be seen that he had an unconformity between f and a, in the region to the east of Glen Logan.[6] Moreover, insofar as "western" gneiss was brought to the surface again, it was supposedly found at Ben Fyn, not Glen Logan. Hicks was very likely the first geologist to have a careful look at the rocks of Ben Fyn. He stated that these rocks definitely resembled those that he had found on the west coast at Gairloch. But this did not make his theory one and the same as that of Nicol.

It may be seen, then, that Hicks's work, though evidently based on a careful examination of the rocks at Glen Logan, introduced as many problems as it solved, if not more. His dependence on the sections of Murchison and Nicol for the ground between Glen Bianasdail (Haasach) and Glen Logan is fairly evident. Further east, where he was (I believe) breaking new ground, his observations allowed him to make only a very inconclusive suggestion for the

5. It must be recalled, however, that Nicol's argument at Durness and Far-out Head relied on an equivocation between schist and gneiss.
6. For modern geologists, Hicks's unconformity is a thrust plane, at the base of the Moine Schists.

structure. The nature of the Logan Rock was not dealt with satisfactorily, and it remained identified vaguely as some kind of syenite. It will be noted that if Hicks's section were correct, it would have the effect of introducing Lewisian (pre-Cambrian) rocks right in the middle of Scotland, thus replacing Murchison's Silurian map colorings. It is not impossible that this consideration weighed with Hicks. After all, much of his reputation was dependent on his discoveries of hitherto unsuspected areas of pre-Cambrian rock, and we may suppose that he would have been pleased to see an extension of "his" territory into that formerly occupied by Murchison.

This, of course, is speculative. But whatever Hicks's motives, he had performed a most important service. He had shown that a respected geologist and observer could see things other than the way in which the "official" view would have them.[7] It was an open invitation for other geologists to travel to the area to see for themselves what all the fuss was about, and it was not long before this invitation was accepted with enthusiasm.

To set the issues in a clearer light, the amateur geologist Wilfred Hudleston (1828–1909) quickly came forward with a review of all the literature on the Highlands controversy up to and including the contributions of Hicks (Hudleston 1879–80).[8] Hudleston, who came from Knaresborough in Yorkshire, had studied under Sedgwick at Cambridge, but he only became actively interested in geology some years later as a result of his association with the palaeontologist John Morris, of University College, London. Hudleston, a wealthy man as a result of his marriage into a landed Cumberland family, was a keen participant in the activities of both the Geological Society and the Geologists' Association, the latter organization meeting more the needs of amateurs, while the former was for the "heavyweights" of the emerging profession of geology. However, though an amateur, he was not a geological "lightweight." He filled the presidency of the Geological Society from 1892 to 1894, having been secretary from 1886 to 1890, and was awarded the society's prestigious Wollaston Medal in 1897. He was secretary of the Geologists' Association from 1874 to 1876 and president from 1881 to 1883.[9] His papers were chiefly palaeontological, but he developed a flair for review articles, bringing before geologists the most salient points involved in any debate or theoretical problem.

Hudleston's paper must have been most helpful to the amateurs in the Geologists' Association, in whose *Proceedings* the paper was published, by reason of the set of sections that he reproduced from the works of the earlier protagonists in the debate, showing all too clearly the great differences that existed in the opinions of the leaders of the geological community respecting the rocks of the Highlands. His talk was delivered on November 1, 1878. It appears, therefore, that Hudleston must have traveled north, to view for himself

7. Hicks had a considerable reputation for finding fossils in rocks where no one else could find them.
8. For reviews of Hudleston's work, see Sollas (1909), Geological Magazine's series on eminent geologists (1909b), and Mineralogical Society (1909).
9. On the Geologists' Association, see Sweeting (1958).

the rocks under dispute, soon after he heard of the new claims being made by Hicks in his paper of May 1878.

Another geologist to take up the challenge offered by Hicks's iconoclastic paper, in both the field and the laboratory, and a man who played a considerable role in the Highlands controversy in its middle years, was Thomas George Bonney (1833–1923).[10] Bonney trained in mathematics and classics at St. John's College, Cambridge. He taught mathematics for a time at Westminster School, where he took holy orders, but in 1859 he returned to take up a fellowship at John's. There he "got up" geology and began to teach the subject in his college. In particular he studied petrography, with the aid of the polarizing microscope, and he soon became one of the leaders in this field in Britain, his teaching yielding a number of distinguished students of the subject.[11] In 1877, Bonney was appointed to a part-time position as professor of geology at University College, London; and in 1881, as a result of his commitments to his duties as secretary of the British Association, he gave up his position in Cambridge to settle in London. He retired to Cambridge in 1905 but still maintained close connections with the geological work being conducted at the university. Bonney was an active member of the Geological Society, and he served as both secretary (1878–84) and president (1884–86), at the time when the Murchisonian theory of the Highlands eventually collapsed. He was definitely a man whose opinion counted.

Bonney's researches covered a wide number of fields, including glacial theory, studies of structures in the Alps, and coral reefs. But his reputation rested chiefly on his studies of basic igneous and metamorphic rocks. His assistance was frequently requested for identifications of rocks of this kind, and wherever possible he made it his business to visit the localities where the rocks were collected, in order to provide better informed opinions. It was in this manner that he became involved in the Highlands controversy.

Like Hicks, Bonney was particularly interested in pre-Cambrian rocks. One of his most important pieces of research, carried out over many years, was to confirm an inlier of rocks at Charnwood Forest in Leicestershire in the English Midlands as pre-Cambrian (Bonney and Hill 1877–80), in opposition to the views of the Survey, which had mapped it as Cambrian. Rocks were brought from Charnwood to Bonney for his consideration by a colleague at St. John's, Edwin Hill, and after extensive petrographical work, using analogies with rocks in Belgium, a decision for a pre-Cambrian identification was eventually reached, which has been sustained by later investigations.[12]

Much the same kind of thing happened in relation to Hicks's investigations of pre-Cambrian rocks. Hicks reckoned that he had field evidence from various places in Britain for rocks being pre-Cambrian, and he would send

10. On Bonney, see his autobiography (Bonney 1921), Watts (1925–1926), and Rastall (1937).
11. For an examination of the role of this important instrument for nineteenth-century petrographers, see Hamilton (1982).
12. At first Bonney was inclined to favor a Silurian (Ordovician) age. On the history of investigations at Charnwood Forest, see Ford (1979).

samples to Bonney for examination and comparison with rocks judged to be pre-Cambrian elsewhere in the world. It was a risky kind of enterprise, since "by definition" the rocks involved were devoid of fossils and lithological analogies might be fortuitous. However, their cooperative work brought success in St. David's, Shropshire, Anglesey, and elsewhere. Naturally Hicks turned to Bonney for assistance in relation to the specimens that he had gathered in the Loch Maree district.

In a review of British pre-Cambrian rocks, Bonney (1879) referred to his examination of rocks from Loch Maree, submitted to him by Hicks, and stated his agreement with Hicks that there did indeed appear to be two kinds of metamorphic rocks at Glen Logan. But he also stated in a footnote that he had subsequently visited the Loch Maree locality, that he had not been convinced by Hicks's interpretation of the structure of the region, and that he still adhered to Murchison's views (ibid., 146). Then on December 3, 1879, Bonney presented an important paper to the Geological Society, in which he described both his fieldwork at Loch Maree and his microscopic examination of specimens at Cambridge (Bonney 1880a). He even provided a sketch map of part of Glen Logan, which will be discussed below. Thus Bonney's paper offered significant changes to the whole character of the Highlands controversy.

According to Bonney's December paper, he spent "some days" in the Kinlochewe area that summer. As might be expected, his main focus of attention was Glen Logan, and perhaps his main objective was to examine the "igneous" rock—variously referred to as syenite, porphyry, diorite, granulite, etc.— to see what he could make of it. It is interesting that Bonney recorded that "immediately on reaching the spot" he formed the view that the mysterious Logan Rock was not igneous, but a *gneiss*. Moreover, it appeared to be a variety of "Hebridean" gneiss—that is, similar to the gneisses of the west coast. In other words, it appeared to Bonney that the rock was Lewisian (or "Fundamental" or "Laurentian") Gneiss. The opinion was subsequently confirmed by examination of thin sections under the microscope, for various rocks in the vicinity. One of his illustrations is reproduced in figure 7.2.[13] The Logan Rock had an unusually high concentration of the yellow-green mineral epidote, which is commonly found in metamorphic rocks (as a result of the alteration of the feldspar plagioclase subsequent to high-grade metamorphism).

Bonney also argued on field evidence against the rock being igneous, for its contacts with the sedimentary rocks in the valley were quite unlike what might be expected to result from a hot igneous intrusion. It did, however, have the appearance (especially in thin section) of having been "crushed," for cracks

13. Bonney did not actually figure "Logan Rock." What is represented in figure 7.2 is a section of a specimen of "Hebridean Gneiss" collected near Ben Slioch. See also plate 6. It is a moot point whether Bonney's figure did or did not represent the section viewed in polarized light. There are, for example, vague indications of the lamellar crystal twinning, characteristic of feldspars, observed in thin section with polarized light. But these appearances are drawn so imperfectly that one may wonder whether Bonney used ordinary light and then gave a representation that might seem to indicate the use of polarized light. (I owe this suggestion to Professor T. G. Vallance.)

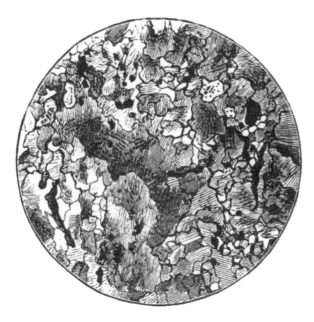

Fig. 7.2. Figure of thin
section of Hebridean Gneiss
(Bonney 1880, 97).

and distorted mineral grains could be discerned; and the contacts with other
rocks suggested faulting. The rock outcropped at various places on the floor
of the valley, particularly near the small but quite striking gorge in the middle
of the valley (see fig. 5.2). It also curved up from the valley floor to the *west*—
that is, up the side of the hill between Glen Logan and Glen Bianasdail.

Now on the higher parts of the *eastern* side of Glen Logan, there was a rock
that Hicks had labeled on his section as "argillaceous, quartziferous, and micaceous flaggy beds," and that he thought were Silurian in age (see fig. 7.1). But
to Bonney, armed with his microscope, the rock was identified as a "micaceous
schist." Microscopic examination assured him that the rock was metamorphic
(it showed foliation), even though in hand specimen it might well appear to be
"flaggy"—which I take to mean sedimentary in this context. Like Hicks, Bonney traced this rock from Kinlochewe up Glen Docharty toward Ben Fyn. But
instead of finding evidence of it lying unconformably on a gneiss, as Hicks
had done in a rather uncertain fashion, Bonney thought that the rock just
continued on to the east, sometimes being more gneisslike in appearance, and
sometimes less so. The microscope revealed that the rock type was essentially
the same, despite certain variations in external appearance. It was, however,
quite different in microscopic appearance from the Hebridean Gneiss. The
rock was, of course, what is now known as Moine Schist.

So Bonney stated quite firmly that there were two eastern metamorphic
rocks: one that resembled the Hebridean ("western") gneiss and that was at a
lower horizon; the other a schist that formed an upper and newer series. The
"syenite" of Nicol and Hicks was reclassified as a gneiss. The only igneous rock
that Bonney found that need concern us was some indications of a diorite by

the track to Glen Bianasdail.[14] But other exposures of Logan Rock in this vicinity now were identified as gneiss, as in Glen Logan.

All this involved substantial clarification. In particular, the conflation of schist and gneiss, which had underpinned a good deal of Nicol's argument, and which Murchison found wholly unacceptable (since it involved, in effect, the equation of Moine Schist and Lewisian Gneiss), was now largely ruled out of court. But as mentioned above, Bonney went a stage further, by exhibiting a sketch map of the floor of Glen Logan. This is shown in figure 7.3 (Bonney 1880a, 95).

It will be recalled that I have previously complained that Victorian geologists had had too swift a recourse to sections rather than maps. To my knowledge, Bonney's map was the first large-scale sketch map of a specific locality to be published in relation to the Highlands controversy, though maps had undoubtedly been displayed at meetings of the Geological Society, the Geologists' Association, or the British Association.[15] Thus, according to my previous strictures, Bonney's map should have provided a significant advance. But it can hardly be said that it did so. There is no scale, no indication of orientation, no indication on the map that it refers to a quite steep-sided valley, and no precise indication of locality—though the reader would know that it referred to somewhere in Glen Logan. Indeed, until the geologist took the map into Glen Logan and examined the ground with its assistance, it would hardly be of much use. But used on location it would, I suggest, help quite a lot in reconstructing what Bonney had seen and what his interpretations were for the geology of the locality. For Bonney also offered two sections, one of which is reproduced in figure 7.4.

Let us consider Bonney's map and section, to see what they tell us. Both map and section, as well as Bonney's written description, indicate that the contacts between the schist and gneiss, and the gneiss and the sedimentary rocks, are faults. But there is a "geometrical contradiction" between the map and the section. For as previously mentioned, a vertical fault should cut through the countryside in a straight line, while a horizontal one does so with a curved outcrop that follows the contours. Now Bonney's section shows two vertical faults, but on the map one of them has a *curved* outcrop. Thus there is indeed a "contradiction."

Furthermore, it appears from the map that the quartzite/limestone overlies the Torridon Sandstone and the Hebridean Gneiss. But actually in the field (this point cannot be gauged clearly from the map, and it is actually contradicted by Bonney's section), one finds that the gneiss curves up the hillside on the western side of the valley and *overlies* the quartzite/limestone. Thus the map does not do justice to the actual field relationships.

14. This is not marked on the modern one-inch maps, and I was not able to find such a rock at this locality in a rather brief search.
15. Other maps that we have looked at were, of course, very general depictions of the geology of large regions—even the whole of Scotland.

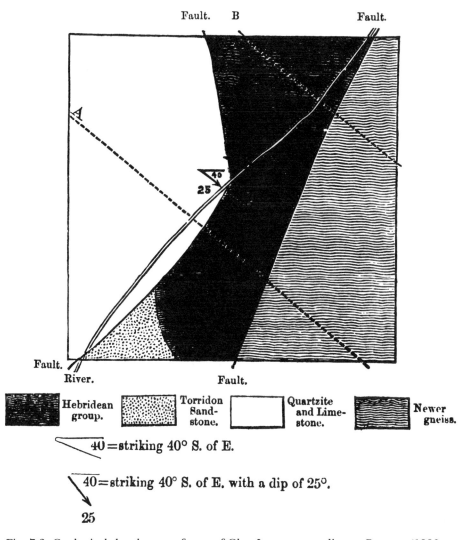

Fig. 7.3. Geological sketch map of part of Glen Logan, according to Bonney (1880, 95). (The sketch refers approximately to the site marked "gorge" in fig. 5.1. See also fig. 5.2.)

For the fault between the schist ("Hebridean group") and the gneiss ("Newer gneiss"), Bonney's section shows a vertical cut, which is consistent with a linear outcrop for the fault. But since the contours (not shown, of course) are straight at that point (they run directly down the eastern side of the valley), a linear outcrop for the fault is consistent with *any* angle of inclination for the fault. Bonney showed it as being vertical in his section. But in fact, this was a wholly arbitrary assumption. Or rather, it was one that was made in accordance with the nature of faults as they were generally understood in Bonney's day. Bon-

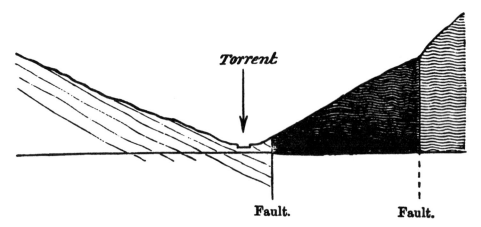

Fig. 7.4. Section of strata at Glen Logan, according to Bonney (1880, 96). (Compare fig. 7.3.)

ney did not have the category of low-angle faults in his theoretical repertoire. Yet this, we now believe, is what one is dealing with in Glen Logan. The boundary between the "Hebridean Group" and the "Newer Gneiss" in Glen Logan is, we hold, the base of the Moine Thrust plane.

It is clear that Bonney had good petrographical reasons for rejecting the large igneous intrusion in Glen Logan advocated by Hicks. But we have also seen that Bonney's petrography prevented him from accepting Nicol's idea that the "western" and "eastern" metamorphic rocks in the Northwest Highlands were identical. It could not be as simple as this, Bonney argued, for there were manifestly two kinds of "eastern" metamorphic rocks, according to the testimony of the microscope. Thus unexpectedly—in view of the significant advances that he made in comprehending the geological situation at Glen Logan—he ended up by siding, in part at least, with Murchison and Geikie as to the geological structure of the district. He supposed that there was an "eastern" gneiss, similar to the Hebridean (or Lewisian) of the west coast, and presumably in fact the same as this, being brought up by faulting. There was also, lying over the gneiss, an "eastern" schist, perhaps becoming more gneissose toward the Central Highlands. By partly ignoring the faults that he showed in his own diagram of Glen Logan, Bonney seemed to accept the idea of an ascending sequence, with a schist at the top of the sequence. "It appears," he wrote, "that, so far as the neighborhood of Loch Maree is concerned, the views advocated by that distinguished geologist [Murchison] and his fellow-labourer Prof. Geikie are fully confirmed by microscopic evidence" (Bonney 1880a, 107).

How could Bonney make sense of the notion of metamorphic rocks lying on top of unmetamorphosed rocks? Metamorphic theory was not well developed at that time.[16] His solution was to suggest that different degrees of metamorphism had been produced owing to the different natures of the rocks

16. For a thoroughgoing treatment by a British author, one has to look as far ahead as the 1930s to find a detailed treatise based chiefly on British samples: Harker (1932). (Alfred Harker

undergoing change, some rocks supposedly being more susceptible to meta-morphism than others (ibid., 106). But why this might be so, Bonney had no satisfactory explanation to offer—indeed, no explanation at all. It was a mystery why quite soft rocks such as the fucoid beds should have apparently escaped entirely the metamorphic processes that had produced the overlying schists and gneisses.

It can be seen, then, that on the one hand Bonney made some highly signif-icant contributions to the Highlands controversy. In particular, he quickly improved the identification of the rocks under dispute with the help of micro-scopical examination of thin sections. The problem was not thereby com-pletely solved, but at least it may be said that the use of improved analytical tools brought about immediate results. The constituent minerals of the rocks concerned could be identified with reasonable certainty, and their approxi-mate proportions determined. Thereby identification followed without undue difficulty. This kind of work was Bonney's forté. He also gestured toward the way of advance by publishing a large-scale map of a small area. But as we have seen, the structure of the area was misconstrued. The map and the accom-panying section were geometrically incompatible with one another, and Bon-ney had no real idea as to how the structures that he thought he could discern had come about—either geometrically, physically, or chemically. On the evi-dence of the paper that we have analyzed, he had expertise in petrography, but not in field geology. His specialty was laboratory analysis. He was not a field surveyor by profession: the professionals for that kind of work were to be found in the Survey.

In the discussion that followed Bonney's paper, Hicks, as we have done, ob-jected to the details of Bonney's map and sections. Hicks was supported by John Blake, another clergyman turned scientist, who was shortly to take up a chair at the University of Nottingham.[17] But Bonney received support from Hudleston. The discussion was apparently fairly evenly divided and was con-ducted without recriminations. Neither Ramsay nor Geikie was reported as being present.

For the moment, then, the debate was conducted between Bonney and Hicks, who by all accounts was a quick-tempered man, averse to criticism just like Murchison. His rock identifications had been queried by Bonney, a recog-nized expert. Not surprisingly, therefore, Hicks needed an expert petrog-rapher to aid him in the debate. For this purpose, he sought the assistance of

---

[1859–1939] was one of the leading members of the distinguished school of earth scientists at St. John's College, Cambridge, following in the tradition established by Bonney himself.)

17. John Frederick Blake (1839–1906) studied under Sedgwick at Cambridge from 1859 to 1862. He taught mathematics and was assistant chaplain at St. Peter's School, York, until 1874, when he gave up teaching for a career in science. He eventually gained a chair in natural science at Nottingham in 1880, where he remained until 1888. Blake published on the Liassic rocks of Yorkshire and did useful bibliographical and review work for the Geological Society, of which he was a council member. He gave much of his energies to the work of the Geologists' Association, and served as its president from 1891 to 1893, also acting as the association's editor. Blake re-appears in our story in chapter 8.

Thomas Davies (1837–92), a mineralogist from the British Museum who has been recorded as having "an eye-knowledge of minerals which has rarely been surpassed" (Woodward 1908, 221). Hicks's reply to Bonney was published in a paper in several parts issued in the *Geological Magazine* (Hicks 1880a). He did not, however, revisit the area between the publication of his two papers, even though there were some significant changes in the interpretations advocated, and alterations in the published sections.

Hicks took his readers once again over the ground of the Loch Maree district, and of Glen Logan particularly, and he incorporated in his account the descriptions of his rock specimens from key localities provided by Davies. Hicks began with what Davies regarded as a "very micaceous schistose rock," found, according to Hicks, "on the shore between Strath Hotel and Gaerloch Kirk." This meshes with what Teall reported to Clough in his letter of February 1890 (see note 2 above), from which I conclude that Teall was probably correct in what he had to say about the starting point for Hicks's work in the Loch Maree area.

Regarding the critical Logan Rock in Glen Logan, Davies's identification was quite at odds with that of Bonney. He regarded the specimen supplied to him as "a much altered hypersthenite" that "had nothing in common with hornblende gneisses," which, it will be recalled, was Bonney's diagnosis. Davies also referred to the rock as a "quartz diorite." Hypersthenite is a term used to refer to a rock somewhat like a granite in appearance, though chemically more closely related to the gabbro family, containing the minerals hypersthene and a kind of a feldspar called labradorite. Hypersthene is an iron/magnesium silicate of greenish, or more usually brown to black, appearance, quite often found in schists. Davies made no mention of the Logan Rock displaying any degree of foliation, and like Nicol and Hicks he seems to have thought that it was igneous in character.

It is difficult to be certain exactly where Hicks's specimen was collected in Glen Logan, but a sample collected by me from near the gorge in the glen, when sectioned and examined under a microscope, was found to contain quartz, feldspar, epidote, and biotite (brown mica) decomposing into the green mineral chlorite. It is conceivable that Davies mistook the greenish minerals for hypersthene, but whether this was so or not, the important thing is that he did not notice the evidence for deformation (cracking and the presence of distorted crystals) and for the action of pressure (foliation) that can be seen under the microscope in a modern specimen, and that provides unequivocal evidence for its metamorphic character (see plate 6). Thus one may have the suspicion, at least, that Davies based his opinion chiefly on the external appearance of the hand specimen. (We note that Woodward recorded that this was Davies's forté.) Whatever the case, the rock was still regarded by him as igneous, not metamorphic. So Hicks felt fully justified, with the aid of his "expert witness," in holding to the opinion that a substantial igneous intrusion was to be found in Glen Logan.

Hicks further provided two sections across Glen Logan: the first (his "Fig.

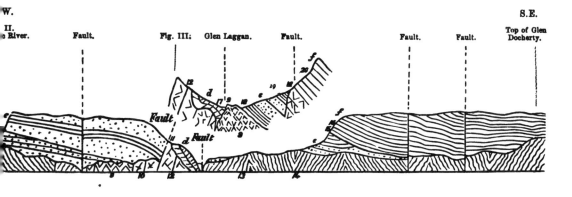

Fig. 7.5. Sections of strata at head of Loch Maree and Glen Logan, according to Hicks (1880a, 159).

II") across the entrance to Glen Logan; the second ("Fig. III") somewhere within the glen, very likely at the point of the small gorge, which is the most obvious locality of geological interest. These are reproduced in figure 7.5.

Davies's identifications of the several units shown in Hicks's sections were as follows:

9. Altered hypersthenite[18]
10. Crushed granitic rock
12. Coarse-grained feldspathic granite
13. Micaceous gneiss
14. Foliated micaceous rock
15. Schist
16. Schist
17. Bluish-grey rock, resembling indurated shale[19]
18. Resembling 17, but with more quartz, some mica, and of somewhat schistose texture
19. Micaceous quartzite
20. Resembling 15

It can be seen from this that there was obviously a good deal of metamorphic rock around, and that the section given was in some ways analogous to what one might expect from Nicol's theory. If we regard 13 and 14 as Lewisian Gneiss, and 15, 16, and *f* as collectively equivalent to Moine Schist, then the

18. This was also referred to as a quartz diorite.
19. I imagine that this was the "fucoid beds" in an unweathered condition.

structure of the eastern side of the glen is not too different from what one would describe today, the chief difference being that Hicks had an unconformity below his schist, whereas we would have it as a thrust plane. The supposed vertical fault between the schist and the gneiss seen in Bonney's sketch and section had no place in Hicks's diagram. The modern observer has as much of a tangle of sedimentary and metamorphic rocks in the middle of the glen as did Hicks. But we don't have his igneous intrusion.

It should be emphasized that Hicks envisaged two distinct types of igneous rock in the locality. There was the altered hypersthenite (9), also called a quartz diorite; and the crushed granitic rock (10 and 12). Hicks thought that these were of different ages. The quartz diorite he regarded as younger than the "Silurian" sediments, for it appeared to intrude into them. The granite, however, seemed much older, and was probably pre-Cambrian, being part of what he regarded as the pre-Cambrian "floor" of the whole region.

As might be expected, Hicks's conclusions were challenged by Bonney (1880b). The main point at issue remained whether the Logan Rock in Glen Logan was igneous or metamorphic. Bonney held to his position that it was metamorphic (a type of gneiss) and claimed that his opinion was supported by (unnamed) "good petrologists," who had examined his slides. He also complained that Hicks had quite altered his estimate of the age of the "granite" from post-Silurian to pre-Cambrian, without adequate justification.

As mentioned above, Hicks was developing some reputation for himself for his identification of pre-Cambrian rocks, and he seems to have been interested in building up a pre-Cambrian kingdom somewhat similar to Murchison's Silurian realm. Thus Hicks published a number of papers in the early 1880s that made use of his claims for the existence of pre-Cambrian rocks in the Northwest Highlands (Hicks 1880b, 1880c, 1881a, 1881b). However, these did not advance the understanding of the northwest stratigraphy to any significant degree, and they need not be given special consideration here. I may mention, though, that Hicks (1881b) was somewhat concerned that the supposed Cambrian rocks of the west coast (red-brown sandstones or Torridonian) contained no fossils, whereas elsewhere Cambrian strata were known to contain quite complex fossil assemblages. Because this rock unit contained pebbles of both gneiss and schist, he was inclined to the view that the supposed distinction between newer and older metamorphic rocks in the Scottish Highlands, as advocated by Murchison and Geikie, was not sustainable. Perhaps most important, Hicks maintained that rocks such as those at Ben Fyn were essentially the same as those at the "western" gneiss, and hence pre-Cambrian. Thus a direct challenge was extended to the old Murchisonian doctrine that all the metamorphic rocks to the east of the Eriboll-Skye line might be colored on maps as Silurian.

The official view of the Survey, however, showed no sign of immediate change, though clearly Geikie, at least, was somewhat concerned by the perceived challenge to the Murchisonian orthodoxy. Thus we find from his field notebooks that he made no less than three field trips to northwest Scotland in 1880 and 1881, presumably to "reassure" himself that his earlier observations and interpretations were sustainable.[20] In May and June 1880, he traveled according to the following itinerary: Edinburgh, Inverness, Garve, Ullapool, Inchnadamph, Scourie, Rhiconich, Lairg, Inverness again, and ending back in Edinburgh. The round trip took nine days. Not long after, in June and July, he traveled by the following route: Edinburgh, Perth, Strath Carron, Shieldag, Torridon, Kinlochewe, Gairloch, Glenelg, and back again to Edinburgh.[21] This journey took about twelve days. Then in April the following year, Geikie traveled to Edinburgh, Perth, Inverness, Lairg, Laxford, Durness, Cape Wrath, back to Durness, on to Scourie, Lochinver, Ullapool, Garve, once more to Inverness and to Perth, then Oban, and finally Edinburgh, the journey taking him ten days.[22] There is no evidence that he went to the critical exposures at Loch Eriboll on any of these trips.

It would appear from these itineraries, and this is confirmed by the notes themselves (especially in Notebook W), that Geikie was chiefly interested in problems relating to the "western" gneiss. Nevertheless, it can be seen that he made a point of visiting well-known sites such as Ullapool, Kinlochewe, Inchnadamph, and Durness. For whatever reason, Geikie did not record any observations relating to Glen Logan. But he remained committed to the concept of an "upper quartzite," recording it at several localities. At Inchnadamph, he figured a regular ascending sequence (lower quartzite, fucoid beds, quartzite, dark limestone, white limestone), much the same as Murchison and Harkness had done. At Loch Kishorn, the sequence was recognized as anomalous (as Murchison had admitted), but it was supposed that a fault cut out the "lower" quartzite as one progressed toward the loch from the north. In this, Geikie was quite correct, but he did not realize that the quartz rock disappeared because it had been overthrust by the limestone. Between Garve and Ullapool, in the territory usually occupied by Moine Schists, he recorded "the most 'Laurentian-like' gneiss" he had yet seen in "the younger crystalline rocks of the Highlands."[23]

In his trip of 1881, Geikie described and figured a regular ascending sequence for the side of Loch Broom at Ullapool.[24] He acknowledged that north of Ullapool, near Strath Kanaird, the limestones seemed to be repeated by faulting, so that they occupied a wider outcrop than would otherwise have

20. The term was Murchison's rather than Geikie's.
21. A. Geikie, Field Notebooks, University of Edinburgh Archives, Gen. 521/1, Notebook W.
22. A. Geikie, Field Notebooks, Haslemere Educational Museum, Notebook Y, 1880–81, Diary of traveling expenses.
23. Notebook W, 43.
24. Notebook Y, 32. Some of the pages from this notebook have been excised. But one of the pages appears to be located in the archive of the Scottish National Museum in Edinburgh, inserted in the back of a bound collection of Geikie's published papers (55[081]/Geikie/G). This shows a sketch, dated 1881, of the southern shore of Loch Broom, opposite Ullapool, with a regular ascending sequence, as evisaged in Murchison's "paradigm," and including both an "upper" and a "lower" quartzite.

been the case.[25] The fucoid beds and the Serpulite grit were now regularly recorded (which had not been the case during Geikie's trip of 1860), and all sorts of odds and ends of rocks were noticed here and there. But the paradigm remained unchanged in Geikie's mind. With three additional trips to the Northwest Highlands, he found no reason to revise his opinions. The visits were all conducted with Geikie's customary haste. He wasn't really surveying the area: it was the usual quick look round to reassure himself of the accuracy of his former observations and interpretations. But to be fair, Geikie's administrative commitments made it impossible for him to deal personally with the problem in the manner he might have wished. It might be argued that in his view the official understanding of the Highlands was still quite sound. The whole matter would, no doubt, be carefully reviewed when the surveyors got round to examining the Northwest Highlands. Meanwhile, there was no reason for concern. It was business as usual: matters could take their course at their normal rate of progress.

But there was rather more to Geikie's journeyings than may appear from the preceding narrative. For we find that on the first visit of 1880 he was accompanied by the director of the Irish branch of the Survey, Edward Hull (1829–1917), by another member of the Irish Survey, Richard Symes (1840–1906), and by "several other friends."[26] Hull subsequently published a full account of the field trip, having given a public statement of his results to the Royal Geological Society of Ireland on December 20, 1880 (Hull 1882–84). The paper presented the official Survey view.

What was the motive for joining with Hull, Symes, and other unnamed persons in the north of Scotland for a brief journey at public expense? According to Hull (ibid., 56), the journey was an official one, made with the sanction of the director general. It was, I suggest, from Geikie's perspective, desirable to gain extra witnesses in favor of the official interpretation of the Scottish stratigraphy, from reputable geologists who would have the appearance of impartiality (being from outside the British "mainland"), yet who might be expected to give support where it was needed, by virtue of their loyalty to the Survey. Hull met these requirements admirably. However, one must not assume Machiavellianism everywhere. It appears that Hull was at that time desirous of making comparisons between the rocks of Donegal and those of the Northwest Highlands of Scotland, and he had in fact requested permission from Ramsay to pay a visit to the northwest (Herries-Davies 1983, 223). Even so, there can be no doubt that this was convenient to Geikie's interests.

As to Hull's capacities to form a sound judgment of the structure of an area on brief acquaintance, it must be said that in his early years on the Survey he

25. Notebook Y, 34.
26. An account of Hull is to be found in his autobiography (1910). See also Geikie (1917), Royal Society of London (1917–19), Harker (1918). His chief work was concerned with the geology of the coal-bearing regions of Britain, but he also wrote voluminously on a wide range of topics and published several textbooks. Besides his Survey position in Ireland, he held a chair at the Royal College of Science in Dublin. He was somewhat distantly related to Murchison in that a cousin of his was married to Murchison's sister.

had something of a reputation for slipshod work (ibid., 197). Moreover, according to Herries-Davies, historian of the Irish Survey, Hull was "not one to be concerned with painstaking investigations of detailed field evidence. Rather was he constantly falling victim to the seductive charms of grandiose schemes of speculative synthesis" (ibid., 208). This seems to mesh with what we find in relation to Hull's Scottish tour of 1880. In any case, he might have been expected to defer to Geikie's opinion concerning the geology of Scotland, seeing that Geikie was director of the Scottish branch of the Survey.

From one point of view, things worked out exactly as Geikie might have hoped. Hull's paper offered full support to the official interpretation of the Highlands strata. In particular, Hull (1882–84, 68) presented a theoretical section through Quinag, Inchnadamph, and Ben More that offered a smooth ascending sequence from lower quartzite to schist, exactly as Murchison had pronounced things to be more than twenty years earlier. But from another point of view, the tactic achieved virtually nothing. Hull stated that the visit was undertaken "under the guidance of Professor Geikie." We can be sure, I think, that Geikie took his party to all the places that might seem to support the Murchisonian theory (especially the hillside of Cnoc an Droighinn at Inchnadamph, and Knockan Cliff), and equally sure that he said nothing of all the peculiarities of structure that the geologist encounters high up in the hills of the Ben More range. Indeed, from what can be discerned from Geikie's field notebooks and his published papers prior to 1880, it is evident that he never made anything but flying visits to Assynt and never got up into the mountain ranges proper. He thus never fully realized the nature of the complexities that he was facing. Certainly, he did not tackle the job of mapping the area properly. As for Hull's published section, there was no warrant in his fieldwork of 1880 for presenting the section as he did. There simply was not time to come to grips with all the details of the structure. The whole business was really a kind of whitewash, and I suspect that most people realized this. Hull's paper achieved no recognition.

Even so, the Survey stuck to its official view. In his presidential address to the British Association in 1880, Ramsay (1881) assumed, apparently without question, that Murchison's ideas about the Silurian age of the great mass of the Central Highlands of Scotland were acceptable. Geikie also made no challenge against Hicks—though it is reasonable to suppose that Hull's paper was Geikie's vicarious response.

The only published contribution to the debate made by Geikie at this time was to suggest—quite plausibly—that the surface of the "western" or Lewisian gneiss, on which the Torridon Sandstone was lying unconformably, was formerly glaciated. Thus the "hummocky" surface of gneiss, so characteristic of the terrain of the west coast of Scotland, could be seen in places like Loch Maree to carry on to the east *under* the Torridon Sandstone.[27] Thus, Geikie suggested (1880b), the characteristic "Archaean" surface was not due to the

27. The phenomenon is well seen from the southern shore of Loch Maree, looking north toward the Torridonian mountain, Slioch, sitting on the curved gneissic basement.

geologically very recent glaciation (i.e., Pleistocene), with which nineteenth-century geologists were by then quite familiar. It marked a very ancient epoch of glaciation, right back in Archaean times. This was a perfectly feasible argument, but it did not touch the main issues at stake in the Highlands controversy. In any case, Hicks (1881b, 75) claimed that the idea had come from him in the first place, though the validity of this claim is somewhat suspect since Hicks first proposed the notion in relation to the pre-Cambrian rocks of Wales, rather than Scotland.

I should now like to take up in more detail the question of the Logan Rock, a term that has already been used in our text as a matter of convenience in a considerable number of places. But the usage has been anachronistic, since the term did not actually appear in print until 1882, in a series of essays on Scottish stratigraphy and mineralogy by the professor of chemistry at St. Andrews University, Matthew Forster Heddle (1828–97).[28] Heddle has an interesting place in our narrative, since he was one of the "amateur" (i.e., non-Survey) geologists who sought to unravel the mystery of the Highlands and made some interesting contributions to the debate, without really coming close to solving the problems.[29] On the whole, he sided with the Survey opinion, but his reasons for so doing are worth consideration. His social role in the affair is also of some interest.

Heddle came from the Orkneys and was educated in Edinburgh, where he studied medicine. On graduation he took himself to Freiberg, the old citadel of Wernerian theory, to study mineralogy and chemical analysis. On returning to Scotland, he practiced medicine for a time and took his M.D., with a very unmedical thesis topic: "The ores of metals." He had always been an avid collector, and in the 1850s he began what came to be his major life-work: the amassing of a very fine mineral collection, particularly Scottish minerals. In 1856, he obtained a junior position in chemistry at St. Andrews, and in 1862 he obtained the chair in that subject at the same institution, where he remained until 1883, when he tried his fortune in South Africa. But this venture was unsuccessful, and he returned to Scotland to spend the remainder of his life in St. Andrews.

Heddle was a great lover of the open-air life and knew the terrain of northern Scotland intimately, to the extent that he was able to bring out a geological map of Sutherland (Heddle 1881), designed to accompany his long series of papers that was appearing in the *Mineralogical Magazine*. The map had an interesting accompanying commentary (Heddle 1882), which asserted that it was almost entirely based on first-hand experience, and where it was not,

28. Of the numerous constituent parts of Heddle's essays (1879–84), the sections that are of particular interest are the descriptions of Sutherland, appearing on pages 135–80, 197–254, 133–89, and 271–324. Heddle's work was also published posthumously in book form (Heddle 1901).
29. On Heddle, see Goodchild (1898) and Thoms (1901). (Goodchild, an officer of the Scottish branch of the Survey, was a son-in-law of Heddle.)

Heddle's indebtedness to other sources was freely acknowledged. Also, the places where there might be uncertainties in his geological boundaries were pointed out with care. Heddle was adept at chemical analyses of minerals, and he certainly knew his mineralogy and crystallography. He helped to found the Mineralogical Society and was its second president. He was also for a time president of the Edinburgh Geological Society. Toward the end of his life, Heddle's splendid mineral collection passed to the Edinburgh Museum of Science and Art, and it currently forms the basis of the collection of the successor of this institution, the Royal Scottish Museum (Macpherson and Livingstone 1982).

Heddle's papers on the "geognosy" of Sutherland make curious reading today, being specimens of what one might call "romantic science." They contain descriptions of the work of authors such as Macculloch as if they were heroes of ancient times. Then there are numerous expressions of Heddle's emotional responses to places that he was investigating, and descriptions in overblown language. Consider, for example, the description of Smoo Cave, on the north coast near Durness (admittedly an awesome spot):

> The southern land-floored end of this great trench is a *cul de sac,* half-domed and roofed by a thin shelf of rock. The light which slumbers within that dome, once seen, is *felt.* The light—what they have of it—of all caves is fine; that of this is surpassingly lovely. Poets would call it a *chastened* light. If by that, is meant that it resembles a character which, through the buffettings and disappointments of the world, shines with a softened sweetness, we partly understand the application; and it would be a fitting one. (Heddle 1882, 172)

Heddle then went on to explain how science might give an explanation of the special qualities of light in caves. The description carried on for several pages with the language becoming ever more flowery. In addition, the text was nicely decorated with Heddle's romantic sketches.

But having brought the reader into a suitably receptive frame of mind, Heddle would gradually change his style, to offer the percentage compositions obtained in his mineral analyses, and technical descriptions of his mineral specimens, with crystallographical diagrams. It was a most unusual approach to the presentation of original scientific information, the like of which I have not seen elsewhere.

In his description of his map, Heddle gave considerable attention to the controversial rock in Glen Logan. Surveying the literature, Heddle (1882, 44), reminded readers that it had been described as "upper gneiss" by Cunningham; "intrusive" by Nicol, Murchison, Harkness, and Hicks; "upper gneiss metamorphosed in situ from the near presence of lime" by Geikie; "old gneiss, faulted up, and partly invaded by extravasated matter" by Hudleston; and "old gneiss, faulted up, its upper portions crushed, especially when near lime, and sometimes granitoid from predominance of felspar" by Bonney. Cunningham, Geikie, and Bonney thought they recognized some indications of bedding. Murchison and Hicks did not, and Nicol did only infrequently. Cunningham, Geikie, Harkness, Ramsay, and Murchison held that the rock did not break the upward succession of rocks; but Nicol, Bonney, Hudleston, Hicks, and Callaway thought that it did.[30] Considering these divergent opinions, Heddle thought it best to renounce theory for the time being. Hence he coined the term "Logan Rock" (Heddle 1884, 144–45, 272, 282–308), which we have already found so useful in our discussion. A specimen from Glen Logan itself, with two photomicrographs, one using ordinary light, the other polarized light, is shown in plate 6.[31]

The Logan Rock seems to have been a sore puzzle to Heddle. He described it minutely and all the many localities where it might be found in the Northwest Highlands (ibid., 285–308). But he could not make out the mineral constituents clearly, even in thin section. Indeed it was, he said, the "only rock mass I have ever examined in which I found no minerals" (ibid., 292). It is true that in many cases the individual minerals of Logan Rock cannot be made out clearly in hand specimens. But as plate 6 shows, the mineral grains show up well with modern techniques of rock sectioning, even though the individual grains often show signs of crushing and alteration.[32] We must assume, therefore, that Heddle's techniques for rock sectioning were not well developed or that he did not examine the sections with polarized light.

Heddle emphatically denied Bonney's interpretation of Logan Rock as altered Hebridean. "It is not," said Heddle,

> the Hebridean gneiss, and it is not even of the same age as the Hebridean gneiss. I hold that stratigraphically it is *impossible* that it is so; geognostically it is *incredible;* lithologically it is *inconceivable;* and mineralogically it is *improbable.* (1184, 287)

In fact, Heddle tended to side with Geikie in thinking that the rock was some sort of gneiss, somehow metamorphosed *in situ.* Heddle had very good rea-

---

30. Callaway will be introduced properly on in chapter 7.
31. I am indebted to Professor T. G. Vallance of the University of Sydney for furnishing me with a thin section of the rock and an identification.
32. Both photomicrographs reveal the foliated structure characteristic of metamorphic rocks. The sample photographed in ordinary light (6b) shows cracking within the rock structure, resulting from pressure. The photograph taken using polarized light (6c) shows the large quantity of epidote (the fine-grained material, much of which appears yellow-orange) in the rock. What appears as a largely undifferentiated mass in ordinary light is revealed as an interlocking mixture of grains of feldspar and quartz under polarized light. The feldspars—some of which have a crosshatched appearance at the top left of 6(c)—show signs of distortion resulting from pressure. So too do the quartz crystals at the center of the picture and the top right-hand corner. A quartz crystal normally has a sharp "extinction angle" when rotated between "crossed" polarizer and analyzer. That is, it appears either dark or light, according to the orientation of the crystal in relation to the polarizer and analyzer, not "in-between." But in strained crystals, such as are to be seen here, the "extinction" is not sharp and a "cloudy" appearance results. This demonstrates that the Logan Rock is metamorphic, not igneous. Such arguments were not advanced, even by Bonney, though he recognized the rock's metamorphic character as soon as he examined it under the microscope, and even before that in the field.

sons for rejecting it as igneous. It always appeared above the limestone. No igneous neck had been found. When it occurred in highly faulted places, it always did so in its proper order, rather than entering the rents of the faults. And it contained no imbedded crystals or mineral veins, quite unlike normal igneous rocks.

Heddle also sided with the Survey doctrine by accepting the notion of there being two quartzites and two limestones—the upper one, he maintained, being dolomitic.[33] It will be realized that in supposing the possibility of the formation of gneiss, without metamorphosis of underlying rocks, he was, like Geikie and Murchison, suggesting something for which there was no plausible explanation in terms of physics or chemistry. But given all the complexities, Heddle was, I think, performing a useful service in suggesting the nontheoretical term "Logan Rock," to be used until the characterization of the rock had been accomplished more satisfactorily.

With these preliminaries, then, we may now turn our attention to Heddle's map. I have selected a portion of this, showing the Loch Assynt district in some detail, for reproduction in plate 7, in order to give an impression of the state of mapping technique during this period. The map had no accompanying section.

Heddle's map did not have the detail that was provided in the maps of the Survey officers (as can be seen by comparison with a later product of the Survey, after the whole problem had eventually been "solved" to everyone's satisfaction: see plate 8). But it must be remembered that the Survey officers were professional mapmakers, who could spend many months in the field and had cartographic assistants in their home bases. Moreover, the map shown in plate 8 took several years for the Survey's most accomplished field officers to produce. By contrast, Heddle worked alone and at his own expense, and he could only go into the field as his university commitments allowed.

The portions of the key to the parts of Heddle's map of Sutherland shown in plate 7 are as follows:

Hebridean Gneiss (as at Lochinver)	Dull green (*A*)
Torridon Conglomerates (as at Coul More)	Orange (*a*)
Quartzite (as to west of Stronchrubie, to southwest of Coniveall, etc.)	Pink (*B′*)
"Igneous" rock of the quartzite (as at Ben Fie) (= Logan Rock)	Green (unlettered)
Dolomite (as at Stronchrubie and Knockan)	Blue-green (*D*)

33. That is, containing magnesium carbonate in addition to calcium carbonate.

| Upper gneiss (forming most of the eastern part of the map) | Dull blue (*B″*) |
| Granite, porphyry, and eruptive rocks | Red (*G, P,* and *E*) |

It is, of course, invidious to make comparisons with modern maps, but as a point of interest it may be said that if we replace Heddle's "igneous rock" (Logan Rock) with Lewisian Gneiss we have a map that looks quite recognizable to modern interpretations. It is least like the modern map in the area round the mountain of Brebag (Breabag), where we show less quartzite, more Lewisian, and more granite. Of special interest are the two localities to the west of Brebag marked with question marks. The modern maps show these spots as being made of Torridon Sandstone lying *over* limestone. This is quite incomprehensible according to the Murchison/Geikie/Survey paradigm, to which Heddle was giving a not-wholehearted support. It is not surprising, then, that he drew a veil over the difficulty with the assistance of the question marks. But he had obviously walked over the spots concerned and had been puzzled—as well he might be. Note that the imbricate structures of plate 8 had no place in Heddle's map.

But while Heddle was giving some support to Geikie's views about the Logan Rock, and the existence of both an upper and a lower quartzite, he was also trying to insert a wedge between Geikie's opinions and those of his former mentor, Murchison. It will be recalled that the "Silurian" fossils had been found by Peach in the limestone at Durness. It had then been assumed that essentially the same limestone occurred all the way down from Eriboll to Skye. But, Heddle emphasized, the limestone at Durness was not connected with the limestones at Eriboll or anywhere else, so far as direct observation showed. So if in fact the limestones were *not* one and the same, then the whole argument about the eastern metamorphic rocks being Silurian (being above, but part of the same system as, the sedimentary rocks at Eriboll, not Durness) might collapse. The fossiliferous sediments at Durness were evidently faulted in, being bounded by a triangular system of faults. But what was the age of the apparently unfossiliferous Eriboll rocks? This had been assumed, rather than proved. As Heddle put it, there is

> *at first sight* a general resemblance,—but "like is an ill mark." And then [he recalled] Peach found the fossils,—and the fossils were pronounced to be Silurian;—and then there was a feast of correlation, and a flood of theory,—and the grand old "gnarled gneiss," which gives their noble outlines to the Highland hills, was declared to be the same as the solidified mud of the hideous hunches [*sic*] of the Lowlands. (1884, 254)

What Heddle was getting at, of course, was the abandon with which Murchison had rushed to paint the whole of the Central Highlands in Silurian colors, even seeking correlations with the rocks of the Southern Uplands. For Heddle, this was a dubious enterprise, for he maintained that the chemical composition of the fossil-bearing limestones at Durness (chiefly calcium carbonate) was different from those at Eriboll and to the south (dolomite, i.e., calcium and magnesium carbonate), which were apparently unfossiliferous: "And if the two 'limestones' be different, then the whole fabric of Murchison's correlation of the rocks which overlie the Dolomite, falls to the ground" (ibid.). Thus while Heddle allowed a continuity of sequence upward from the sediments of Eriboll, Assynt, and elsewhere into the metamorphic rocks, he denied the claim that the "eastern" metamorphics had been shown to be Silurian. Rather, he judged them to be "Archaean."

So much for Heddle's theory.[34] But were there any personal questions that might have swayed him one way or the other? Here I enter the realm of speculation. But some facts bearing on the issue are worth bringing to light. Heddle and Geikie were friends of very long standing. They had first met as students in Edinburgh when they attended the classes in mineralogy and analytical chemistry of the private teacher Alexander Rose (Geikie 1924, 30). Moreover, it appears from a footnote in Heddle's paper of 1882 that the two of them had actually traveled together, or had at least met, in the Northwest Highlands not long before its publication (Heddle 1882, 230). There are also some letters from Heddle to Geikie, held at the Edinburgh University Archives, which show that Heddle was extremely beholden to Geikie.[35] We only have one half of this correspondence, and even that is far from complete, but nevertheless it brings out features that are of interest to our inquiry.

Back in 1873, Heddle had written to Geikie, in his capacity as chief of the Scottish branch of the Survey, seeking some occasional work performing mineral analyses, and it appears that Geikie was able to offer something in this line.[36] Heddle's problem was simply one of money. He had a wife and seven children to support and could barely manage on his meager salary at St. Andrews. This was a recurring theme in his letters, and by 1878 things had got so bad that in desperation Heddle offered Geikie the use of his house for summer fieldwork, for the sum of £50.[37] But nothing came of the suggestion. There was some exchange of letters that year relating to the chair vacated in Ireland as a result of Harkness's death.[38] But nothing came of that either.

Geikie did manage to help his old friend, however, by securing for him a grant from the Royal Society, to which Heddle responded with unrestrained joy:

> Hooray!!
> Geikie you are a trump—king of trumps. . . . your persuasive eloquence has been enough. . . . God bless you my boy—as the venerable

---

34. The Survey officers later showed that Heddle's doubts about the identity of the limestones of Durness and Assynt were not warranted. There are some units of the Durness Limestone that are dolomitic and others that are not, both at Durness and at Assynt.
35. Edinburgh University Archives, Gen. 525.
36. Heddle to Geikie, Dec. 22, 1873; Jan. 1, 1874.
37. Heddle to Geikie, Aug. 9, 1878.
38. Heddle to Geikie, Oct. 21 and Dec. 27, 1878.

Fathers say—"When things are at their worst they mend"—question is
whether are they at their worst. This thing may turn the luck for me,
very likely will, for it may make those Classical duffers here think twice
before they push me and science further back.[39]

But later that year, Geikie and Heddle seem to have had a theoretical disagreement, in which Heddle spoke somewhat slightingly of Geikie's "Sutherland work" and gave some credence to the controversial views of Thomas
Sterry Hunt (1826–92), who worked for the Canadian Survey, and of whom
Geikie had a low opinion because of his supposed "neo-Neptunism" or "neo-
Wernerism."[40] Exactly what passed between Heddle and Geikie in 1879 one
cannot tell because of the incompleteness of the correspondence, but it appears that Geikie had reacted with anger to whatever it was that Heddle had
written; and that Heddle had then largely climbed down, trying to pass off
what he said as a kind of joke.[41] Remembering that Geikie was in some measure Heddle's patron, and that with his large family Heddle would still have
been anxious to continue to be the beneficiary of Geikie's patronage, we can
well imagine that Heddle would not have wanted to take issue with Geikie in
public, considering the strong response he had apparently received to a critical comment in private.

Obviously, such a line of argument cannot be pushed very far. But it is to be
emphasized that the "chemical" features of Hunt's stratigraphy would natu-

39. Heddle to Geikie, Jan. 1, 1879.
40. Hunt was trying to work out a stratigraphy for the Archaean rocks on the basis of their
chemical composition. In a sense his ideas lay in the old tradition of the Wernerians or Neptunists,
in that he believed mineral composition to be a guide to relative ages. Since the work of Smith and
Murchison, this methodology had long been rejected by British geologists, though for the unfossiliferous rocks of the pre-Cambrian a chemical approach to stratigraphy was warranted. But
Hunt went further: he sought to establish a theoretical scheme according to which different rocks
and minerals would necessarily be formed as the earth cooled out of its primeval chaos and the
original primeval ocean cooled and deposited crystalline matter. Indeed, he wished to establish a
whole new branch of geology: "chemical geology" (as opposed to geochemistry), which would
construct models for the processes occurring during the very early history of the globe. Such a
science could not be developed satisfactorily during the nineteenth century, and Hunt's views
were generally rejected as wildly speculative. Nevertheless, he had some followers, especially his
colleagues in the Canadian Survey. His work had a marginal impact on the Highlands controversy, in the sense that he appealed to Scottish examples in support of his general theory. But
these views were unacceptable to Geikie since they would disallow a "Silurian" schist or gneiss.
Such rocks, on Hunt's view, should be much deeper in the whole stratigraphical column. Thus
Hunt's "chemical geology" was in flat contradiction of the whole Murchison-Geikie theory of the
Highlands. For Hunt's work, see his *Chemical and Geological Essays* (1875). In this book (p. 271), he
specifically queried the stratigraphical relations of the Highlands as construed by Murchison and
Geikie and suggested that the chemical evidence required that the "eastern" metamorphic rocks
be much older than Murchison and Geikie had supposed. See also Brock (1979).
41. Heddle to Geikie, Nov. 22, 1879.
42. We may note his emphasis upon the importance of the chemical composition of the various
limestones.

rally have appealed to Heddle—a professor of chemistry.[42] His early training at Freiberg no doubt made him receptive toward Hunt's "neo-Neptunist" doctrines. In fact, in the Heddle interpretation of the Highlands, outlined above, we can see that he was going against the Murchisonian doctrine of a "young" gneiss or schist, while seemingly keeping faith with the *structure* that Murchison had espoused, and that Geikie continued to sustain more than twenty years after the field trip of 1860. But to do this landed Heddle in an incoherent explanation. What was an ancient metamorphic rock doing on top of quite young-looking sediments? As to the "Logan Rock," that was acknowledged to be a puzzle by the very name that Heddle had chosen to attach to it. Evidently, the mystery of the Highlands rocks was far from solved in Heddle's work. But while a reader of his papers might think that he was giving credence to Geikie's doctrines, it can be seen that there were substantial differences between Heddle's ideas and the official views of the Survey. Heddle was quite subtle in appearing to agree with Geikie in some measure, while in fact putting forward ideas that were incompatible with the old Murchison paradigm. Whether it was an intentional subtlety, designed to placate and to ensure Geikie's continued patronage, or whether Heddle's long series of papers was based wholly on his field observations and mineralogical analyses, cannot now be determined. But the thought that Heddle leaned toward Geikie's doctrines more strongly than would otherwise have been the case cannot be wholly discounted.

One of the most interesting papers to emerge from the whole Highlands controversy was published by Wilfred Hudleston (1882), who spent about a fortnight of the summer that year in the Assynt region in the company of Heddle. The great interest that attaches to Hudleston's account of the excursion, entitled "First Impressions of Assynt," is that he frankly and in a non-partisan manner explained what he saw and where the difficulties lay in interpretation. He had no particular axe to grind (being a palaeontologist), but he knew more than enough geology to be able to appreciate the problems and their several proposed explanations. He had also read all the previous literature on the topic, as is evidenced by his earlier review paper. Such papers as "First Impressions of Assynt," expressing puzzlement and a personal response to bewildering data, rather than facts or theory or point of view, are rare in the literature and of great value to the historian.

Hudleston and Heddle approached Assynt from the rear—that is, from the east (from Glen Cassley), rather than from the southwest along the Ullapool road as do most travelers—and crossing a saddle to the north of Ben More they then descended into Glen Dubh (the "Black Glen") and down the track to Inchnadamph. Coming this way, Hudleston was immediately introduced to the peculiarities of the Logan Rock—the "everlasting 'Logan' rock," as he put it. In the next few days, the two geologists made a careful examination of the district, and Hudleston began to understand some of the many problems that it presented. Eventually, he boiled the problems down to three crucial questions:

1. Is there an Upper Quartzite?
2. What is the nature and geological position of the "Logan" rock?
3. Is the Upper Gneiss really a newer formation properly overlying the Quartzite-dolomite? (The term "Upper Gneiss" was used by Heddle and Hudleston to refer to what we call Moine Schists. By "Quartzite-dolomite" they meant what we call Durness Limestone, or a subunit thereof.)

At the section at Knockan Cliff near Stronchrubie, the sequence was evidently:

Upper Gneiss [Moine Schists]
Dolomite
Fucoid Beds, etc.
Quartzite

Here, then, it would appear that there was no "upper quartzite." But the sequence seen at Cnoc an Droighinn (close to the Inchnadamph Hotel) was

Quartzite (with intercalated igneous beds [A])
Quartzite (turning over and reversing its dip)
Quartzite (with intercalated igneous beds [B])
"Pale grey dolomites"
"Dusky dolomites"
["Lower Quartzite"—to the west of Loch Assynt]

The igneous beds (A and B) were thought to be in reverse order. Therefore, although there *appeared* to be an "upper quartzite," as Murchison had for long insisted, this was doubtful to Hudleston, with the evidence for a roll-over of the quartzite. At Loch Maoloch (Gillaroo Loch—the famed trout loch), inspection of the dip of the claimed upper quartzite suggested to Hudleston that it dipped below the limestone and was therefore actually one and the same as the "lower quartzite."

So on the whole, and with some further arguments that will not be detailed here, Hudleston concluded that the evidence was against the existence of an upper quartzite. The district was "chopped-up." There might, he suggested, be "a roll up of the Lower or western Quartzite in a series of convolutions to the east of the Dolomite [limestone] which has [formerly] been assumed to pass beneath it" (ibid., 394). He also ventured an interesting generalization, namely, that "where there is no 'Logan' Rock, there is no Upper Quartzite" (ibid., 393). That is, wherever claims had been made for an upper quartzite one could always find Logan Rock in the vicinity, and vice versa. (At Knockan Cliff, for example, there was no Logan Rock and no indication of an upper quartzite.)

What then of the Logan Rock? Hudleston gave a very nice description of it, which may be compared with the photograph of a specimen from Glen Logan shown in plate 6:

> I hold in my hand at the present moment a thoroughly typical speci-men of "Logan" rock obtained from the somewhat isolated exposure to the west of Ledbeg. It is striped something like a tiger, with bands of a hackly pinkish felspar, partially relieved by a dull white quartz, alter-nating with thick or thin bands of a dark green mottled hornblende. The cracks and backings are lined with abundance of pale green epi-dosite,[43] which is so characteristic of the rock in the Logan valley. Al-together this is a fair specimen, though there are others far richer in quartz. (ibid., 396)

Hudleston decided, like Bonney, that if anything the specimen had to be clas-sified as a form of Hebridean Gneiss, though he reached this conclusion with some surprise that "so acute an observer as Nicol" should have thought it a "granulite," a "syenite," or a "diorite," in different localities. Evidently, it var-ied considerably from place to place. But Hudleston found that if he replaced the "syenite" in Nicol's sections with Logan Rock or gneiss, he could reconcile his own observations with those of Nicol reasonably satisfactorily.

In Hudleston's view, the Logan Rock was in fact the central feature of the whole of the Assynt region, and perhaps the key to the whole problem, for as he said, "Wherever there is 'Logan' rock, trouble is sure to ensue" (ibid., 392). Might it not be, he suggested, "the framework or core round which the newer formations are folded, sometimes in great winding sheets of white quartzite" (ibid., 397)? Here he was leaning toward ideas once put forward by Nicol (1861, 99), who had thought of "granitic gneiss and mica slate, with intrusive rocks" as forming the core or nucleus of the mountains of Assynt. In a very loose sense, Hudleston was right, as can be seen by comparison of his hypoth-esis with the structure that was eventually produced when the stratigraphy of the Assynt region was regarded as "solved" (see figs. 9.2 and 9.3 below). But it can further be seen from this section that Hudleston's metaphor of a winding sheet of quartzite offered no suggestion of any kind of thrusting process, and he had to acknowledge that in a few places the quartzite seemed to lie below the Logan Rock. Nevertheless, he thought that his generalization usually held.

In reference to his third question, Hudleston regarded the evidence for the Murchison-Geikie doctrine as stronger than that for Nicol's. Looking at the Knockan Cliff section, he found it to be unlike the way Nicol had figured it

---

43. This term is usually taken to refer to a rock rich in the mineral epidote. But here Huddleston seems to be using the term for the mineral rather than the rock.

(see figs. 6.3 and 6.4 above), and similar to the way Murchison regarded it. At Knockan there was no upper quartzite (and no Logan Rock either), and the upper gneiss (or Moine Schist) did indeed seem to lie regularly on top of the limestone, and there was not the slightest evidence of inversion at that site.

How the whole problem should be resolved, Hudleston could not say, though he made the valuable point that the strike of a stratum could not, in his opinion, be a certain guide to the group to which it belonged. (Murchison, it will be recalled, had always argued that the "western" and "eastern" gneisses were different from one another for they differed in strike as well as in lithology.) Hudleston's conclusions were stated in the following remarks:

> It is not without feelings of regret that I cannot see my way to an Upper Quartzite, and if I have been wrong about the position of the "Logan" rock, I still think that this monster will, in most places, have to be dealt with on the basis of a fold over of some of the lower beds. The sequence at Craig-a-Knockan, showing the superposition of the Upper Gneiss, seems to be unshaken.
>
> There are workers now in the country, skilled in the interpretation of the older rocks. May it be given to them effectually to pierce the mists of Assynt, and to raise the veil which yet hangs over portions of the North-West! (1882, 399)

Hudleston did not succeed in raising this veil. But he did show others more clearly the nature of the problem that faced them—or what the veil was made of.

Among the "workers now in the country, skilled in the interpretation of the older rocks," was Dr. Charles Callaway (1838–1915), who was in fact with Heddle and Hudleston for part of their field trip to the Assynt region in 1882.[44] Of all the workers who grappled with the Highlands problem before it was essentially solved by Lapworth and the Survey itself, Callaway came by far the closest to achieving a satisfactory understanding. We shall therefore look at his accomplishments in some detail.

Callaway was born in Bristol and received his early education in that district. In his early life, he planned to enter the Church, and he attended Cheshunt College (a theological college) with that end in view. However, he also attended London University, where he studied philosophy, economics, and psychology, eventually turning to geology, in which subject he graduated with first-class honors in 1872. By 1878 he held a D.Sc. in geology, which would have been one of the first degrees of this kind to be held in Britain. During the course of his life, Callaway held a number of museum curatorial posts, namely, at the Bradford Philosophical Society, in the New York State Museum at Albany (under the distinguished Professor James Hall), and at the

44. On Callaway, see Richardson (1915) and Smith Woodward (1916a).

Sheffield Public Museum. Thus Callaway was one of the first men in Britain to have a professional career in geology outside the Survey or the universities. He was fully trained and a man to be reckoned with, though he did not serve on the Council of the Geological Society.

In 1878 Callaway married and settled for a time at Wellington in Shropshire, near the Wrekin, and he worked extensively on the geology of this region. Through careful examination of the Cambrian rocks in the district, and the rocks that underlay them, Callaway played a major role in establishing the presence of pre-Cambrian rocks of Shropshire (Callaway 1879–82; Sarjeant and Harvey 1979). He later turned his attention to Anglesey, and likewise demonstrated the presence of Archaean rocks in that island (Callaway 1884). Thus he was engaged in a research program rather like that of Hicks. Happening to live in a district where pre-Cambrian rocks might be found, he became interested in the topic and spent much time and energy endeavoring to extend the pre-Cambrian domains through more of Britain, in a manner somewhat reminiscent of Murchison's efforts on behalf of the Silurian system. It is not at all surprising, therefore, that Callaway interested himself in the northwest of Scotland, and we first find him in the field there in the summer of 1880. His first Scottish results were presented to the Geological Society on January 5, 1881 (Callaway 1881).

Callaway's initial attention was directed to the limestones of Durness and Inchnadamph. He took fairly small localities and mapped them with some care, even producing sketch maps of small portions of each region—though the one for Inchnadamph was very crude. The special feature of his work was the close examination, and recording on his sketch maps, of the various dips and strikes of the strata, to see whether they were compatible with the Murchisonian doctrine. Callaway's sketch for the Durness region and Far-out Head is shown in figure 7.6.

On the basis of this survey of the terrain, Callaway argued cogently that the Murchisonian doctrine was unsustainable in the Durness region. On Murchison's view, the "flaggy gneiss" (Moine Schist) of Far-out Head, like all the schistose rock, was supposed to overlie the limestone, and both were believed to have a general dip to the southeast.[45] But clearly, from the dips (marked with arrows in the figure) and from the presence of the fault at the southern margin of the "flaggy gneiss," this claim was mistaken. Callaway rejected the idea that the gneiss of Sango Bay could possibly be said to be higher in the stratigraphical sequence than the limestone—for it was impossible to see how it might have been metamorphosed while the limestone escaped metamorphism. (This kind of point had been made by Nicol before, of course.) Rather, suggested Callaway, the limestone might have been deposited on the gneiss, which might then have been brought to the surface by the faulting. Again, this

---

45. Murchison did, of course, allow for local deviations from this rule. It will also be recalled that the schist of the Far-out Head is today regarded as an isolated "western" fragment—a klippe—of the "eastern" metamorphics, or Moine Schists.

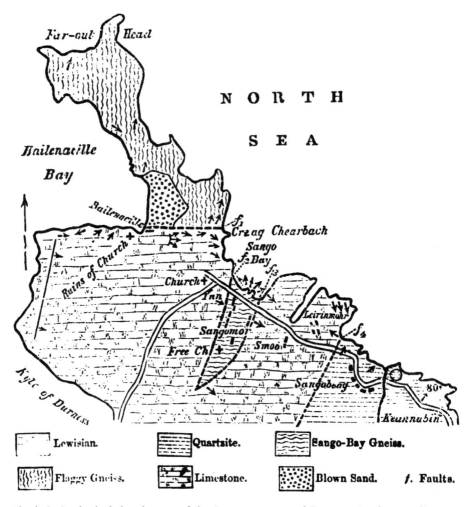

Fig. 7.6. Geological sketch map of the Durness area and Far-out Head, according to Callaway (1881, 240).

was the kind of suggestion that Nicol favored. However, Callaway agreed with Murchison that the gneiss of Sango Bay was quite different in appearance from the characteristic "fundamental" gneiss of the west coast.

Callaway also noted that the quartzite of Sangobeg dipped to the northeast and could be seen to be faulted against the neighboring limestone with evidence of crushing, and the two units had completely different strikes. He did not actually state that the orthodox view was nonsense, but contented himself by saying that the way in which "the limestone can [be said to] hold a conformable relation to the altered groups is a problem which the advocates of the received view may be fairly called upon to solve" (1881, 242).

For the Assynt region, Callaway carefully examined the hill of Cnoc an Droighinn, noting the strata and their dips, and indications of faulting. His

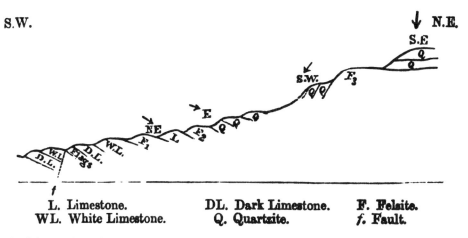

L. Limestone.   DL. Dark Limestone.   F. Felsite.
WL. White Limestone.   Q. Quartzite.   *f.* Fault.

Fig. 7.7. Section of strata at Cnoc an Droighinn, according to Callaway (1881, 242).

resulting section is shown in figure 7.7. It is difficult to assess this section, since one cannot be certain where it was taken. I am puzzled why Callaway called the intercalated igneous rocks "felsites," since these are generally light colored and commonly porphyritic, while modern maps have the igneous rocks of Cnoc an Droighinn as lamprophyres, which are dark in color. Perhaps he was using the term to refer to texture rather than mineral composition. Be that as it may, his "felsite" (F_3) seems to govern the whole structure of the hillside— which is in actuality far from the case. The igneous rocks there are really only of subsidiary importance structurally, though as we have seen, Hudleston not long after was to make good use of them in analyzing the structure of the hill.

But whatever we may think of Callaway's Cnoc an Droighinn section, with one of his statements I would heartily agree: "It is obviously unsafe to base a succession on such a broken structure as this" (ibid., 243). Yet Cnoc an Droighinn was one of Murchison's most favored sections. If it were seriously questioned, his whole scheme would be in jeopardy. The reported discussion following the presentation of the paper indicated no unanimity of opinion. It also revealed that several other geologists who had not yet published papers on the district—or who never did—were interesting themselves in the geology of the northwest. The president of the Geological Society and Survey palaeontologist, Robert Etheridge (1819–1903), had apparently visited the area twice, but he said he "felt difficulty in tracing the succession of the rocks." Likewise, Judd said that "after several visits to the district, he felt great difficulty in offering an opinion as to the succession of beds." Evidently, the clearest thing that might be said was that geologists at last knew that they didn't know the structure of the Northwest Highlands. They were, according to the dictum of Socrates, becoming wiser.

Nevertheless, the Survey thought it appropriate to offer a semi-official rejoinder to Callaway's skepticism. This came in the form of a letter to *Nature* from Hull, expressing his "entire concurrence in the interpretation of the structure of the country given by my late chief [Murchison], whose elaborate and graphic descriptions in the pages of the *Quarterly Journal* of the Geological Society . . . will, I feel sure, never be invalidated" (Hull 1881, 289).[46] However, the evidence Hull offered to controvert Callaway was based on the observations that Hull had made during his journey with Geikie the previous year— and it chiefly referred to Ullapool. It wasn't likely to reassure people about the structure of Durness, where (to the best of my knowledge) Hull had never been![47] It is not the least surprising that the geological community was little satisfied with the Survey's continued support for the Murchisonian doctrine.

In another paper delivered in 1881, on December 21, Callaway (1882) sought to query a point on which there had been consensus for some years, namely, that there was a distinct unconformity between the "Silurian" and the underlying Torridon Sandstone.[48] He claimed that the dips were generally one and the same, that a smooth passage from one unit to the other could be discerned, and that the "same conditions of deposit prevailed through both formations."[49] However, he did not succeed in gaining any agreement for his suggestion, and it is perhaps fortunate that this was so, in that it would probably have generated yet more confusion if it had been taken up seriously. The proposal would not have been made, I believe, if the Highlands had been well mapped at that time, for there are many places where the quartzite is now known to lie both on the western gneiss and the Torridon Sandstone at adjacent points, which is a clear indication of unconformity. But if Callaway used Heddle's map of Sutherland, as doubtless he did, this point would not have shown up clearly. (See plate 7, which does not suggest unconformity at the base of the quartzite.)

Though we may not think that Callaway's paper on the Torridon Sandstone–quartzite boundary brought matters forward to any marked degree, it was another thing altogether with his remarkable paper to the Geological Society on May 9, 1883, in which he detailed his extensive fieldwork of the previous summer in the Northwest Highlands (Callaway 1883b). This was one of the most important documents pertaining to the Highlands controversy, and it must claim our attention for several pages.

Callaway stated that the fieldwork leading to the presentation of his paper

46. The letter was dated January 18. Whether Hull attended the presentation of Callaway's paper is not known.

47. Hull did see something of the rocks at Inchnadamph, and Geikie would undoubtedly have taken him up the hill of Cnoc an Droighinn. But as we have said, it was a very rushed visit, and the main focus of concern at that time was the "western" gneiss.

48. Callaway now called it Ordovician, rather than Silurian. This geological system had been proposed by Charles Lapworth a couple of years earlier for the epoch about which Murchison and Sedgwick had fought bitterly for many years, Murchison regarding the rocks as Lower Silurian while Sedgwick wished to claim them for the upper part of his Cambrian system. See Lapworth (1879) and Secord (1986a). It was to be a few more years before the Survey could bring itself to accept the notion of an Ordovician System, and this did not occur during the period of Geikie's directorate.

49. He referred to what is called "cross-bedding" in both deposits. But in itself this is no guide to the presence or absence of unconformity.

was carried out in the summers of 1881 and 1882. He spent about a fortnight on the job the first year, and two months in 1882. We learn that he met Heddle several times in the field, and he was undoubtedly guided by him on the first excursion. He also met a number of other geologists up in the Highlands during the course of his work. Three were mentioned in the paper: Professor L. C. Miall of the Yorkshire College at Leeds; R. H. Tiddeman, a Survey officer who specialized in work in Yorkshire, and a Mr. Eccles, F.G.S. Callaway reported as well that in 1881 he was accompanied by a friend, G. H. Bailey. We know that quite a few other geologists visited the area at about this time, including Blake, Bonney, Etheridge, Geikie, Hicks, Hudleston, J. W. Hulke (president of the Geological Society in 1883, when Callaway read his paper), Hull, Jolly, Judd, J. E. Marr (a friend of Tiddeman, student of Bonney, and later professor of geology at Cambridge), Symes, J. J. H. Teall, and most important, Charles Lapworth, whose work will be discussed in detail in later chapters. There were doubtless several others who did not leave their mark on the historical records. It can be seen that the Highlands controversy was really "on the boil" at this time. There was a clear recognition that there was something badly awry with the official interpretation, and sterling efforts were made to try to find a solution to the problem of the tangled geological structure. This was eventually accomplished by the largely independent investigations of Callaway, Lapworth, and the Survey's own officers, led by Benjamin Peach and John Horne, who began their work in the northwest in 1883.

The areas that Callaway investigated were Ullapool and its environs, the Assynt region, and the eastern side of Loch Eriboll. His account was so detailed that I cannot give a complete analysis of it here, but some of the salient features were as follows. First, he offered the following revised stratigraphical column:

The Assynt Series
  Dolomite:
    Dark Grey Dolomite
    White Dolomite
  Salterella Grit [Serpulite grit] and Quartzite
  Brown Flags [fucoid beds]

  Quartzite:
    Annelidan Quartzite [pipe rock]
    Seamy Quartzite
  Torridon Sandstone and Ben More Grit[50]

Caledonian
  Hope Series [schistose: approximately the same as Moine Schist]
  Arnaboll Series [gneissose]

50. Callaway used this term to designate exposures of Torridon Sandstone that he found high in the hills of the Ben More Assynt range.

Hebridean
[This included the "western"/Lewisian/Laurentian/Fundamental
Gneiss and Heddle's Logan Rock.]

Needless to say, this classification was the end of Callaway's work, not its beginning. Nevertheless, it may be useful to put it down at the outset, to facilitate exposition. It will be seen that Callaway introduced some useful new nomenclature, and he retained his earlier assertion that the quartzite and the Torridonian belonged essentially to the same system.

Starting at Ullapool, Callaway did not make just two traverses as Geikie had done in his journey of 1860. Callaway took a number of diverging pathways, as if walking along the spokes of one quadrant of a wheel, with Ullapool as its center. The results obtained were quite different from those obtained by Geikie, for the sequence was manifestly different according to the path one followed. So where was Murchison's "clear ascending sequence"?[51]

Even where Callaway followed the same track as Geikie (i.e., along the northern shore of Loch Broom), things didn't appear the same way to the two men. Thus where Geikie observed a gradual transition of conglomerate into gneiss and thereby inferred that the gneiss was somehow a metamorphosed feldspathic grit, Callaway found "not the slightest evidence of a gradation." On the contrary, he saw "slickensiding," indicative of the presence of a fault, with the gneiss "slightly overhanging the conglomerate" (Callaway 1883b, 364). Then walking over the hill east-northeast from Ullapool—a route followed neither by Murchison nor by Geikie—Callaway encountered gneiss faulted over crumpled-up limestone, with "Caledonian" (Moine Schist) faulted over the gneiss. And in that area, he found the ground to be "literally a pavement of fragments." Considering all his Ullapool sections, it appeared to Callaway that "the zone between the Quartzite and the Caledonian [Moine Schist] is sliced by several sub-parallel faults."

Moving on to Assynt, we find Callaway's description to be so specific that it was almost as if he were giving a map "in writing." I cannot provide all the details here, but some of the major features should be noticed. Considering the famous Cnoc an Droighinn section, Callaway did not simply walk up the hillside. Rather, he examined the sections in the streams to the north and south, the former of which displayed to Callaway's eye a section such as is shown in figure 7.8.[52] This kind of observation encouraged Callaway to think in terms of *lateral* forces.

51. It may be recalled that Murchison relied on Geikie for the account of the Ullapool region in their joint paper of 1861, and that Geikie was only in the Ullapool district for a very brief time.
52. If the reader cares to examine the map of this area, later produced by Peach and Horne (see plate 8), it will be seen that the stream where this section was taken is marked as "Chalda Mòr." (It runs up from C[h]alda House by Loch Assynt.) The dips shown by Callaway in his sketch section are much the same as those marked by the Survey officers. But they have located a thrust fault in the position where Callaway had the quartzite folding over. The red bands in the Survey map mark lamprophyre intrusions that were not shown in Callaway's section. Also, the Survey officers distinguished two kinds of quartzite here (a_2 and a_1 in plate 8).

W.   E.

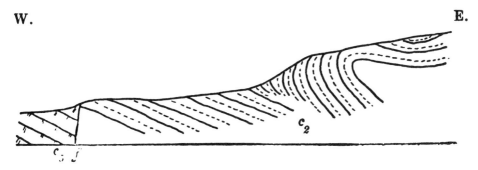

Fig. 7.8. Section of strata along the line of Allt a Chalda Mor, just to the north of Cnoc an Droighinn, according to Callaway (1883, 368). c_2 = Quartzite; c_5 = Dolomite; f = fault.

It will be seen from the stratigraphical sequence proposed by Callaway, detailed above, that he was beginning to recognize subdivisions in both the quartzite and the limestone (dolomite). This process of subdivision was carried much further in the work of Lapworth and the Survey officers Peach and Horne, as we shall see. But even with the fairly simple subdivisions suggested by Callaway, much progress could be made. For having established a general rule for the initial order of deposition (prior to disturbance by earth movements), it was then possible to recognize whether any given sequence that one might encounter was the right way up or inverted. Logically, of course, this procedure did not in itself yield certainty since there could be doubt as to which of the two possible sequences, one of which is inverted, represents the original order of superposition. One had to take the balance of evidence, starting with what had the best appearance of being the bottom of the sequence. In the given case, one would look for the kind of quartzite that seemed to be in most frequent contact with the Torridon Sandstone. Working in this manner, Callaway achieved considerable success in establishing which rocks were the normal way up and which were inverted; and then he could begin to envisage what the overall structure was like. Reasoning in such a fashion, and examining the exposures in detail, Callaway came to the firm conclusion that Murchison's notorious "upper quartzite" was a chimera. Likewise, there was no "upper limestone."

A good idea of the general kind of structure that Callaway was envisaging for the Assynt region is revealed in his section across the mountains of Brebag (Breabag) and Coniveall—two of the main peaks of the Ben More range. This section is shown in figure 7.9. It will be seen that the structure offered conveys in some measure the notion of a force exerted from an easterly direction. However, Callaway did not invoke a low-angle fault, and the gap in the structure at the middle of the diagram is itself an indication of some inadequacy in his theory and understanding of how the structure might have been brought into being.

Even so, the figure is roughly reconcilable with the modern maps, so far as

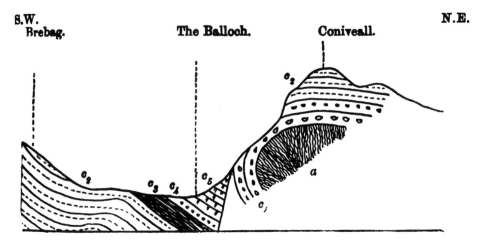

Fig. 7.9. Section of strata forming the mountains of Breabag and Coniveal and
the intervening saddle (Ben More Assynt range), according to Callaway (1883, 383).
a = Archaean Gneiss; c_1 = Torridon Sandstone and Ben More Grit; c_2 = Quartzite;
c_3 = Fucoid beds; c_4 = Salterella grit; c_5 = Dolomite.

the outcrops and dips are concerned. Later investigators, however, have envisaged several thrust planes along the particular line of section of figure 7.9.
Notably, it is thought that there is a thrust fault below the Torridon Sandstone. One might therefore describe Callaway's section as a kind of "first approximation."

As for Heddle's Logan Rock, Callaway agreed on mineralogical and petrographical grounds with Bonney's diagnosis that it was a version of the Hebridean (Archaean) Gneiss, with a generally similar direction of strike.[53] It was
not, Callaway decided, conformable with rocks either above or below it. Bands
of Logan Rock were not intercalated into the quartzite, as Heddle had supposed. The contacts were marks of "dislocation": they were faults. Where
repetitions of the sequence occurred, this was due to repeated faulting. The
junctions always displayed signs of alteration and crushing, and often of
weathering as a result of water getting into the fractures.

As for the section at Knockan Cliff, which had seemed to Murchison and
Geikie to provide such good evidence for a smooth upward sequence from
limestone into eastern metamorphic rock (Moine Schist), Callaway rightly contended that to judge the matter satisfactorily one needed to examine a section
that cut across the dip slope, rather than one running parallel with the strike.
For this purpose, therefore, he recommended the sections provided by Loch
Glencoul and its feeder rivers, to the north of Assynt.

This locality, which is today recognized as one of the best places to view the
so-called Glencoul Thrust plane, had hitherto not been the subject of very

53. In fact, Bonney was now acting as Callaway's petrographer and he supplied petrological information in an appendix to the paper.

Fig. 7.10. "Pipe Rock," collected from below Stack of Glencoul, showing distortions of the "pipes" due to lateral pressure.

detailed description, perhaps because it is somewhat difficult of access unless one travels by boat. There are two valleys at the head of Loch Glencoul, and between them there is a bluff somewhat like the prow of a ship, known as the Stack of Glencoul. Glen Coul itself runs southeast from the (sea) loch, on the northern side of the Stack, being the northern of the two valleys. It provides a fine section of the strata.

At first, it seemed to Callaway that things were as the Murchisonian theory would require, but close examination revealed that the quartzites below the "prow" (which is formed of Callaway's "Caledonian," or Moine Schist) were folded back on themselves. Moreover, the annelid tubes ("pipes") in the quartzite showed clear indications of distortion, suggesting that the rock had been subjected to lateral forces. (An example of this is shown in figure 7.10, which is a specimen that I have myself gathered from just below the overlying "prow" of schist.) Moreover, careful examination of the dips of the quartzite (35 degrees) and the overlying schist (15 degrees) revealed to Callaway that they were not conformable with one another. The sum of the evidence seemed to be that the schist had been forced over the quartzite by lateral forces. In the valley of Glen Coul, the schist could also be found overlying the Hebridean Gneiss (Logan Rock). Thus Knockan Cliff was recognized for what it was: an illusory section, which gave the appearance of a conformable upward sequence because one was merely examining the exposures along the strike, rather than a section across the dip slope. For a proper picture, one should

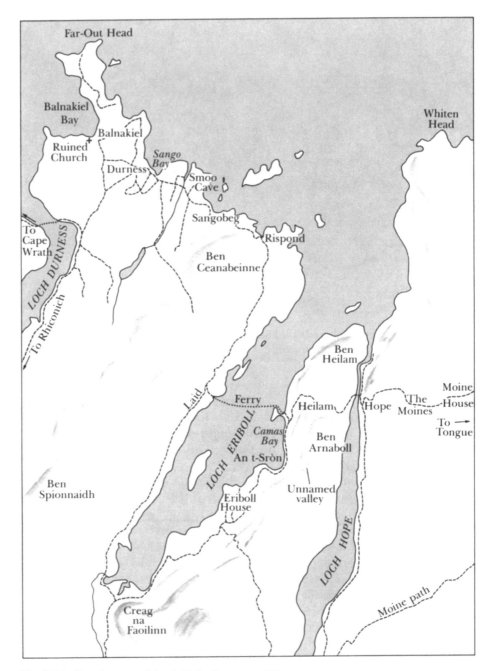

Fig. 7.11. Sketch map of Loch Eriboll area and Durness.

have recourse to the much more extensive evidence furnished, for example, by Glen Coul.

Finally, Callaway took his readers round the sections on the eastern shores of Loch Eriboll, where the evidence was probably the most convincing of all. Indeed, he claimed, "nothing but the hastiness of most previous workers [could] account for their failure to perceive the true interpretation" (Callaway 1883b, 396). Only Nicol could be said to emerge with any credit; and Callaway was glad to have the opportunity, "however humbly, to vindicate his reputation."

Some of the salient features of the topography of the Eriboll region are shown in figure 7.11. Again one of the most important facts to be established was whether there was or was not an "upper quartzite." Callaway settled the matter to his satisfaction by walking round the hills of Druim na Teanga (which he called Druim an Tenigh and which may be translated as "tongue ridge"). He found that, starting at the northeast of this hill and moving southward and westward, he could—walking all the time on their outcrop—trace quartzite beds in such a way that they revealed a kind of S-shaped structure, and thus presented clear evidence of overthrow.

Another section that proved to be of great importance, and was also utilized by Lapworth, was that along the southern side of Camas Bay (Camas an Dùin, or "enclosed bay"). It has now become a classic geological site, like those at Knockan Cliff and Cnoc an Droighinn, and is usually known as the An t-Sròn section. It was so named by Lapworth, after the headland at the south of Camas Bay. Callaway's interpretation of the section is shown in figure 7.12.

It is not easy to reconcile this section precisely with modern maps, since the strike of the beds is slewed round in the southwest corner of Camas Bay, and Callaway's section does not bring this point out clearly. However, there can be no doubt that he construed the major features of the locality satisfactorily, notably the S-shaped folding of the quartzite at the eastern end of his section, and the associated crushing. We may notice also that the idea of a low-angle thrust fault was beginning to be included in the diagram, arising, it would seem, from the same lateral forces from the east that folded the quartzite into an S-shape. However, Callaway did not go the whole hog on this, for we see that he proposed a high-angle thrust fault between the Caledonian Gneiss and the folded quartzite onto which it abutted. This point is of significance, for it suggests that Callaway arrived at his section on the basis of the field evidence before him, not on the basis of some higher-level theory that he might have encountered in foreign literature.[54] Indeed Callaway concluded his paper by stating that he "had invented no new method of investigation."

At Camas Bay, Callaway apparently arrived at his notion of an S-shaped folding of the quartzite on the basis of examining dips and strikes, and as many sections as he could find, and then piecing together a composite section,

---

54. The case may have been different for Lapworth, who several times mentioned his familiarity with the texts of authors such as Heim, Brøgger, and Rogers.

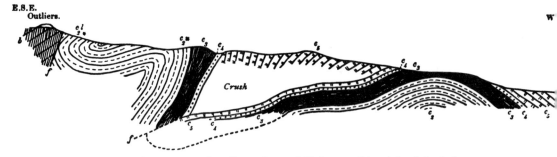

* There are several small overthrown folds between this point and the fault.

Fig. 7.12. Section of strata at Camas Bay, according to Callaway (1883, 399).
c_5 = Dolomite; c_4 = Salterella grit; c_3 = Fucoid beds; c_{2u} = Pipe-rock
quartzite; c_{2l} = Seamy quartzite; b = "Caledonian" Gneiss; f = fault.

which was partly imbued with theory. However, at the southern end of Loch
Eriboll, he examined the hill called Creag na Faolinn, and there he could see
an S-shaped structure directly exposed to his view. This is so striking that anyone examining it might have been disposed to think in terms of overfolds for
the rocks of the Eriboll district. As I suggested in chapter 3, I suspect that
Murchison missed this interesting structure as a result of taking the ferry
across Loch Eriboll to Heilam, instead of going by the longer route round the
southern shore.

As for Murchison's favorite section in front of and behind Eriboll House,
Callaway showed that the sequence had been misconstrued. There *appeared* to
be an upper limestone and an upper quartzite because the strata were inverted there. A wedge of dolomite was faulted in between the quartzite and
the gneiss, and the quartzite was all folded back on itself, as in other places in
the vicinity.

How did the gneiss arrive on top of the strata of the Assynt Series? Callaway plausibly suggested that the same force that doubled the Assynt Series
back on itself forced the gneiss over the younger rocks. Thus in large measure
the mystery of the Northwest Highlands was solved in its essentials by Callaway—a point that is often forgotten, for most of the credit is usually allotted
to Lapworth as an independent investigator, and to Peach and Horne, leading
the Survey team that eventually mapped the region and unraveled the structure in its entirety. For all of these men, it was only possible to sort matters out
when the stratigraphical sequences were deciphered. I have indicated in an
approximate fashion how Callaway would have done this. But it was done in a
much more thoroughgoing manner by Lapworth and by Peach and Horne; so
a detailed account of the mode of reasoning deployed will be deferred to the
next two chapters.

There are first some points that should attract our attention where Callaway differed from the other workers. In particular, he did not agree with
Bonney that the gneiss found on the eastern side of Loch Eriboll was one and

the same as the "western" Hebridean Gneiss. (This was the rock that Nicol had called a "granulite," which was Eriboll's version of the "igneous rock": he thought it was always to be found at the locality of the major Northwest Highlands faulting. It was also what Heddle and Hudleston called Logan Rock.) Callaway acknowledged that the Arnaboll Gneiss, as he called it, was almost identical in both composition and strike to the pocket of gneiss faulted in at Sango Bay, between Durness and Eriboll. But it seemed grossly different from the gneiss of the far west coast, as indeed Murchison had long claimed. And just as Nicol had been inclined to conflate gneiss and schist, so too did Callaway. That is, he regarded the Arnaboll Gneiss and the overlying Hope Schist as belonging to the same system (the "Caledonian"), and not really distinct from one another. By contrast, Lapworth, Peach and Horne, and modern maps depict these as two distinct units—the Arnaboll Gneiss and the Moine Schist. It is also supposed that there is a thrust plane between them—as well as a separating band of mylonite in the Eriboll locality, extending up to Whiten Head.

Finally, then, Callaway offered his interpretation of the history of the Northwest Highlands. It went as follows:

1. Caledonian rocks were laid down on the basement of the Hebridean rocks.
2. The Caledonians were faulted down to the east, along a fault line running south to southwest.
3. The rocks of the Assynt Series were deposited on the Hebridean, but thinning out to the east, they only partly overlay the fault between the Hebridean and the Caledonian rocks.
4. The direction of the faulting reversed, and the Caledonian rose up, driving its cover of Assynt Series upward, folding the rocks, and thrusting them in a westerly direction. At the same time, some Hebridean was supposedly brought up so as to be forced over Assynt Series rocks at Durness.
5. The Assynt rocks east of the great fault were denuded so as to leave only patches here and there.

This historical reconstruction was not without its difficulties. In particular, it seems to me that by including the Torridon Sandstone in the Assynt Series, considerable problems were raised, since this sandstone exists in massive quantities on the western margins of Scotland and then somehow rapidly disappears to the east. Callaway's theory handled the sudden disappearance of the quartzite/fucoid beds/Serpulite grit/limestone reasonably well, but it was hardly able to deal with the disappearance of the Torridon Sandstone in the same way. There was also the problem of where the Arnaboll Gneiss actually belonged in the succession. Lapworth, who was present at the reading of Callaway's paper, stated that he believed that it *did* belong to the Hebridean.

However, most members of Callaway's audience seemed to be well satisfied with his performance. Hicks "expected to hear some one rise to defend the views maintained by the Geological Survey" (Callaway 1883b, 422). But apparently no one was forthcoming. Or if anyone was, the fact was not recorded for the benefit of posterity. It would appear that the Survey had finally realized that it could maintain its dogmatic attitude no longer, and that new work would have to be done to polish up its tarnished reputation. We shall investigate its activities in chapter 9. Meanwhile, we may note that Lapworth listened to Callaway's paper with "great pleasure," since he was unfamiliar with the Assynt region. So far as the Eriboll district was concerned, he found that his views largely coincided with those of Callaway. We are told that Lapworth brought with him to the meeting some of his own sections and a map that he had prepared of the Eriboll region. What these may have contained, and how Lapworth's results were obtained, I shall endeavor to recount in the next chapter.

8

Charles Lapworth: Digressions and Diversions to the Southern Uplands and the Alps

So far in my narrative I have been able to treat the geological controversies concerning the rocks of the Northwest Highlands as a matter that was to a large degree independent of developments elsewhere. Also, the debates were, to perhaps a surprising degree, little involved with major problems of high-level theory. To be sure, they related to issues raised by theorists such as Sterry Hunt, to Murchison's territorial ambitions for his Silurian system, or to the causes of metamorphism. But by and large the problems were "standard" for stratigraphical research in the second half of the nineteenth century—only more difficult than British geologists were accustomed to handling. In this chapter, we must introduce accounts of certain new techniques and theoretical principles that were worked out elsewhere—particularly in the Southern Uplands of Scotland and in the European Alps. In both the Southern Uplands and the Northwest Highlands it was the "amateur" geologist Charles Lapworth (1842–1920) who was largely responsible for the events that occurred. His photographic portrait, reproduced from the *Geological Magazine* for 1901, is shown in figure 8.1.[1]

Lapworth was born in Berkshire and reared in Oxfordshire where his father worked a farm. He trained as a teacher and obtained a post at Galashiels, situated in the Gala Valley that runs north-south through the middle of the Southern Uplands of Scotland, about thirty miles south of Edinburgh. There he lived and worked for eleven years until 1875, when he took up a position at a school (Madras College) in St. Andrews. Lapworth became keenly interested in geology as a result of finding fossils in the strata near Galashiels, and from 1869 he began an intensive study of the geology of the Southern Uplands. This led to discoveries and theoretical results of fundamental importance, and he also devised and deployed methods for stratigraphical work that were unequaled in his time. He attended scientific meetings, was elected a Fellow of the Geological Society in 1872, and achieved such renown in his work that he obtained a chair of geology and mineralogy at Mason's College Birmingham in 1881.[2] Lapworth served in that institution with great distinction until his

1. On Lapworth, see Geological Magazine, Eminent Living Geologists series (1901); Watts (1921a, 1921b, 1939); Watts and Teall (1921); Boulton et al. (1951); Bailey (1952, 90–94).
2. This college was the forerunner of the University of Birmingham, which was established in 1900.

Fig. 8.1. Portrait of Charles
Lapworth (*Geological
magazine* 1901).

retirement in 1913, helping to build his department into one of the leading
centers for geological teaching and research in Britain.[3]

The fieldwork that Lapworth carried out in the Northwest Highlands was
performed at Durness and Eriboll in the years 1882 and 1883. It was his first
major project after his election to the chair at Birmingham. In what follows,
I want to argue that Lapworth's Highland researches had a good deal in
common with the work that he carried out in the Southern Uplands so far as
technique and theory were concerned. In addition, the Southern Uplands
were part of the territory that Murchison had for long claimed for his Silurian
kingdom. There is therefore good reason to look at this Southern Uplands
work before we examine what Lapworth achieved in the Highlands. We shall
also examine briefly some of the ideas on mountain building that were being
developed at that time in Europe and America, for Lapworth is known to
have been familiar with this work and was the first to make important use of it
in British stratigraphy.

3. Lapworth's papers are held in the geology department of the University of Birmingham. They
have been examined by Dr. Beryl Hamilton, who has kindly made a copy of her catalog of them
available to me. I understand that she is currently writing a biography of Lapworth, a work that is
much to be desired.

The hills of the Southern Uplands stretch across the south of Scotland from one coast to the other. They consist chiefly of unfossiliferous grey-colored grits (coarse sandstones), commonly called "greywackes" after the German term *Grauwacken*. In the early years of the nineteenth century, these Scottish rocks had been classified by Robert Jameson (1805) as belonging to Werner's "Transition Series." It was, of course, Murchison's lasting contribution to geology to begin the elucidation of the stratigraphical sequences for rocks of "Transition" type in Wales and the Welsh Border region.

Though most of the rocks of the Southern Uplands are unfossiliferous, there are occasional dark shale bands within the greywackes containing numerous curious fossils called "graptolites." But these were not well known at the time that Lapworth began his investigations. A typical set of graptolites of various kinds is shown in figure 8.2. Graptolites, which have long been extinct, are thought to have been colonial animals, with one organism living in each "pocket" (theca) on the "stem" (common canal) of the colony. They probably lived floating on the surface of the sea, or attached to seaweed. A typical graptolite might be an inch or so in length.

It will be recalled that Murchison published his major claim for the existence of the Silurian system, with its own characteristic fossil assemblages, in 1839 (Murchison 1839). It will be further recalled that it was James Nicol, in 1848, who suggested that the rocks of the Southern Uplands might be correlated with the Welsh Silurians (Nicol 1848, 207–8). The following year, Nicol suggested a more precise correlation, namely with the Llandeilo Series of Wales, which Murchison had in his Lower Silurian (Nicol 1850).[4] Nicol suggested further a broad anticlinal structure for the Southern Uplands. During the 1850 season, he and Murchison worked together in these rocks, accompanied at times by Harkness; and Murchison published his results the following year (Murchison 1851). The repetitions of the dark fossiliferous shale bands at various places parallel to the strike were suggested by Murchison to be due to a complex set of folds, whereas earlier Harkness (1851) had ascribed the repetitions to faulting. Harkness (1856, 245) later deferred to Murchison's opinion.[5]

By 1863, the officers of the Survey were beginning their work in the Southern Uplands, and it was not long before beds that appeared to belong to the upper part of the Silurian system were revealed. In a review of progress to 1867, Geikie (1868)—soon after he had assumed the directorship of the Scottish Survey—summarized the supposed succession among the older rocks of the Southern Uplands into eight units, ranging upward from equivalents of the Welsh Llandeilo Series, through Bala or Caradoc, to Lower Llandovery, all at that time being regarded by the Survey as belonging to the Lower Silurian. But, as Geikie acknowledged, the structure could not be made out

4. Nicol's paper was read in May 1849.
5. Harkness offered no particular reason why he changed his mind. The text suggests that he simply deferred to Murchison.

Fig. 8.2. Figures of graptolites, reproduced from Pringle (1948, 13). Ordovician, Arenig series: (A) *Didymograptus extensus* (Hall); Glenkiln Shales: (B) *Nemagraptus gracilis* (Hall); Hartfell Shales: (C) *Diplograptus* (*Orthograptus*) *truncatus* (Lapworth), (D) *Pleurograptus linearis* (Carruthers), (E) *Dicellograptus complanatus* (Lapworth), (F) *Dicellograptus anceps* (Nicholson). Silurian, Lower Birkhill shales: (G) *Akidograptus acuminatus* (Nicholson), (H) *Mesograptus modestus* (Lapworth), (J) *Monograptus cyphus* (Lapworth), (K) *M. fimbriatus* (Nicholson); Upper Birkhill Shales: (L) *Cephalograptus cometa* (Geinitz), (M) *M. sedgwicki* (Portlock), (N) *Rastrites maximus* (Carruthers), (O) *M. turriculatus* (Barrande), (P) *M. crispus* (Lapworth), (Q) *M. griestoniensis* (Nicol); Wenlock series: (R) *Cyrtograptus murchisoni* (Carruthers). By permission of the Director, British Geological Survey: British Crown copyright reserved.

satisfactorily, and the proposed subdivisions were "still confessedly vague and generalised."

By 1871 Geikie thought that he was beginning to understand the sequence and structure properly, as shown by the explanation that was published that year for Sheet 15 of the Survey's maps of Scotland, covering the area in the central parts of the Southern Uplands in the counties of Dumfriess, Lanark, and Ayr (Geikie 1871b). The interpretation offered apparently rested chiefly on the work of one of the younger Survey officers, Robert Logan Jack (1845–1921).[6] Jack mapped an area that contained some black shales swarming with graptolites, which were allotted to Upper Llandeilo age; and in a neighboring locality he found some coarse grits and conglomerates containing shells that seemed to belong to the Caradoc age (recognized in Wales to be the unit overlying the Llandeilo beds). Furthermore, it seemed to Jack that the area that he was mapping was synclinal in structure, and on this basis it appeared that the succession could be understood, and a general sequence for the Scottish Llandeilo and Caradoc beds could be established.[7]

A couple of years later, the explanation to Sheet 3, for the western extremity of the Southern Uplands, was published (Geikie 1873), and an attempt was made to apply the sequence established for Sheet 15 to the area of Sheet 3. But now it appeared that the structure was not as might have been expected. The Caradoc beds did not always lie on the same unit or subdivision of the Llandeilo series. Hence Geikie concluded that there had to be a gap in the sequence of the deposition—an unconformity at the base of the Caradoc, such as had not been recognized elsewhere in Britain. Thus, instead of throwing doubt on the sequence and structure that had been proposed by Jack for the area of Sheet 15, and the extrapolation of this to other areas of the Southern Uplands, Geikie announced with some apparent satisfaction that the

6. My account here of what went on in the Southern Uplands Survey work relies heavily on an uncataloged document in the Lapworth Archive, University of Brimingham. This undated manuscript, in Lapworth's hand, bears the title "Recent Discoveries among the Lower Silurian Rocks of the South of Scotland." There is also a pencilled addition to the title page, not in Lapworth's hand, saying "?Address to the Geological Society of London, The Moffat Series." But this appears to be incorrect, for the manuscript can be correlated with the published version of a paper that Lapworth read to the Geological Society of Glasgow in 1878 (Lapworth 1882a). The manuscript is of great value, for it displays the Survey's thinking on the question of the Southern Uplands much more clearly than one can find it revealed in the official publications. In particular, the role of Jack as the person responsible for the Geikie's ideas about the structure of the Southern Uplands is shown here in a way that is not found in any other document that I have consulted. It may be mentioned that *if* Lapworth's manuscript was read to the Geological Society of Glasgow in the form that is preserved in the Birmingham archive, we can well understand why his relations with Geikie were so bad: the Survey's thinking is represented as being thoroughly muddled, with Geikie relying on the mistaken ideas of one of his subordinates.

For an obituary of Jack, see Oldham (1922). Jack was appointed to the post of goverment geologist of Queensland, Australia, in 1877. He also spent some time in China. For his work in Australia and a photographic portrait, see Cribb (1976). Also, see figure 6.11.

7. This was (in descending order): Caradoc; Black Shale Group, Lowther Group, Haggis Rock Group, Dalveen Group, Daer Group, Hartfell Shale Group, and Queensberry Grit Group (all Llandeilo).

supposed unconformity marked "a new feature in the geology of Britain" (ibid., 5). He also "suspected" an unconformity between the Upper and Lower Silurian rocks of the Southern Uplands. The trouble was, however, that the whole argument rested upon an assumption that the synclinal structure of Jack's area of Sheet 15 (the Leadhills district of the north-central region of the Southern Uplands) was properly understood. Yet that area was greatly disturbed by often unsuspected faults, overturns, and repetitions of strata due to faulting or folding; and as events later showed, the classificatory scheme there established was unreliable. Thus the foundation of the argument for the analysis of the structure of the Southern Uplands as a whole was insecure, though at the time it appeared to Geikie that the Survey was making considerable progress with its researches in southern Scotland. The Survey's interpretation was publicly applauded in a review by one of its supporters, Robert Harkness, in the pages of *Nature* (Harkness 1873).[8]

It is believed that Lapworth began his investigations in the Southern Uplands almost as soon as he started teaching at Galashiels in 1864 (Boulton et al. 1951, 436). His serious systematic work began in 1866, and he was working in top gear by 1869, so that by 1870 he was issuing the first results of his investigations (Lapworth 1870). During the next twelve years, after which he turned his attention to the rocks of the English Midlands and the Northwest Highlands of Scotland, he published twenty-four further papers on graptolites and on the stratigraphy of the Southern Uplands (Watts 1921b, 49–50), revolutionizing the understanding of the structure of the region, even though he was but a local schoolmaster by profession, without formal training in geology. In these several publications, Lapworth offered a number of reviews of the literature on the topic, but he did not state in detail exactly what it was that he had against the Survey's interpretations.[9] Fortunately, however, the manuscript in the Lapworth archives at the University of Birmingham entitled "Recent Discoveries among the Lower Silurian Rocks of the South of Scotland" sets out his objections in detail, from which we are able to see the difficulties that the Survey's doctrines had to face. We need not suppose that the arguments put forward in this lecture were all present in Lapworth's mind when he started his work. Nevertheless, they are worth our examination in more detail, in order to show why he thought it necessary to revise the Survey's work. We may then examine what it was that Lapworth actually did, so that we may understand the techniques he developed and subsequently applied in the Northwest Highlands.

8. Harkness's puff for the Survey probably attracted some private criticism, for in a letter published a week after the review Geikie made the point that he was not personally responsible for the new results: the fieldwork had been done by Jack, Horne, Irvine, Skae, and Peach. Somewhat unctuously, perhaps, Geikie thanked Harkness publicly for "his most valuable and welcome papers," and he expressed his "gratification that the labours of the Survey [had] found so courteous an exponent, . . . whose knowledge of the country [was] so minute and extensive."

9. Much later, in a review of the Survey's memoir on the Southern Uplands, Lapworth (1899) did reveal some of his objections.

The Survey had adopted (Geikie 1863a, 8) the broad structure for the Southern Uplands originally proposed by Nicol, namely, that the hills were anticlinal, so that one moved from older to younger beds as one walked from the central axis to the margins, either to the north or to the south. If the sequence were to parallel that established in Wales, it should, according to the Murchisonian interpretation favored by the Survey, have appeared as follows, each unit (except the Longmyndian—which is today regarded as pre-Cambrian) being defined by its characteristic fossils:

Ludlow	Upper Silurian
Wenlock	Upper Silurian
Llandovery	Upper Silurian
Bala/Caradoc	Lower Silurian
Llandeilo	Lower Silurian
Arenig	Lower Silurian
Tremadoc	Lower Silurian
Lingula Flags	Lower Silurian
Longmynd	Cambrian

In fact it appeared as if the axis of the Uplands might consist of the ancient Longmyndian rocks, and perhaps parts of the lower levels of the Lower Silurian. The various black shale beds within the greywackes, with their rich fauna of graptolites, seemed to belong to the Llandeilo unit; and to the south of Girvan (on the west coast) the more sandy and shell-containing beds to be found there were thought to betoken Caradoc beds, with upper units also appearing, up as far as the Wenlock unit. As we have seen, there appeared to be an unconformity below the Caradoc beds, not found in Wales. Despite the discrepancy, this was the broad theory that the Survey espoused—even though the officers well knew that the whole region was folded and faulted in a highly complex manner, so that it was everywhere very difficult to be sure whether one was ascending or descending the stratigraphical sequence at any given point.[10]

As mentioned, the rocks of the Southern Uplands are for the most part unfossiliferous, with only occasional highly fossiliferous bands—chiefly containing graptolites—in the black shales. These graptolites, however, had not been adequately examined and characterized for the different units of the "Silurian" in Wales—and certainly not in Scotland. Indeed, following an argument of the French-Bohemian geologist Joachim Barrande (1799–1883), the graptolites were not thought to be reliable stratigraphical indicators,

10. The Survey made no secret of the difficulties faced in the interpretation of the stratigraphy of the Southern Uplands. In 1863, Geikie had written "It is very difficult to decide in any particular district whether one set of beds is actually higher or lower than another. At the best the decision in most cases rests only upon a combination of probabilities" (Geikie 1863a, 8). But as I say, after the work of Jack, it seemed to Geikie that his officers were gaining an understanding of the stratigraphy.

compared with well-known fossils such as trilobites and brachiopods.[11] The graptolites were nevertheless used as a very rough rule of thumb: where they were numerous in dark shales, these bands were equivalent to the Llandeilo of Wales. But this was much too crude a generalization. So the belief, based on this mistaken rule, that the shell-bearing beds of Girvan and elsewhere were in fact above the black shale bands, and were therefore Caradocian in age, was not well founded.

A further problem had to do with the thicknesses of the rocks. If one added up the *apparent* thicknesses of the greywacke beds associated with the

11. Some remarks on the theories of Barrande should be interpolated here, since they formed a significant "background noise" to the disputes that were conducted concerning the structure of the Southern Uplands. Barrande was a palaeontologist and stratigrapher who made a major contribution to geology by his studies of Palaeozoic fossils in Bohemia, especially trilobites (Barrande 1852–1911). But in some places in Bohemia, he found that assemblages of younger fossils were apparently intercalated among older rocks. (For example, in an apparently synclinal structure the fossils of the young rocks at the axis of the syncline might also be found in isolated localities toward the margins of the syncline.) Barrande's explanation (1859–60) was that migrations of organisms had occurred from one region to another in advance of the arrival of the "main party," as it were. He called these precursory forms "colonies." Such a hypothesis in effect allowed the stratigraphical principle of William Smith to be disregarded, where this seemed to be required by the empirical evidence, for the identification of strata of a particular age with fossils of a particular kind was not insisted on. Barrande's ideas on colonies were hotly contested in Europe, particularly by Professor J. Krejci of Prague, who realized that an alternative and perhaps preferable explanation was that the Bohemian strata had been dislocated by earth movements, so that the confused stratigraphical sequence was due to a disturbed physical structure rather than the abnormal "behavior" of the organisms sending out "colonizing" precursory forms. Barrande's theory was found increasingly attractive by Murchison, who (I suggest) was always keen on any kind of military or colonial metaphor in geological doctrine. Thus we find that while there was no mention of Barrande's colonies in the first edition of *Siluria* (1854), the theory was mentioned with respect though not with entire acceptance in the second edition (1859, 401), and it was given considerable prominence and was largely accepted in the third edition (1867, 380). Murchison's acceptance of the theory did not sit well with his usual insistence, as in his debates with De la Beche and Sedgwick, that fossils should be the arbiters of stratigraphical identifications and correlations. Yet it was convenient to deploy Barrande's ideas whenever stratigraphical correlation was in difficulty because of structural complexities. Further, Murchison was glad to have his Silurian empire fully established in Bohemia as a result of the magnificent palaeontological work of Barrande. Reciprocally, Barrande was glad to have his ideas concerning colonies accepted by the king of Siluria. So the debate about colonies came to form a kind of background assumption to the work of the Survey in the Southern Uplands. Wherever the fossil sequence seemed confused, one might invoke a "colonial" invasion as a kind of ad hoc hypothesis and the fieldworker would be spared the laborious task of working out the sequence and structure rigorously. It was this carefree attitude that, I believe, earned for the Survey Lapworth's contempt in the Southern Uplands. Geikie still spoke favorably of the doctrine of colonies in his *Text-book of Geology* (Geikie 1882a, 627–30; 1885b, 618–21), with specific reference to the strata of the Southern Uplands. He supposed that the graptolites flourished when the conditions were such as to lead to the deposition of the black shale deposits. They departed when the conditions became unfavorable—that is, when the greywackes were being deposited—only to return when conditions became favorable once again. Looked at in this way, there was nothing inherently implausible about the hypothesis. It was just that it was too facile and conflicted with the results of Lapworth's very close examination of the rocks and fossils. It should be noted that the Barrande hypothesis of colonies was not specifically invoked by the Survey in its memoirs accompanying the maps of the Southern Uplands. On Barrande, see further Horny (1980).

black shale bands, there appeared to be some 20,000 feet of them. This made the Llandeilo of the south of Scotland about as thick as the rest of the Silurian system elsewhere in the world. Worse still, the graptolites of Scotland, from the point of view of a steadily ascending sequence in an anticlinal structure, did not mesh at all with the sequences elsewhere, insofar as they were understood. Then there was the problem of the unconformity in the sequence suggested by Geikie. How and why did it come about? It was extraordinary that the greywackes above the supposed unconformity looked much the same as those below, although the production of the unconformity would have involved considerable disturbances and considerable changes in the conditions of deposition. Elsewhere in the stratigraphical column, unconformities were usually marked by substantial differences in lithology—as, for example, between the Old Red Sandstone and the underlying Silurian greywackes of Scotland.

These were some of the reasons that led Lapworth to question the stratigraphical analysis for the Southern Uplands proposed by the Survey. But what had he to offer in its place, and how did he achieve results that were so dramatically different from those suggested by the Survey? Lapworth's "alternative" stratigraphy emerged gradually during the 1870s, and in his published work he never gave a detailed account of the methods that he employed. However, there is a manuscript in the archives of the geology department at the University of Birmingham, of two lectures given by Lapworth at Edinburgh, which tells us what we need to know on this head— though it is possible that they present a somewhat idealized version of what Lapworth actually did.[12]

Lapworth needed to locate a spot where he could find an unambiguous stratigraphical sequence where the correct order of superposition might be established, with fossils identifying each unit satisfactorily. We do not know how much time and trouble it took him to locate such a place, but it was probably not long before he lighted upon the spot called Dobb's Linn, already known to local geologists, and which has subsequently become one of the classic sites in British geology.[13] Lapworth's work there was chiefly carried out between 1872 and 1877, though he had already made considerable progress toward unraveling the structure of the Southern Uplands before he began serious study of the strata at Dobb's Linn.[14]

The road between the towns of Selkirk and Moffat in the Southern Uplands runs for some distance approximately parallel with the strike of the greywackes. It crosses a pass at the county boundary between the shires of Selkirk and Dumfries, at a small, lonely shepherd's house called Birkhill Cottage, and it was here that Lapworth stayed during his vacations during the period of his

12. C. Lapworth, "The Silurian Age in Scotland," undated manuscript in Lapworth archive at the Department of Geological Sciences, University of Birmingham (uncataloged in 1987): typed transcript prepared by Beryl Hamilton.
13. See, for example, Strachan and Toghill (1975).
14. See, particularly, Lapworth and Wilson (1871, 1872). (Wilson was a Galashiels journalist with whom Lapworth made friends and with whom he conducted his fieldwork.)

studies at Dobb's Linn.[15] Dobb's Linn itself is a small valley to the north of the road, into which feed several minor streams forming part of the headwaters of the river, Moffat Water, leading down to Moffat.[16] The little valley of Dobb's Linn—a kind of gash in the hillside—branches, and there are quite substantial waterfalls in one of these branches. The walls of the valley are steep, not much vegetated, and display obvious bands of dark shale. Figure 8.3 shows a portion of the valley, with the cottage of Birkhill by the road in the background. Figure 8.4 shows the interior of the valley and some of the exposures of shale, which on examination are easily found to be crowded with graptolites.[17]

On walking into the entrance gorge to Dobb's Linn, one may notice a repetition of the beds on either side, dipping away from the axis of the gorge. The beds can be seen to match one another in terms of their lithologies on either side of the glen. Thus it becomes apparent that the entrance gorge is an eroded arch or anticline, with the line of the gorge parallel to the axis of this anticline. Lapworth found that the shales may be subdivided into five bands according to their different lithologies. His next step is perhaps best given in his own words:

> Having already finished our arrangement of the shales into five successive groups, we next busy ourselves by collecting as many as possible of these fossil graptolites. In our cabinet we have five corresponding sets of drawers. We place all the graptolites we procure from the highest grey shales in the first set—those from the second set of beds in the second set of drawers, and so on. In our spare time of an evening we name and identify these graptolites but we are always careful to replace them in the proper drawer. Long before we have collected our hundred specimens, we shall have made another and more startling discovery. We find that our drawers contain three distinct sets of graptolites. The two upper sets of rocks contain one set of graptolites. Those answering to the two middle sets of shales contain a second set and those answering to the flinty beds at the bottom of the section contain a third set. None of these species of graptolites found in one set passes into another.[18]

Lapworth's procedure seems simple enough. Whether it happened just as described here, or whether his description is something of an idealization, I cannot say, but his account offers a very plausible statement of how he

15. A plaque at the cottage bears the following inscription: "Birkhill Cottage where between 1872 & 1877 Charles Lapworth recognized the value of graptolites as a clue to the Geological Structure of these hills—erected by Scottish Geologists 1931."
16. A "linn" is either a waterfall or a ravine.
17. It is a romantic spot, and in his Glasgow lecture Lapworth told his audience that it had once been a hiding place for hunted covenanters and the waters of the stream had sometimes been tinted with blood!
18. C. Lapworth, "The Silurian Age in Scotland," 2d lecture, 6.

Fig. 8.3. View from Dobb's Linn, looking south to Birkhill Cottage.

deployed (or perhaps came to understand in use) the notion of fossil *zones* at Dobb's Linn. The recognizable anticlinal structure allowed Lapworth to establish a small portion of the sequence in terms of lithologies, knowing which way up the sequence was. In addition, he now knew three distinct specific combinations of fossils and the correct order in which they occurred in time.

Fig. 8.4. View of vertical graptolitic shales, Dobb's Linn.

From this starting point, he was then able to map the Dobb's Linn area in detail and work out some sections, establishing in the process some further fossil zones. Lapworth's beautiful and meticulously drawn map of the area is shown in figure 8.5. The section of the (faulted) anticlinal arch should be noted particularly.

Working on the basis of the sequence established at Dobb's Linn, then, Lapworth examined other localities in the neighborhood where the black shales could be found outcropping in the greywackes, and he provided maps and sections in each case.[19] Also, using his fossil zones, he established where his three black shale bands belonged in the standard Silurian sequence of Wales.[20] It appeared that the Upper Black Shale (which he called the Birkhill Shales) was equivalent to the Lower Llandovery; the Middle Shale (Hartfell Shales) belonged to the Bala; and the Lower Shale (Glenkiln Shales) belonged to the Llandeilo. The great mass of greywackes, now shown to be overlying

19. The localities were Craigmichan Scar, Selcoth Burn, Crossleuch, Muckra Burn, Riskinhope Burn, Black Grain, Fala Grain, Thirlstane Score, Morry Syke, Earl's Hill, Yellow Mire, Whitehope Burn, Berrybush Burn, Cowan's Croft, Scableuch, Shiel Syke, Cossar-Hill Burn, Mount Benger Burn, Edinhope Burn, Sundhope, Ettick River, Entertrona, Frenchland Burn, Glenkiln Burn, Black Linn, Belcraig Burn, Garple Burn, Rittonside, Headshaw Linn, Hartfell Spa, Hartfell Score, Syart-Law Score, Craigierig, Boar Cleuch, Palmoody, Carrifran, Bodsbeck, and Shortwoodend. This long list is given simply to provide an indication of the thoroughness of Lapworth's work. He furnished sections for most of these localities.
20. Lapworth also utilized Swedish analogies.

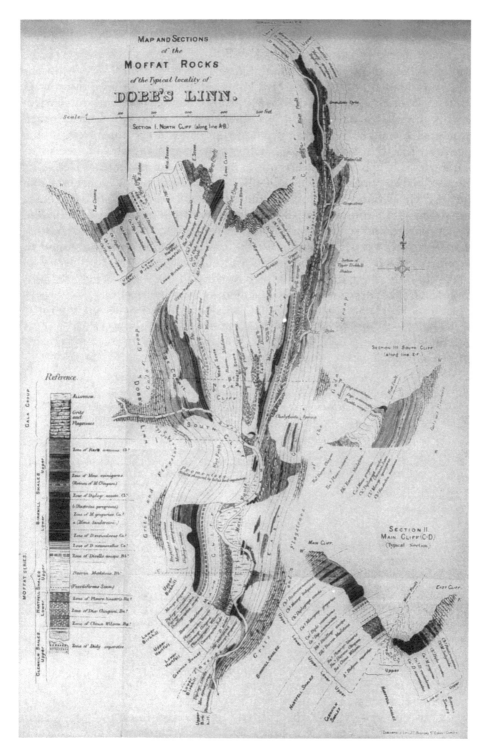

Fig. 8.5 Geological sketch map of Dobb's Linn area, according to Lapworth (1878, plate 12).

the shales, could not be older than Middle Llandovery. On the basis of the argument that has just been outlined, Lapworth was able to show how to work out the general structure of the Moffat region. For he now knew the correct sequence of the fossil zones, and from this he could tell whether any particular sequence that he might encounter was the right way up or inverted. It then appeared that there were indeed many *repetitions* of the shale beds in the Southern Uplands, and that the whole region was folded and faulted in an extraordinarily complicated fashion. However, it was now also evident that the greywackes were not abnormally thick. The thickness was apparent, being due to the repetitions arising from faultings and foldings. The Llandeilo of the Moffat area, far from being unusually thick, was in fact unexpectedly thin. It was represented by the Lower Shales, or Glenkiln Shales, only about fifteen feet thick at Dobb's Linn. All this, and much more, was published in Lapworth's classic paper "The Moffat Series" (Lapworth 1878).[21]

Turning his attention to the western outcrop of the rocks of the Southern Uplands, Lapworth made a similar detailed study of the celebrated exposures near Girvan—a fishing village on the Ayrshire coast to the west. Here there were numerous shells to be found, as well as graptolites. Lapworth recognized four major lithological groups at Girvan:

> Purple and green shales and flagstones
> Shelly sandstones, grits, limestones, black and grey shales
> Flagstones
> Pebble beds and limestones with corals.

These appeared, by their fossil shell contents, to have the following Welsh analogies:

> Tarannon[22]
> Llandovery
> Bala/Caradoc
> Llandeilo

The lower three of these had already been shown to have their equivalents in the three black shale units at Moffat. Thus Lapworth was able to correlate rocks of different lithologies at coastal Girvan and inland Moffat. The analogy was supported by the occurrence of some black shale bands at Girvan having graptolite contents in agreement with the proposed correlations.

The sequence at Girvan was several thousand feet thick and looked like material that had been deposited near a coastline, whereas the temporal analogues at Moffat were only about three hundred feet thick and consisted of fine black shales. This was so, Lapworth proposed (Lapworth 1882b), because

21. The paper was read on Nov. 21, 1877.
22. An upper unit of the Llandovery series.

the material had been derived and transported from the west, and the finer sediments had been carried further—to the Moffat region—while the coarse sediments had been dropped in the Girvan locality.

This, then, was Lapworth's solution to the problem of the Southern Uplands. It was an extraordinary achievement. Virtually single-handed he had shown how to analyze a complex stratigraphical problem by minute attention to detail. The key to the whole problem was the old principle of William Smith, now given a tighter interpretation in terms of geological age—that rocks of a particular geological time were characterized by fossil organisms of particular types. For Lapworth's purposes, the graptolites, far from being useless for stratigraphy, were peculiarly well suited, since they were free-floating organisms, so that a given type might turn up in a range of rocks of different lithologies but the same age. Thus there was no need to invoke the notion of Barrande's "colonies" to explain apparent anomalies in the stratigraphical sequences.[23] The fossils could be found to keep their order perfectly satisfactorily, provided that they were collected with due care, and the exact localities of collection properly recorded.[24]

All this was a severe embarrassment to the Survey. They were offering a structure for the Southern Uplands that was not properly controlled stratigraphically, and they had a massive thickness of Llandeilo beds as a result— and a complete muddle of the graptolites, so that the earliest forms sometimes appeared at the top of the sequence (as it was ordered according to Lapworth's interpretation). The Survey also invoked an unconformity (below the Caradoc) that was hypothetical and essentially ad hoc in character. Nor did Survey officers appreciate the extent to which faulting and folding could make it appear that there were additional black shale bands high in the sequence. In contrast, Lapworth's method, based on fossil zoning, showed that these seeming "higher" beds were in fact repetitions of the lower formations, reappearing because of faulting.

The Survey tried to defend itself. In the discussion recorded after the presentation of Lapworth's Moffat paper, Ramsay pointed to the occurrence in Lanarkshire of black shale bands elsewhere in the sequence than those that Lapworth had investigated at Moffat (Lapworth 1878, 344). But this simply restated the Survey's view, rather than vindicating it. Geikie gave credence to Barrande's theory of colonies in his *Textbook* (Geikie 1882a, 627–30), apparently having written to Barrande not long before, offering a section of the "Silurian" rocks of the Southern Uplands, which Barrande wanted to include in his *Défense des colonies V*, then in preparation.[25] One of the subordinate Survey staff, the fossil collector Arthur Macconochie (1850–1922), publicly

23. See note 11 above.
24. This was one of Lapworth's chief complaints against the Survey's collecting practices. The stratigrapher was not offered an exact locality for each specimen collected. He found, however, that a Girvan resident, Mrs. Robert Gray, had amassed a large collection of fossils from the Girvan district, which were tagged with exact collection sites and thus proved invaluable for his work.
25. British Geological Survey, Edinburgh Archives, Gen. 1425/18, Barrande to Geikie, Jan. 5, 1882. Barrande's volume was never published, since he died in 1883.

attacked Lapworth's interpretation at a lecture in Glasgow on April 10, 1884 (Macconochie 1881–84).

Nevertheless, the Survey eventually came round to the view that Lapworth's interpretation was to be preferred. According to the Survey historian, Sir Edward Bailey, it was John Horne who "first saw the light" (Bailey 1952, 93).[26] Horne was glad to be able to postpone the publication of the maps of the Moffat area and Loch Doon, for which he was responsible, until the ground had been resurveyed according to Lapworth's method of zonal collection and analysis. Indeed, nearly the whole job of surveying the Southern Uplands had to be done again, in the light of Lapworth's work. But this was not accomplished immediately, for in the meantime the Survey had to give all its attention to the far northwest, where again it was Lapworth who finally convinced them that the official view of the structure was all wrong. However, it must be said that when the Survey's memoir on the Scottish Silurians was eventually published (Peach and Horne 1899), it was a magnificent piece of work and met with the full approval of Lapworth (1899).

As a further thorn in Geikie's side, Lapworth (1879) proposed the concept of the Ordovician system as a compromise solution to the long-standing battle between Murchison and Sedgwick (and their respective followers) concerning the disputed territory in Wales, which Murchison wanted for his Silurian kingdom, and which Sedgwick claimed for his Cambrian (Secord 1986a). As is well known, Lapworth's suggestion was beautifully simple: namely, to take the rocks in the disputed area and erect a new system for rocks of that age. They were well characterized by their own fossil fauna—chiefly graptolites. But Barrande, we recall, had believed that graptolites were unreliable palaeontological criteria for characterizing formations; and the Survey, following Murchison, was for a long time under the sway of this opinion. So we find that in Geikie's time as director general, the Survey never officially recognized the Ordovician as a valid system. No doubt, this had to do with Geikie's continued devotion to everything that Murchison stood for and his unwillingness to let outsiders make decisions about the main stratigraphical subdivisions of British rocks.[27] But it was a somewhat Canute-like gesture, in my view.

From what has been said above, it may appear that Lapworth's unraveling of the structure of the Southern Uplands was more or less a single-handed achievement, of a David-and-Goliath character. But this is not entirely correct. Hamilton (1984) has shown that the zonal collecting methods used by Lapworth at Moffat were made known to him, in part at least, by his correspondence with the Swedish geologist Gustaf Linnarsson (1841–1881), chief palaeontologist to the Swedish Geological Survey, who was engaged in work-

26. On Horne, see chapters 9 and 10.

27. Geikie's formal argument for the rejection of the Ordovician as a valid system was based on the fact that Murchison had priority for his term; the Lower Silurian. Geikie wrote: "To wipe out Murchison's accepted designation from half of the system which he was the first to define and describe, is a change quite unwarranted by any discoveries that have been made since his time" (Geikie 1903, 2: 917).

ing out problems in Sweden analogous to those of the Southern Uplands.[28] Further, Lapworth received help and encouragement from Professor H. Alleyne Nicholson of Aberdeen University, an old friend of James Nicol, and for a time virtually his sole supporter.[29] And according to an anecdote originally recounted by Benjamin Peach, Lapworth first started collecting graptolites non-zonally at Dobb's Linn, only changing to a zonal method when he heard of Linnarsson's work (Hamilton 1984, 190).[30]

Be this as it may, by the end of his work in the Southern Uplands, Lapworth was a world authority on graptolites; he had proposed the solution to the Cambrian-Silurian dispute that eventually received general acceptance; he had driven the Survey to reconsider its Southern Uplands work; and in time the structure that he proposed was accepted.[31] His achievements soon became widely recognized (though only reluctantly by Geikie), and Lapworth was appointed to the chair at Birmingham in 1881. But from the present perspective, perhaps the most important thing was that Lapworth had learned and applied the cardinal rule: if one is going to unravel the stratigraphy of a

28. On Linnarsson, see Lapworth (1882) and Hamilton (1984). It may be noted that Lapworth stated (page 171) that when he read the manuscript of a paper of 1876 by Linnarsson on Swedish graptolites he was delighted to find that the results obtained in Sweden harmonized with the results that he had himself been obtaining in Scotland. As a result: "[Linnarsson] followed with the keenest interest every stage of my later work among the Palaeozoic rocks and fossils of Britain, [and] applied the same methods of research to the Graptolite-bearing rocks of Sweden." It is not exactly clear, therefore, whether it was Lapworth or Linnarsson who first deployed the "zonal," as opposed to the "formational," method of studying strata.

29. Nicholson (1844–99) was professor of natural history at Aberdeen. He was an experienced palaeontologist and stratigrapher among fossiliferous rocks, and published widely used texts in zoology and palaeontology. His specialism was graptolites. See Nicholson (1872).

30. According to Hamilton, the anecdote was recounted by Sir Edward Bailey in a letter to W. W. Watts (Lapworth's successor at Birmingham), dated July 30, 1921, and held in the Birmingham archive.

31. It prevailed until the 1950s, when substantial revisions of Lapworth's structural ideas for the Southern Uplands were initiated. It was not Lapworth's work in the northern part of the Southern Uplands (from Girvan to Moffat and at Dobb's Linn) that was modified, but his ideas on the general structure of the folds of the main body of the Uplands. A situation developed such that the written descriptions in the various publications did not mesh with the figured sections. The Survey (in its memoir of 1899) followed what is now regarded as a mistake on Lapworth's part, the mistake having arisen because one of the main stratigraphical units (the "Hawick Rocks") is unfossiliferous, so that the rocks were not amenable to Lapworth's methods of stratigraphical analysis by graptolite zones. It may be suggested that the enormous later reputation of Lapworth made it difficult to effect changes in the structural interpretations, so that they remained unchanged for some eighty years. However, these revisions in the structural interpretations of parts of the Southern Highlands have not undermined either Lapworth's methodology (his establishment of the stratigraphical column for the region of the Southern Uplands by zonation of strata with the help of graptolites) or his general determination of the stratigraphical column for Lower Palaeozoics on that basis. For his developed views on the general structure of the Southern Uplands, with illustrative sections and a tabular comparison of the strata in Scotland and England and Wales, see Lapworth (1889). For a modern criticism of Lapworth's interpretation of the structures, see Craig and Walton (1959). In recent years, under the influence of ideas about continental drift, it has been suggested that the rocks near Ballantrae, to the south of Girvan mark the position of a former "subduction zone."

district, it is essential to establish the correct order of superposition for the strata of that district. He was able to do this with the help of fossils in the Moffat district, starting from the recognition of the anticline or archstructure at the entrance to Dobb's Linn. We shall be interested to see how he proceeded in the Northwest Highlands, which are for the most part barren of fossils.

There was, perhaps, another thing that Lapworth learned at Dobb's Linn that may have been useful to him in the northwest. As we have seen, he viewed the apparent great thickness of the greywackes of the Southern Uplands as due to folding and faulting. The folding, however, was in many cases not "normal" but "reverse." The Southern Uplands were cut up by numerous thrust (reverse) faults. A fine example, shown in figure 8.6, may be seen in the hillside just above Dobb's Linn, within easy view of Birkhill Cottage. This is a manifest thrust fault, arising from lateral forces in the earth's crust. Admittedly, it is not a huge, low-angle thrust fault, such as Lapworth was to encounter in the Highlands, but it must surely have awakened him to the possibility of reverse faulting in Scottish rocks.

We also know, however, that Lapworth had further theoretical, as well as empirical, information at his disposal, for he was reading the recent publications from Continental and American authors on mountain formation, and he was to deploy their theories in his elucidation of the Northwest Highlands. It is needful, therefore, that we digress still further from the discussion of the Northwest Highlands to consider the developments that had been taking place in Europe and America concerning the theory of mountain building.

A valuable account of early investigation and theorizings about the formation of mountains in Europe, particularly in Switzerland, has been given by Sir Edward Bailey (1935); and a more recent publication by Mott Greene (1982) provides an excellent overall view of the history of theories of mountain building in the nineteenth century.[32] In the Alps, the prevailing view in the earlier years was similar to that preferred elsewhere—namely, that the dominant motions of the earth's crust were vertical rather than lateral. However, at the Zurich Polytechnic, Arnold Escher von der Linth (1807–72), recognized the phenomenon of overthrusting as early as the 1830s at the northern side of the Aar Massif, and somewhat later in the mountains of the Canton of Glarus, to the southwest of Zurich. In Switzerland, the structures could to some extent be perceived "directly" by simple examination of the well-exposed mountainsides. Also, because the mountains were much younger than those in Scotland, and quite often fossiliferous, it was not so very difficult to demonstrate an inversion of the normal stratigraphical sequence, and Escher did just this ([Escher von der Linth] 1841).[33] Subsequently—the date is

32. For another useful earlier account of the geological researches in Switzerland, see Heritsch (1929).
33. In discussing the geology of the area near Mount Säntis, Escher referred to the "well-known" "overpushing" (*Ueberschiebung*) of the Secondary limestone mountains by the Tertiary molasse deposits, which were themselves overturned ([Escher von der Linth] 1841, 55).

Fig. 8.6. High-angle thrust fault at Dobb's Linn.

not known exactly since he did not publish his views, though he spoke about them freely to other geologists—Escher developed the notion of a massive *double*-fold for the Glarus district, entailing the pushing, from both north and south, of older rocks over younger ones. The manifest evidence of an inversion of the standard stratigraphical sequence was sufficient even to persuade Murchison (Bailey 1935, 46–49), who looked over the sections with Escher during an extended tour of Europe that Murchison undertook with his wife in 1847–48 (Geikie 1875, 2: chap. 19). Murchison (1849) gave full expression to his understanding of the inversions in a major paper with numerous figures (though he did not refer to the idea of the "double-fold," which may not then have been developed by Escher). Given this ready acceptance of Swiss rocks being inverted, it may seem strange that Murchison was so vehemently opposed to the notion of analogous structures in Scotland. However, the dimensions of Silurian territory were not at issue for the Swiss rocks that Murchison examined with Escher, the structures of the Scottish rocks are by no means so easy to construe, and Nicol was not proposing a Swiss-style inversion.

When Escher died in 1872, he was succeeded in the Zurich chair by his pupil Albert Heim (1849–1937), who inherited Escher's manuscripts and was thus well placed to promulgate his teacher's ideas.[34] This he did to great effect in 1878, with his publication of *Untersuchungen über den Mechanismus der Ge-*

34. On Heim, see Bailey (1935, 49–54 and passim; 1939); Greene (1982, chap. 8), Arbenz (1937), Geological Society of London (1938), de Margerie (1946), von Franks and Glaus (1987).

birgsbildung (Investigations on the mechanism of mountain building) (Heim 1878).[35] Besides giving graphic accounts of the structures of the Swiss mountains, beautifully illustrated in color in the accompanying atlas, this publication also sought to develop a general theory of mountain building. It was this work, which was also based on investigations in the Canton of Glarus, that Lapworth found of greatest use in his attack upon the structure of the Highlands and that he described in some detail in his paper on his work in that area, which we shall be examining below.

The essential features of Heim's theory were really quite simple: beds of rocks, when squeezed by lateral forces, will fold into S-shaped structures, possibly asymmetrical; and at even higher lateral pressures these may begin to shear, so that we have the beginnings of a thrust structure, as shown in figure 8.7.

Then, with even further lateral pressure, the whole structure may "give way," so to speak, and the rocks of one limb of the overfold become thrust over the rocks of the other limb. In this manner, a low-angle thrust fault develops; and as can be seen from the figure, it is possible under these circumstances for strata to become inverted and for older rocks to lie on top of younger ones. Also, one may anticipate that all sorts of metamorphic changes occur in the rocks as a result of the gigantic pressures exerted.[36]

In the nineteenth century, the general explanation as to why all these things might occur slowly over time was that the earth was cooling and shrinking, and hence the crust of the earth might become wrinkled like the skin of a dried-out apple—a favorite simile (e.g., Geikie 1877, 13). This was indeed a perfectly satisfactory hypothesis at the time (and might be deployed to account for both normal and reverse faulting), but the theory has not continued to find favor in the twentieth century, partly because it is no longer believed that the earth is cooling to any great extent, since the radioactivity within the body of the earth sustains its heat. However, with the advent of plate tectonic theory, the notion of lateral movement of portions of the earth's crust is considered more important than ever. One might say that the nineteenth-century discoveries concerning the structures of mountains—such as the work of Callaway, Lapworth, and others in the Northwest Highlands—provided the empirical basis for subsequent grand theoretical interpretations involving lateral movements. The reason why such movements might occur was uncertain. Meanwhile, the "wrinkled apple" theory was quite serviceable.

As Bailey (1935, 50–51) has pointed out, there were all sorts of other diffi-

35. For a review in English, see Renevier (1879).
36. It is worth noting that although Heim had the idea of thrust faulting, his chief theoretical concept was that of the fold, as can readily be seen from the figures in the atlas accompanying his *Mechanismus der Gebirgsbildung*. It is interesting, then, that although Lapworth and the surveyors (especially Peach and Horne) obtained essentially the same results at Durness and Eriboll, Lapworth placed greater emphasis on folding, while Peach and Horne displayed the importance of thrust faulting with particular skill. It may be that the emphasis on folding in Lapworth's work was derived particularly from Heim.

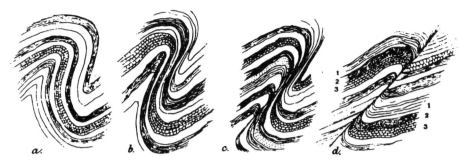

Fig. 8.7. Faulting and folding of strata under lateral pressure, according to Heim (1878, Atlas, plate 15, fig. 14).

culties associated with the theory of the double-fold of Glarus.[37] It involved a lateral displacement of folded strata southward from the north, and conversely from the south. Yet all the evidence in the Canton of Glarus seemed to suggest lateral displacement from the south northward. Rather remarkably, the French geologist Marcel Bertrand (1847–1907), suggested in 1884 that the structure might better be explained in terms of a single thrust of material from the south, the central portion of the thrust mass being absent because of erosion (Bertrand 1883–84; Mather and Mason 1939, 631–36).[38] Never having been to Switzerland, Bertrand suggested the idea on the basis of his reading of Heim's work and by analogies with thrust structures known to exist in the coal-mining regions of the south of Belgium and the north of France. He followed ideas that had previously been propounded by the great Austrian geologist Eduard Suess (1831–1914), who argued in his *Die Entstehung der Alpen* (The formation of the Alps) (1875) that mountain structures in any given area tend to show movement from one direction only.[39] For the Belgian strata, Bertrand drew on the work of Jules Gosselet (1880).

Interestingly, Heim, the great theoretician of the structures of mountains and the theory of mountain building, was at first quite unwilling to contemplate the huge one-sided thrust that Bertand wished to invoke—which was equivalent to the two very large arms of the double-fold that had been proposed for Glarus by Escher, combined together. Likewise, Suess at first ignored Bertrand's suggestion in his magnum opus, *Das Antlitz der Erde* (The face of the earth), which he began to publish in installments in 1883. However, Suess became converted by Bertrand's reasoning, and in 1892 he demonstrated his views to Heim with the help of the latter's large relief model of the district. Even so, it was not until 1902 that Heim eventually came round to

37. The hypothesis did, however, offer an explanation of the observed reversals of the usual stratigraphic order while minimizing lateral earth movements.
38. On Bertrand, see Geikie (1908).
39. Suess argued the case with a wealth of empirical evidence, but without really offering a theory as to why it should be so.

accepting the idea of a single major thrust for the Glarus district, and it was as late as 1919, in his *Geologie der Schweiz* (The geology of Switzerland), before he finally published his ideas (Heim 1919–21).[40] But even though Heim himself was not applying it satisfactorily to one of the most important geological sites in Switzerland, by 1878 he had developed a theory of folding that was both sound and useful to Lapworth.

Besides his bold hypothesis concerning the structure of the Glarus region, Bertrand was also responsible for introducing the useful term "nappe" into the geological literature. In his paper of 1884, he spoke of the thrust mass of Permian rock in the Glarus district as a *nappe de recouvrement,* since he envisaged it as being analogous to a bent-over (recumbent) fold such as may be seen on a wrinkled tablecloth or sheet (*nappe*). A good deal of other new terminology was introduced at about this time, as the new ways of thinking about the processes of mountain building were developed and displaced the older notions emphasizing vertical movements, which derived from authors such as Hutton, Lyell, and Darwin. Of particular importance in codifying this new terminology in both French and German was the volume by Emmanuel de Margerie and Albert Heim entitled: *Les dislocations de l'écorce terrestre/ Die Dislocationen der Erdrinde* (The dislocations of the earth's crust) (1888). This provided a glossary of terms such as: "nappe" (*Decke* in German); "klippe" (an erosional remnant of a nappe—such as that envisaged by Bertrand at the northern side of the Glarus, or that portion of the Moine Schists exposed at Far-out Head near Durness, which we have mentioned previously); and "fenestre" (*Fenster,* or window), which we have encountered at the Ord Window in Skye. By thus introducing this new nomenclature, the authors were performing a useful service toward the acceptance of the theories in which the terms found application.[41]

In his paper setting out the preliminary results of his investigations at Durness and Eriboll, Lapworth (1883c) mentioned, besides Suess and Heim, two American geologists, the brothers Henry Darwin Rogers (1809–66) and William B. Rogers (1805–81), who had mapped considerable areas of the eastern United States in the 1830s and 1840s and had developed also a theory to account for the formation of the Appalachian Mountains. In addition, Lapworth mentioned the Norwegian geologist Waldemar C. Brøgger (1851–1940).

The work of the Rogers brothers that Lapworth probably had in mind was concerned with the structures of the Appalachian Mountains. It had been made available to British geologists by an abstract published in the *Report of the British Association for the Advancement of Science* in 1842 (Rogers and Rogers 1843a, 1843b). Henry Rogers had also published in the *Transactions of the*

40. Heim declared for Bertrand's hypothesis in a letter included in a publication of his student Maurice Lugeon (1901).
41. Simiarly, during the "Chemical Revolution," toward the end of the eighteenth century, a new nonmenclature was proposed. But the terms were not "neutral": one could hardly deploy the new chemical terminology without giving support to the new chemical doctrines.

Royal Society of Edinburgh (Rogers 1857). In mapping the Appalachians, the Rogers brothers noted the passage of inverted or folded flexures into faults, which extended over great distances; and their published sections illustrated such structures in an eye-catching way. Henry Rogers suggested that such structures might be accounted for in a "catastrophist" manner, according to which the sudden escape of large quantities of subterranean vapors and gases might give rise to wavelike motions in the earth's crust, which would spread outward from the locus of the explosion. These waves might, then, turn over, like surf on a beach (Rogers 1857, 432), and thus the overfolding and associated faulting might be explained.

The theory was attractive to Murchison (1849), usually willing to endorse "paroxysmal" notions, and he referred to it in his paper on the Alps. It may be that it was by this route that Lapworth's attention was drawn to the Rogerses' theory, though Lapworth very likely read the papers in the British Association *Report* and the *Transactions of the Royal Society of Edinburgh,* which would have been easily accessible to him. By the time he began his investigations in the Highlands, the "catastrophist" modes of explanation such as one finds in the Rogerses' theory were largely outmoded. Even so, the Americans' empirical work certainly demonstrated the possible occurrence of overfolds and associated large-scale faults, and concomitant inversions of strata. Thus the kind of structure that Lapworth wished to suggest for the Northwest Highlands was rendered more plausible by the American analogies.

The work of Brøgger—who was professor of geology and mineralogy at the University of Stockholm—to which Lapworth referred was almost certainly a volume that Brøgger published in 1882, dealing with the details of the geology of the Oslo region and intended as a textbook for his course of lectures at his university that year.[42] The book was remarkable for a university textbook, in that it provided a very detailed account of the geology of a local region, besides being a substantial contribution to the zonal stratigraphy of part of the Silurian system, with numerous figures illustrating the relevant fossils (chiefly trilobites). It included also some general theoretical ideas about tectonics and metamorphic phenomena.[43] Lapworth's work was mentioned in several places, and it is reasonable to assume that he was sent a complimentary copy of the book, which could easily have reached him in Birmingham before he began his work in the north of Scotland in August 1882.[44]

The feature of Brøgger's book that would undoubtedly have appealed to Lapworth would have been the numerous large-scale maps and accompanying sections—showing, for example, the stratification in the rocks in the area of a single street, or even under a single house. Several of these sections showed tight folds passing over into faults, as envisaged in the theory of Heim, and Brøgger specifically discussed how folding might occur in such a

42. At that time, the institution was called the Stockholm Høgskola.
43. The publisher was Brøgger's father, which might account for the elaborate physical character of the book.
44. The foreword of Brøgger's book was dated March 1882.

way as to give rise to inversion and associated faulting, according to the ideas of Heim (Brøgger 1882, 177).[45] The book offered an exemplar for the kind of detailed map work that Lapworth was to undertake in the Highlands, though without the advantage in Scotland of numerous types of fossils to assist in the establishment of the stratigraphical horizons in the area under investigation.

With knowledge of these preliminaries concerning the various theoretical notions that were available to Lapworth when he began his researches at Durness and Eriboll, and which I feel confident that he knew quite well, for he was a skilled linguist as well as geologist, we may now turn to examine in some detail just what he did in the northwest that soon led to the destruction of the Murchison-Geikie theory of the Highlands.[46]

On obtaining his chair at Mason College, Birmingham, in 1881, Lapworth first turned his attention to the Lower Palaeozoic rocks of Shropshire and Warwickshire, where he began to bring order into the subject with the help of his new stratigraphical category, the Ordovician system; and he began preliminary studies of the Cambrian strata with the help of his technique of zoning by fossils. His first publication in this area of inquiry appeared in 1882 (Lapworth 1882c, 1883a), and he continued with work in the Birmingham district, westward into Wales, and in the latter part of his career, eastward to the pre-Cambrian rocks of Charnwood Forest in Leicestershire.

But as early as 1882, Lapworth felt the call to the north, to see whether his meticulous techniques might finally solve the great question of the Northwest Highlands. It must be emphasized, I think, that by this time Lapworth had formed an unflattering view of the work of the Survey, and in particular of Archibald Geikie (as his correspondence, which will be considered below, reveals). Lapworth well knew what a botch had been made of the Survey's investigations of the Southern Uplands, and he was obviously aware that a heated controversy was developing over the Northwest Highlands and the official Murchison-Geikie interpretation of its structure.[47] He would naturally have suspected that the official interpretation was mistaken and would have known that there was, in the northwest, a grand opportunity to display once again the results that might be obtained with the help of detailed mapping—an opportunity to take the Survey down a further peg or two, for Lapworth knew the scornful attitude that some of the Survey officers (particularly Geikie) held toward "amateurs." There was a feeling that the Survey was be-

45. Brøgger stated that he did not entirely agree with Heim's ideas, but that his work was based on his system nonetheless. In particular, he utilized Heim's system of "fault-folds" (*Faltenverwerfungen*).
46. Lapworth worked out the principles of the structure of the Southern Uplands well before he got to know Heim's work. See Lapworth Archive, University of Birmingham, letter A29, Lapworth to Bonney, Dec. 28, 1884.
47. It should be remembered that at this stage the Survey officers had not begun their detailed mapping of the Northwest Highlands.

coming arrogant, with a lordly belief that *it* was the final arbiter on how the British stratigraphical column was to be subdivided. By virtue of the fact that the Survey was responsible for the government's official maps, it seemed to some of the Survey staff that their classificatory decisions had to be respected. Naturally, this was a view that Lapworth, who had revealed grievous mistakes in the Survey work for southern Scotland, was keen to challenge, especially as he had recently attained a chair at a university college—albeit one built of the very reddest of bricks.

In the Lapworth archives at Birmingham there is extensive correspondence from 1882 onward pertaining to Lapworth's Highlands work, and his original field maps for Durness and Eriboll are also preserved there. However, we do not have Lapworth's field notebooks for his Highlands investigations, and there is no diary of any sort. Thus one has to rely on miscellaneous documents, such as obituaries written by his acquaintances, and so on, to attempt to reconstruct what Lapworth did in the northwest. We may be confident that he began with the assumption that the problem would yield to detailed mapping of the terrain. As mentioned above, he was bringing with him the theoretical armory provided by texts such as Heim's *Mechanismus der Gebirgsbildung,* and he would be able to utilize analogies from the structures of the Appalachians and the Alps. A letter from Lapworth to Bonney, dated September 9, 1882, written shortly after Lapworth's first period of fieldwork in the northwest, also provides valuable information about his early activities at Durness and Eriboll.[48] It informs us that Lapworth obtained a Royal Society grant for his work in the Northwest Highlands, and that he headed north immediately on hearing of his award on August 3. It also appears that he had largely come to understand the structure of the region by the time the letter was written. Thus he broke the back of the problem in about a month's work, though to achieve this, as he said in his letter to Bonney, he labored more or less continually, night and day.

I suggest that Lapworth chose Durness and Eriboll, rather than Loch Assynt, Loch Maree, or Ullapool, because the northern coastline provided him with the opportunity to see as complete a section as possible of the strata; and also because it was only at the Kyle of Durness that any significant fossil remains had been found with any promise for the application of zonal techniques. For a palaeontologist, Durness would have been the place to begin. In the summer of 1882 we therefore find him beginning his mapping at Durness, furnished with accurate six-inch topographical maps issued by the Ordnance Survey. According to W. W. Watts (1921b, 20), Lapworth also utilized maps of twenty-five inches to the mile (which he must have made for himself).

In an obituary notice of Lapworth by J. J. H. Teall (Teall and Watts 1921,

---

48. Geological Society Archives, Bonney Correspondence, LDGSL 776, Lapworth to Bonney, Sept. 9, 1882. (A typed copy of this important letter was made by Lapworth's former colleague and biographer, W. W. Watts, in March 1940, and this is also preserved in the archive.)

xxxv), we learn that Lapworth did not immediately find the clue to the High-lands structure at Durness, despite the occurrence of fossils there, and so he soon moved on to Eriboll, where he took up residence in a shepherd's cottage at Heilam. Nearby, at a locality at the southern side of Camas Bay, he found a section that met his needs at the place called An t-Sròn ("the promontory," or "nose"), where he was able to establish a satisfactory sequence for the sedi-mentary rocks of the region. It will be recalled that Callaway also examined this section in detail. Whether it was Callaway or Lapworth who first did the fieldwork at An t-Sròn, I have not been able to determine, neither have I found any record of their having met in the field.[49]

Perhaps one of the first things that Lapworth would have done was to try to settle once and for all whether there was indeed an "upper" quartzite as Murchison had for long stoutly maintained. I take it that Lapworth would have determined this question in 1882, but his "proof" was demonstrated to Teall and Professor J. F. Blake of Nottingham University College, as is re-vealed in the following passage from Teall's remarks in Lapworth's obituary (ibid., xxxvi). Teall, Blake, and Lapworth met in the field in 1883.

> Leaving Arnaboll hill we worked northwards towards Whiten Head, and, to remove any doubt that might linger in our minds as to the identity of the "Upper Quartzite" of Murchison with the "Lower Quartzite," [Lapworth] placed me on the latter with strict instructions to walk along it, making sure that I never left it. He and Blake took up their positions on the "Upper Quartzite," and moved in a direction roughly parallel with me with the same care. Progress was slow, for in such a disturbed area it was necessary to examine every inch of the ground. Finally we met on quartzite, shook hands, and declared that beyond all shadow of doubt the "Upper Quartzite" was merely the "Lower Quartzite" brought up again by the disturbances of which we had already seen such striking evidence.[50]

This is an interesting passage. It suggests, I think, that Lapworth had al-ready performed this exercise for himself and was simply *demonstrating* to his fellow geologists a conclusion that he had already reached. He was inviting them to repeat his own experience as a means of persuading them of his con-

49. Callaway went to Assynt before Durness and Eriboll. Both Callaway and Lapworth were at work in the northwest in 1882.
50. Some account of the joint observations made by Teall, Blake, and Lapworth is to be found in a paper published by Blake (1884), describing his northern excursion of 1883. The paper is some-what reminiscent of Hudleston's "First Impressions of Assynt" (1882). It shows a geologist strug-gling to make sense of the confused and confusing evidence before him. But whereas Hudleston's guide was Heddle, Blake had the advantage of the assistance of Lapworth, and it is clear from his report that he had managed to attain at least a partial understanding of the structure. However, his sections were in some cases little more than rough sketches, rather than representations of the three-dimensional structure of a district based upon the results of field mapping; and while Blake had some success at Durness and Eriboll, where he had Lapworth's guidance, he was much less successful further south at Assynt.

clusions. They were to be *witnesses* to the truth of Lapworth's opinions (cf. Shapin and Schaffer 1985). However, it must not be thought that Lapworth had invited Teall and Blake to the north simply for that purpose. Teall recorded that it was by chance that he and Blake met Lapworth at Rhiconich (in gneiss country on the west coast), and they then went on together to Durness and Eriboll. Nevertheless, it was highly convenient to Lapworth to have geologists such as Blake and especially Teall as witnesses to his theory concerning the northwest sequence and structure.

The point about the "upper" quartzite being settled, Lapworth needed to determine the true sequence of sediments at Durness and Eriboll. Once this was established, he could then know whether a particular set of beds that he was mapping was or was not inverted. It will be recalled that at Dobb's Linn the clue was provided by the archlike structure of the beds at the entrance to the linn. Then the sequence of graptolite zones could be established satisfactorily. But at Durness and Eriboll, the palaeontological method failed. The fossils were too rare, were insufficiently varied in type, and ones that might conceivably be used for zoning were only found in the Durness Limestone, and were difficult to extract. There were virtually none of these fossils at Eriboll, where, however, the structures seemed most promising for establishing a sequence. So Lapworth was forced to turn to lithological characters as a basis for his sequencing. I imagine he did this with considerable regret, for the method with fossils had proved such a resounding success in the Southern Uplands.

Before the work of Lapworth and Callaway, not much had been done toward attempting subdivisons of the quartzite, the fucoid beds and the Serpulite (or Salterella) grit, and the limestone. Yet there were distinctive lithological variations of color, texture, and mineral composition that persisted over wide areas and that could be recorded consistently on maps. This made subdivision on a lithological basis feasible. Lapworth examined the rocks on the western and eastern shores of Loch Eriboll, and he came to the conclusion that the following sequence held:

IV. Durness Limestone
    a Grey and white mottled limestone

    b Dark and flaggy limestones with cleaved calcareous shales

    c Impure dolomitic limestone, with *Orthoceras, Linguloid* shells, and *Salterella maccullochi*

III. Salterella grit with quartz pebbles, empty wormholes, and casts of *Salterella maccullochi*

II. Fucoid Beds
    c Quartzose flags (Upper or Heilam Flags), wormholes, and an upper band of dark blue shale

    b Calcareous shales and flags (Fucoid Limestone)

    a Flaggy grey shales with "fucoids" (actually branching worm casts)

I.   Quartzite
    d Pipe Rock (containing the well-known in-filled vertical worm tubes)

    c Massive white quartzite

    b Tinted quartzite (thick-bedded, flaggy, and weathering to a faint pink or buff color)

    a Basal quartzite, either a breccia or conglomerate.[51]

As for knowing which was the top of the foregoing sequence, Lapworth reckoned that the arrangement was fairly clear to the west of Loch Eriboll up as far as layer I(d), with layer I(a) lying unconformably on the gneiss beneath, and the sequence proceeding regularly thereafter up to I(d). I(d) could also be found on the eastern shore of the loch near Heilam, and so the sequence on the western and eastern shores could be connected. For the further sequence on the eastern shore, Lapworth relied particularly on the well-exposed shore section at An t-Sròn, which (like the arch at Dobb's Linn) was anticlinal—though this time Lapworth was dealing with a much broader fold than the tight arch at Dobb's Linn. On the two limbs of the An t-Sròn fold, he could see the ascending and the descending sequence—from IV(c) to I(d)—and make out the structure satisfactorily.

Having sorted out the stratigraphical sequence, Lapworth could then begin to map the whole area thoroughly, onto sheets of the printed six-inch maps of the Durness and Eriboll regions. These working sheets have been preserved and are held in the Lapworth archives at Birmingham. The area covered by Lapworth in his two seasons of fieldwork was as follows. At Durness, he mapped from Far-out Head southward and eastward, so as to cover the whole area of the faulted inlier of the sediments of the Durness region between the bands of gneiss to the west and east of Durness. He showed the gneiss lying between Durness and Eriboll and mapped the boundary of the quartzite lying on it unconformably on the western side of Loch Eriboll. Much attention was given to the areas round Heilam, and south of this toward Eriboll House (Murchison's favorite section it will be recalled), and at Ben Arnaboll, which particularly interested Callaway and where the demonstration to Blake and Teall was made. Lapworth then mapped in detail the terrain all the way north of Loch Eriboll up as far as Whiten Head, where there are no roads and no habitation, but much bog on gneiss terrain. Roughly speaking, then, he mapped the whole coastal strip from Far-out Head to Whiten Head. The detailed character of Lapworth's mapping may be gauged from figure 8.8, which

51. Conglomerates contain rounded pebbles; breccias have angular fragments.

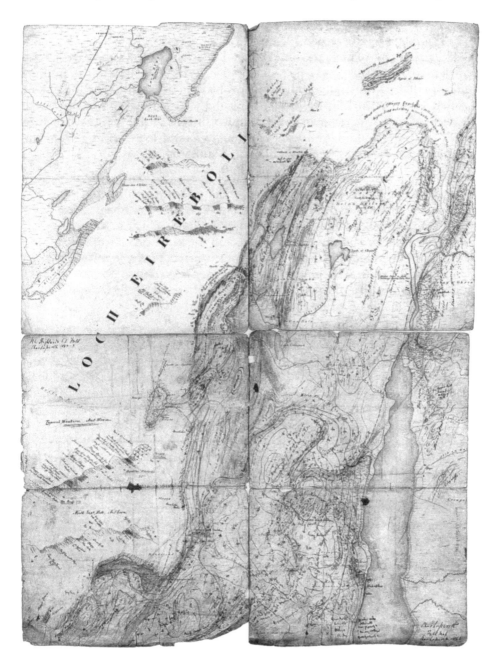

Fig. 8.8. Portion of Lapworth's field map for the Durness-Eriboll region, 1882–83, showing the area adjacent to Camas Bay, Loch Eriboll. Courtesy of School of Earth Sciences, University of Birmingham.

reproduces the portion of his field map for Camas Bay, An t-Sròn, et cetera.[52] (The original is hand-colored.)

The field maps (which are dated 1882–83) do not reveal which areas Lapworth mapped in 1882 and which the following year. But his letter to Bonney indicates that Lapworth had solved the problem in its essentials by the end of his first month's work in the field. Using the ideas of Heim and the Swiss, Norwegian, and American analogies, Lapworth recognized that the structure might be understood as resulting from huge lateral forces acting from the southeast. These forces were described to Bonney as follows:

> Conceive a vast rolling & crushing mill of irresistible power, but of locally varying intensity, acting not parallel with the bedding but obliquely thereto; & you can follow the several stages in imagination for yourself. Undulation, corrugation, foliation & schistose structure— slaty cleavage are all the effects of one and the same cause—the local result is dependent upon the local power exerted. Shale, limestone, quartzite, granite & the most intractable gneisses crumple up like putty in the terrible grip of this earth-engine—and are all finally flattened out into thin sheets of uniform lamination and texture.[53]

Thus Lapworth (1886, 1026) proposed a new petrological category—"mylonite"—a rock produced by the pulverization of rocks in regions of intense folding, shearing, and thrust faulting.[54] Such rocks are found exposed in a band extending from near Loch Assynt to Whiten Head, particularly on the eastern side of Ben Arnaboll between Loch Eriboll and Loch Hope. Figure 8.9 shows a detached block of this rock.[55]

Examination of figure 8.8 will show that Lapworth was mapping the outcrop of a thrust plane at the base of the gneiss to the east of Camas Bay (the bay south of Heilam), and this junction is shown much as it appears on modern maps. Further east, however (not shown in figure 8.8), there are significant deviations from modern maps, and the boundary Lapworth drew between the gneiss and the overthrust Moine Schist does not appear in quite the manner that one might expect according to modern maps. However, the distinctions between the various types of metamorphic rocks were very difficult to establish in the field, and Lapworth's problems in this are not at all surprising.

The way Lapworth construed the situation in the Durness-Eriboll region was never described by him in detail, for reasons that will be explained shortly. However, the general nature of his thinking was made known to his fellow geologists in a brief paper published in the *Proceedings of the Geologists' Associa-*

52. I am grateful to Peter Osborne of the School of Earth Sciences at the University of Birmingham for providing me with the reproduction of part of Lapworth's map.
53. Lapworth to Bonney, Sept. 9, 1882 (Watts transcript, p. 4).
54. The term was coined from the Greek word *mylon*, meaning "mill." A mylonite consists of comminuted fragments that have been welded together by friction-generated heat. In the Highlands, it was also sometimes known as "flinty crush rock."
55. This fine specimen was photographed at the Stack of Glencoul, not in the Eriboll area.

Fig. 8.9. Block of mylonite loose on surface below the Moine Schist at Stack of Glencoul.

tion (1883–84).[56] He distinguished between the gneiss found at Ben Arnaboll (which he called the Arnaboll Gneiss) and schistose rocks of the "Moen" (Moine) and other parts of Sutherland. The Arnaboll Gneiss, he suggested in agreement with Callaway, was (as Nicol had long held to be the case) the "western" (Hebridean, Laurentian, or Archaean) gneiss brought up to the east of the Assynt Series (quartzite, fucoid beds, Durness Limestone, etc.). But whereas Nicol had attributed this configuration to faulting, Lapworth spoke of "gigantic overfolds." The "Sutherland Schistose Series" (Moine Schists), on the other hand, was a "complete intermixture of Archaean and Assynt rocks, the two series being so interfolded and interfelted together that . . . they can never be separated in the field, but must be mapped simply as 'metamorphic'" (Lapworth 1883–84, 438).

As to the planes of foliation and schistosity in the "Sutherland Schistose Series," these were, in Lapworth's view, "gliding-planes, along which the rocks have yielded to the irresistible pressure of the lateral Earth-creep during the process of mountain-making" (ibid.). This interpretation was, of course, totally at odds with that of the Murchison-Geikie doctrine, which construed the planes of schistosity of the gneisses and schists as indications of the original bedding. In that way Murchison and Geikie had been able to speak of a conformable upward sequence into the eastern gneisses and schists. But exactly what was mylonite and what was schist, what the schist was formed of,

56. This paper was read on July 4, 1884.

and what its age was, Lapworth did not clearly indicate, either in his field maps or in his published papers. He was, however, definite that the schists were the most recently formed rocks in the region.[57]

Whereas the details of the petrological relationships of the metamorphic rocks and their ages were not worked out fully by Lapworth, he seems to have settled the question of the geometrical structure of the Assynt Series at Eriboll satisfactorily in his mind, though one cannot be certain even of this. In fact, he never published most of his sections or any of his maps of Durness and Eriboll. In the letter to Bonney of September 9, Lapworth mentioned some sections that he enclosed, which he wished Bonney to pass on to Judd, with the letter. The diagrams were to be returned to Lapworth, and hence they would have become separated from the letter; unfortunately I have not been able to locate them, if they still exist. Thus I have no information from that source. However, Lapworth's first published statement of his results, entitled "The Secret of the Highlands" (1883c), which was read in three parts to the Geologists' Association in March, May, and August 1883 (Watts 1921b, 21) did show a section at An t-Sròn, and also one near the village of Durness, and these are included in figure 8.10, which is reproduced from "The Secret of the Highlands."

In figure 8.10 (top) we see the Northwest Highlands sequence according to the orthodox doctrine of Murchison and Geikie. The middle section shows the stratigraphical sequence at Eriboll, as worked out at An t-Sròn. It may usefully be compared with its source in Lapworth's field map (fig. 8.8). The bottom section shows An t-Sròn with its overfolding and thrusting. The small klippe, labeled "Outlier of flaggy quartzite," is to be noted, and the manner in which Lapworth mapped it may be seen in figure 8.8. We notice also the "syenitic gneiss ('Igneous Rock') of Ben Poll" to which Nicol had drawn attention. This was Heddle's Logan Rock, or Hebridean Gneiss according to Callaway and Bonney. Figures 7 to 10 of Lapworth's sections (which are not reproduced here) offered English-language readers synoptic views of Heim's theory of mountain building. Of these sections, figure 9 provided a diagram demonstrating how an overall synclinal structure might appear, with cursory fieldwork, to be anticlinal; and a large-scale anticlinal structure might appear to be synclinal. (This was what had been happening with the surveyors in the Southern Uplands, particularly R. L. Jack.) The map of figure 8.8 shows Lapworth's careful marking-in of all the dips and faults of the sedimentary rocks of the Assynt Series, and there is ample evidence to show that he was construing the general shape in much the same way as had Callaway.[58]

In 1883 Lapworth had started fieldwork in April and had begun presenting his results as outlined above, but toward the end of the season he was so

57. Lapworth's views on this point have not found favor among later geologists, who regard the Assynt Series as Cambrian, while the Moine Schists are regarded as pre-Cambrian.
58. Callaway, did not publish his maps, however. According to Bailey (1952, 114), Callaway mapped on a one-inch scale, and this led to his work receiving less regard than that of Lapworth. Also, Callaway was regarded as "cantankerous."

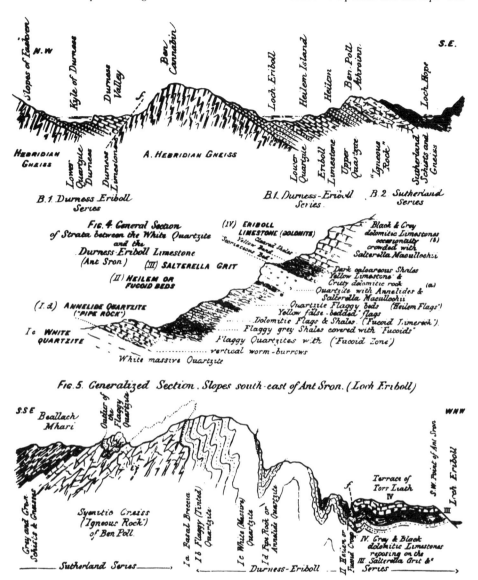

Fig. 8.10. Sections of strata in the Durness-Eriboll region according to Lapworth (1883, part of plate 5), including his characterization, in the top section, of the Murchisonian theory.

exhausted both mentally and physically that he suffered a severe illness, which was partly psychological in origin.[59] There is a well-known story, recounted by Sir Edward Bailey, as to what happened:

59. This was unusually early in the season for fieldwork by a university teacher in the north of Scotland, and in itself is indicative of the urgency that Lapworth felt toward completing and publishing the results of his investigations.

> While in his field quarters, [Lapworth] broke down under the excitement of discovery, feeling the great Moine Nappe grating over his body as he lay tossing on his bed at night; and he had to see others bring the work to full fruition. (Bailey 1952, 114)

But Lapworth had been thinking along these lines, in a sense, since 1882, as may be seen from his letter to Bonney, where he refers to the "terrible grip of this earth-engine."

In his discussion of Lapworth's work, Mott Greene has suggested that Lapworth's breakdown was caused by the expectation of "professional humiliation" (Greene 1982, 201). But I do not believe this to be the case. Though his work had not always been well received by the Edinburgh establishment, his efforts in the Southern Uplands had stopped the surveyors in their tracks and, according to the published comments, had been well received at the Geological Society by a number of auditors when the paper on the Moffat region was presented. Lapworth was in active correspondence with respected geologists such as Judd, Bonney, and Teall, and so far as the latter was concerned he had apparently convinced him in the field of the merits of his views. Best of all, perhaps, the success of his Uplands work had enabled him to gain a chair at Birmingham, so that he now had an institutional base and some intellectual support from the Birmingham scientific community. It was obviously with the greatest enthusiasm and confidence of success that Lapworth entered the field in the Highlands. Moreover, so far as the structure of the Assynt Series was concerned, it seems that he had already broken the back of the problem by 1882. But by 1883, he was entering what was the more difficult phase of the work. He would have been working with the complex petrological problems of the schists and gneisses, and he would have been extending his mapping into physically more remote and demanding terrain, out at Whiten Head, and so on. By contrast, the work needed to sort out the Assynt Series had been done in relatively gentle country, along the coast, and not far from human habitation.

One thing would, however, certainly have been giving Lapworth concern—the forthcoming battle with Geikie. Geikie was a proud and powerful man, who had already been seriously bested by Lapworth in the matter of the structure of the Southern Uplands. The loss of face that Geikie would have to endure, at the very beginning of his tenure as director general of the Survey (which began in 1882), would be immense if he finally had to acknowledge the inadequacy of the Murchison theory of the Highlands because of the efforts of amateurs such as Callaway and Lapworth. There was also the question of the Ordovician to consider. If Lapworth's suggested compromise solution for the Sedgwick-Murchison dispute were accepted, it would mean a radical loss of territory for the old Silurian kingdom and would, if one were to take a Murchisonian view of geology, allow Lapworth to begin building his own (Ordovician) empire. As mentioned earlier, Geikie never permitted the use of the term "Ordovician" in official publications of the Survey. Until the end of his tenure as director general (in 1901) it remained "Lower Silurian."

It is highly likely that it was *this* sort of encounter with Geikie that Lapworth feared, either consciously or unconsciously. But I do not think his worry was that of being ignored, or that his views would not ultimately prevail. Indeed, before his illness, Lapworth seemed quite willing to challenge "authority." He wrote a lengthy and by no means entirely complimentary review of Geikie's *Text-book* (1882) in the *Geological Magazine* (Lapworth 1883b). While praising Geikie's exposition of general geological theory, Lapworth had some harsh things to say about the stratigraphical portions of the work. He condemned Geikie for his adherence to Barrande's doctrine of colonies, and for his "timid little section" where he still expressed support for the old Murchisonian idea of there being a metamorphic rock lying conformably on top of a sedimentary sequence—in some ways the central question of the Highlands controversy. At one point Lapworth became quite vehement in his condemnation:

> Among the older rocks [Geikie] is heavily handicapped by his intense anxiety to defend the peculiar position he has adopted with respect to the age of the Metamorphic rocks of the Highlands, and the sequence in the Southern Uplands, and by his amusing reluctance to admit that any British-born subject who has not been, or is not, a member of H.M. Geological Survey, can possibly produce anything worthy of notice in geology. But once he gets fairly out of the Silurian maze, his work is excellent. (ibid., 84)

Obviously, such public remarks cannot have endeared Lapworth to Geikie. But it may be observed that Geikie (1882, 672n) had made a point of ridiculing Lapworth's views on the Moffat succession. Lapworth was thus not afraid to challenge Geikie publicly, but he appeared to have little taste for the staking of personal claims in the disputed territory.

In the event, of course, Lapworth found that he could not carry on, and he had to retire from the field for six months in 1883–84. During this period of his forced rest, we can follow other interesting developments in the area, with the assistance of the surviving archival material.

On examining Lapworth's correspondence, it becomes immediately clear that besides Bonney he was on intimate terms with the geologists Judd and Teall, and it is therefore appropriate to say a few words about these men, though they have already entered the narrative from time to time and will do so in important ways further on.

John Wesley Judd (1840–1916), who was born at Portsmouth, was reared in London and trained as a teacher at a training college in Westminster.[60] He obtained his first employment at a school in Lincolnshire, where, inspired by his reading of Lyell, he began his early efforts in geological fieldwork. He sat for an examination as a science teacher with the Department of Science and Art, and his examiners, Ramsay and Smyth, were so impressed with his scripts

60. On Judd, see Chambers (1896, cvii–cviii); Geological Magazine's Eminent Living Geologists (1905); Harker (1917); Reeks (1920, 133–37); Cole (1914–22); Bonney (1916); Kendall and Wroot (1924, 346–47).

that he was invited to take a course at the Royal School of Mines, which he duly did. After graduation, Judd worked for a time as an industrial chemist in Sheffield, where he met the mineralogist and petrologist Henry Clifton Sorby (1826–1908) and was introduced by him to the techniques of examining rocks in thin section under the microscope.[61] Judd nearly lost his life in a railway accident in Sheffield, and while recuperating he began fieldwork in Yorkshire. His researches in that part of the world led to an appointment at the Survey in 1867, where he worked for four years, being chiefly engaged with work in the Midlands.

In 1871, Judd accepted a position as a science inspector in London schools, but he continued geological research in an amateur capacity. He surveyed Secondary (Mesozoic) rocks on the eastern and western coasts of Scotland and then began detailed investigations of the Tertiary volcanic rocks of Skye and elsewhere in the Inner Hebrides. As will be recounted in chapter 10, Judd became involved in a heated controversy with Geikie in relation to this work. At the Geological Society, he came into contact with distinguished geologists such as Lyell, Scrope, and Darwin, and under Scrope's auspices he made journeys to Italy, Hungary, and Bohemia to study volcanic phenomena.

In 1876, Judd obtained Ramsay's former chair at the Royal School of Mines, which later became part of the Royal College of Science (now Imperial College). As we shall see in due time, Judd played quite an active part in the Highlands controversy, and he became a severe critic of the Survey under Geikie's directorship. Indeed, he was instrumental in bringing about the Wharton Inquiry into the Survey's work, which terminated Geikie's period of office. Judd retired in 1905, but still continued his researches and did some useful early work on Darwin and the history of the theory of evolution. As can be seen, there are clearly certain parallels between the careers of Lapworth and Judd.

Jethro Teall (1849–1924) came from a well-to-do family in the Cotswolds.[62] He attended St. John's College, Cambridge, where he studied mathematics and science and came into contact with Bonney, from whom he received early instruction in petrography. After graduating, though possessing private means, Teall acted as a University Extension lecturer, and gave geological lectures in Nottingham, among other places, which brought him into contact with J. F. Blake. After his marriage, Teall settled in Kew (where Judd also lived) and devoted himself to the private investigation of British rocks and minerals, with the help of the polarizing microscope. Though not initially employed by either the Survey or a university, he nevertheless soon established himself as one of the country's leading petrographers, and his reputation was confirmed with the publication of his *British Petrography* (Teall 1888).

In 1888, Teall accepted an invitation from Geikie to join the Survey as petrologist, Geikie well realizing that his organization's work was being seri-

61. On this technique, see Hamilton (1982).
62. On Teall, see Geological Magazine's Eminent Living Geologists series (1909a); Marr (1924); Flett (1924–25, 1925, 1937); Bailey (1952, 115–22, 144–71 and passim).

ously hampered by its lack of an expert in hard-rock analysis able to handle the problems that were emerging in Scotland.[63] Also, Geikie needed the services of a first-rate petrographer to bring to completion his own researches on the volcanoes of Great Britain (Geikie 1897). As will be shown, Teall made important contributions to the study of the rocks of the Northwest Highlands, but he never became publicly involved in the associated controversy.[64] When Geikie retired in 1901, it was Teall who was appointed successor, with the difficult task of effecting a reorganization of the Survey's activities according to the requirements of the Wharton Inquiry. Teall proved fully equal to this demand, and was judged a considerable success in his role as scientific bureaucrat, being knighted in 1916. He was for many years a close friend of Lapworth, and at the time when Lapworth was finally demolishing the Survey's theory of the structure and stratigraphical sequence of the Highlands, Teall acted as a kind of confidant to the man, seven years his senior. He also assisted in the publication of Lapworth's work in 1884.

As we have seen, Lapworth had made known his views to Bonney concerning the Durness-Eriboll area soon after his return from the northwest in September 1882, and the letter was quickly passed on to Judd. It appears that Lapworth also wrote to Judd at about the same time, and this letter was passed on to Bonney.[65] Thus they were in three-way communication. We also find that early the following year Judd was setting up a dinner party at his home in Kew, with Bonney and Lapworth, to talk over the new discoveries in the Northwest Highlands.[66] Another letter reveals that the circle was being extended, with a letter from Lapworth to Bonney being sent on to Hicks by Judd.[67] We can here discern the crystallization of a group of "amateur" geologists who, for one reason or another, were opposed to the views of Geikie and the Survey.[68]

The reasons that may have led Judd to be in opposition to Geikie will be considered in chapter 10. Hicks and Geikie were in a state of open warfare as a result of a difference of opinion concerning rocks in the St. David's area in the southwest corner of Wales, which Hicks considered to be pre-Cambrian in age. This was, in fact, part of a wider battle between the "amateurs" and the "surveyors." As we have seen, hitherto unexpected pre-Cambrian rocks were being suggested at about that time by a number of authors—for example, by Bonney at Charnwood Forest in Leicestershire and by Callaway in Shropshire and Anglesey. These suggestions were not entirely welcome to the Survey. Ramsay, for example, supporting fieldwork done by the surveyor William Aveline (1822–1903), had supposed the St. David's rocks to be metamorphosed

63. A comparative youngster, Frederick Hatch (1864–1932), was already doing petrographical work for the Survey.
64. Teall's work as a petrographer came to be so highly regarded by his colleagues in the Survey that he was, in the words of his obituarist, Sir John Flett (1925, lxiv), thought "almost infallible."
65. Lapworth Archive, University of Birmingham, letter A4, Judd to Lapworth, Sept. 20, 1882.
66. Letter A6, Judd to Lapworth, Jan. 22, 1883.
67. Letter A7, Judd to Lapworth, April 4, 1883.
68. See Hicks (1873–86) and Geikie (1882a, 643; 1883).

Cambrian, with associated igneous intrusions. In the case of the St. David's exposures, the accuracy of some of the earlier maps of the Survey was now being questioned, and this was one of the reasons that Geikie was rising to the defense of his former colleagues, especially Ramsay. The published report of the discussion at the Geological Society of Geikie's paper (read on March 21 and April 11, 1883) shows that feeling was running high on the issue, and that most of those present felt, for one reason or another, that the official version was unsatisfactory. Lapworth, it may be noted, entered the discussion (remarks accompanying Geikie 1883, 329) with the suggestion that "the supposed intrusions appeared to be nothing but overturn faults." In his correspondence, he referred to the group of geologists opposing the Survey's views as the "Archaeans," by reason of their advocacy of the pre-Cambrian status of some of the rocks that the Survey held to be Cambrian.

Lapworth's three-part presentation of his ideas in 1883 in the *Geological Magazine* had unfortunately been interrupted by his illness later that year. However, a further brief paper from him was read to the Geologists' Association in July 1884 (Lapworth 1883–84), and there were two later papers that should also be noted: one published in the *Geological Magazine* in 1885 (Lapworth 1885); the other in the *Report* of the British Association meeting held in Aberdeen in 1885 (Lapworth 1886). Thus Lapworth did recover his health sufficiently to play some public part in the conversion of the geological community to his way of thinking concerning the disputed rocks of the northwest. But a good deal of his action was carried out "behind the scenes" through his correspondence with Teall, Judd, and Bonney.

In the following chapter, I shall describe the activities of the officers of the Survey, as they followed close on the heels of Lapworth into the Highlands, as well as Geikie's *volte face* that soon followed. Without going into all the details here, it may be mentioned that following the work of Benjamin Peach and John Horne at Durness and Eriboll in 1883 and 1884 and their coming to a view of the structure essentially the same as that advocated by Callaway and Lapworth, Geikie visited the northern regions in 1884, and was eventually convinced by his own officers in the field of the inadequacy of the Murchisonian doctrine, to which Geikie had for long given such loyal support. He then took a somewhat unusual step. Instead of publishing the results in the customary fashion in the Survey's official publications, he issued a brief paper on November 13 in the pages of *Nature* (Geikie 1884), introducing a somewhat longer paper by Peach and Horne (1884), which set out their recent findings and published a general section of the Durness-Eriboll region, showing thrust faulting. Geikie made a frank admission of his previous errors and of the inadequacy of the old Murchisonian doctrine. But he made no acknowledgment at all that other geologists such as Callaway and Lapworth had been earlier in the field and had worked out the structures independently and in advance of the officers of the Survey. This caused intense annoyance to the amateur party, even though they were glad enough to see Geikie's recantation.

Lapworth's paper to the Geologists' Association, printed in the *Proceedings* of the association for 1883–84, was in fact no more than a letter that Lapworth had written to Teall on July 4, 1884, and that was read to the association in London by Teall on Lapworth's behalf the very same day.[69] It is probable that it would never have been published in the form in which it appeared had it not been for the events that followed on the publication of the papers of Geikie and of Peach and Horne in *Nature* in November 1884. These papers gave rise to a tremendous amount of epistolary activity among the members of the amateur party, and although the correspondence is by no means complete the events can largely be reconstructed with the help of the Lapworth archive in Birmingham.

Geikie's paper was published on November 13. The day after, Lapworth wrote to Teall expressing his pleasure at Geikie's recantation. It appeared to Lapworth that "the matter ha[d] ended beautifully altogether, and Geikie ha[d] done the gentlemanly thing."[70] But others were less satisfied with the outcome. Bonney, it seems, was incensed with the director general's failure to acknowledge the work of the "Archaeans" and of Lapworth, and he sent two letters about the matter to Lapworth, one of which was apparently a copy of an intended letter of protest to *Nature*.[71] Unfortunately, these letters appear not to have been preserved, but we have Lapworth's references to them in another letter of his to Teall.[72] Lapworth was extremely anxious to avoid another major confrontation, such as had occurred in relation to the pre-Cambrian rocks of St. David's. He certainly didn't want Hicks and Callaway rushing in with priority claims. Rather, he urged Teall to discuss the matter with Judd and to "play the game best for British geology and independence [of the amateurs]." We may presume that Bonney's chief objection to Geikie's action arose from the well-known fact that Lapworth had been unable to complete his series "The Secret of the Highlands" in 1883 because of his illness. The Survey had moved in and published without even mentioning Lapworth's prior investigations, merely promising that a detailed account of prior work would be made in the full report that would be published subsequently.

Though apparently chiefly concerned that the truth should emerge no matter who was responsible, Lapworth was not entirely oblivious to his own interests. He suggested that Teall might consider publishing the paper that he had read for Lapworth at the Geologists' Association on July 4. On the other hand, Lapworth telegraphed Bonney requesting that the intended letter to *Nature* not be posted. Presumably Bonney withdrew it, on receipt of Lapworth's request (it had in fact been sent, but was never published). The only public indication of the feeling developing within the geological community was a notice in *Nature* recording a meeting of the Geologists' Association at

69. Lapworth to Teall, Lapworth Archive, letter H70, July 4, 1884. The date of the meeting of the Geologists' Association is also given as July 4, 1884, in its *Proceedings* (p. 419).
70. Lapworth to Teall, letter H71, Nov. 14, 1884.
71. By "Archaeans" Bonney presumably meant himself, Hicks, and Callaway, and possibly some others.
72. Lapworth to Teall, Letter H72, Nov. 18, 1884.

which Hicks made very clear just who had been doing the work in the north-west. Perhaps it showed a little concealed glee in the enforced climb-down of Geikie. But the statement was couched in guarded terms: "The author expressed gratification at the candid manner in which the whole question had been dealt with by the Director-General and the Surveyors in their recent report."[73]

Letters from Lapworth to Teall dated later in November suggest that Lapworth was vacillating and in a state of considerable nervous tension. First he wanted to see his letter of July 4 published; then he didn't. At one stage he urged that Teall should publish his own account of the matter. Teall never did this, but he did prepare an introductory statement concerning Lapworth's views, to accompany the published version of the July 4 letter.[74] This did not appear either. Fortunately for Lapworth's peace of mind, however, a meeting of the Geologists' Association determined that it would be appropriate to publish the July 4 letter in its *Proceedings;* and, as we have seen, this was eventually done.[75]

On Christmas Day, 1884, Lapworth wrote a long and gracious letter to Peach, congratulating him and his colleagues on the results of the Survey's work in the Northwest Highlands.[76] He spoke approvingly of Geikie's change of heart (his "gentlemanly modest and straightforward introductory remarks"), and expressed the hope that "your report has sounded the death-knell of that *odium geologicum* that has been growing rather than dying out amongst us for the last few years." Lapworth also remarked that "if the Survey's officers concede the fact that Geology is a progressive science and act accordingly amateurs to a man will refuse to countenance such an absurdity as the cessation of the Geological Survey."

This remark opens up another dimension to our narrative, namely, that the Survey itself was by no means impregnable and had its critics in the political sphere and in the ideas of the general public, as well as in the view of amateurs such as Bonney, Hicks, or Judd. The preliminary survey was nearing completion for England, Wales, and Ireland. There was obviously a great deal more to do in Scotland, but already the question was beginning to be asked whether the Survey should eventually be established on a permanent footing, as a national institute of geological research, or whether it should be wound up when all the maps of the United Kingdom were published. This question will be addressed more fully in chapter 10, but it is worth noting here that Lapworth should choose to mention the matter to Peach—perhaps as a re-

73. See "Societies and Academies, London: Geologists' Association, Jan. 2," *Nature* 31 (1885): 258–59.
74. Teall, "On Lapworth's views as to the physical structure of the Durness-Eriboll district," Lapworth Archive, H88, undated.
75. Lapworth to Teall, letter H78, Dec. 8, 1884. There was a new president of the association at that time, William Topley (1841–94), who was then taking over from Hicks. Topley was a Survey officer, but one who played no significant part in the Highlands controversy. His support at that juncture would have been welcome.
76. British Geological Survey Archives, GSM 1/520, Lapworth to Peach, Dec. 25, 1884.

minder that the Survey had to realize it was not a law unto itself, and that in a sense it was beholden to the goodwill of the community as a whole. So far as Lapworth was concerned, as an increasingly influential amateur, the Survey had not come up to scratch, since it was failing to keep up with the latest developments in geological work that were appearing overseas. It was thus Lapworth, not the professionals in the Survey, who was responsible for introducing and applying in British geology ideas such as those of Linnarsson, Heim, Brøgger, and Lehmann.[77]

The other interesting matter in Lapworth's letter to Peach concerns Blake's work. When Blake met Lapworth with Teall in the northwest in 1883, he was by his own admission still a "Murchisonian." To be sure, Blake was soon fully convinced by Lapworth's arguments and demonstrations that the Murchisonian theory was hopeless, and that a folded and faulted structure was required in order to make sense of the field evidence. But this had more or less been established by then anyway, through Callaway's work and through Lapworth's own "Secret of the Highlands," which was then appearing in the *Geological Magazine*. In fact, Lapworth was now less concerned with such relatively trivial matters as whether there was or was not an "upper quartzite." The real question for him was the difficult relationship between the different kinds of metamorphic rocks in the Northwest Highlands, in terms of age, provenance, and mineral-chemical composition.

Yet according to his letter to Peach, Lapworth chose deliberately not to unburden himself of his thoughts on these matters to Blake.

> [Blake] took sketches and copious notes at the time, & I expected that his object was a future paper upon the subject, and therefore carefully abstained from giving him any clue to the origin and relations of the Sutherland schists.[78]

This did not betoken a wholly candid attitude on Lapworth's part. Like nearly all scientists, he was clearly not unconcerned with matters of priority. It appears from Lapworth's letter that left to his own devices Blake had misconstrued the section at Far-out Head, thinking that the rocks there were an "upward continuation" of the Durness Limestone—very much a Murchisonian doctrine. Also, Blake (1883–84, 427) gave a most peculiar structure for Stronchrubie (near Assynt), where he went without the help of Lapworth's eyes. On the other hand, Blake managed to offer quite a coherent account of the structures to the east of Loch Eriboll, where he had had Lapworth's personal guidance. And by the time his paper was published, Blake found himself able to construe the structure at Far-out Head as being one that arose from thrusting (ibid., 429).

Peach replied to Lapworth on January 4, 1885, thanking him for his "very generous letter" and stating that he had shown Geikie parts of a letter from

---

77. Lehmann author has not been mentioned so far in our account, but his work will be discussed below.
78. Lapworth to Peach, Dec. 25, 1884.

Lapworth to a colleague of Peach, H. M. Cadell, which had opened Geikie's eyes to Lapworth's character.[79] However, while public niceties were being observed, there appears to have been a good deal going on behind the scenes that even the correspondence does not reveal fully, let alone the formal publications. Lapworth was in Edinburgh on January 6, 1885, to address the Edinburgh Philosophical Society, and he spent about a week in the north. Not surprisingly, he took the opportunity to call in at the Edinburgh office of the Survey to have a chat with some of the staff, Geikie being well out of harm's way in London. We do not, of course, have a verbatim account of the conversations in the Edinburgh office, but some record survives in the form of a letter from Lapworth to Teall, written shortly after Lapworth's visit.[80]

It may be that what Lapworth had to recount to Teall from Edinburgh was no more than rumor and office gossip. Nevertheless, it gives an idea of what the surveyors thought was going on, even if it was far from the truth of the matter: Geikie was believed to fear the publication of a "mighty paper" by Teall and Lapworth on the Eriboll district that would entirely forestall the surveyors' work; all the work in the north was to be taken out of the hands of the local staff down to London; the group of Highlands surveyors was to be broken up and Peach put back to work in the Girvan area; an essay on metamorphism was to be published in *Nature* to enhance the glory of the director general. Only Geikie, it seemed, had it in for "amateurs." So whatever the amateurs might do, as long as it was right and fair, would be supported from within the Survey.

It would appear that there was no friction between the rank and file of the Survey and the amateurs—or, at least, not with Lapworth. Looking back over the events that I have been expounding, it is evident that the friction had always been with the directors of the Survey, first Murchison, then Ramsay (though only to a minor degree with him), and then Geikie, with Hull adding his weight to the official party from Ireland. Further, it was the directors who had chosen to nail their colors to the mast of the official Survey doctrine concerning the structure of the Highlands.

I suggest that this reflects the rather peculiar intellectual situation of the directorate. The director general could not personally supervise all the work going on in his domains at any one time. He had to rely on the competence of his subordinates. But he also had to speak for the Survey and try to defend its interests, with the public and with the government. If a director was led astray by his subordinates—as appears to have occurred to Geikie with the work of Jack in the Southern Uplands—he might be led into all sorts of embarrassments. Yet he was expected to defend the integrity of the Survey and not let it be the object of carping criticism. The situation was such that it was scarcely possible for a single man to accomplish, and Geikie signally failed in the Southern Uplands and in the Northwest Highlands, despite his manifold gifts and ceaseless labors. But then, it would have been more prudent for the

79. Lapworth Archive, letter H48, Peach to Lapworth, Jan. 4, 1885.
80. Letter H79, Lapworth to Teall, Jan. 13, 1885.

Survey not to have had such firm views on a region such as the Northwest Highlands, which it had not yet properly surveyed. In Geikie's case, his fatal error was not taking seriously the work of Lapworth, despite its manifest excellence in the papers such as those of the Moffat and Girvan successions. It was sheer foolishness on Geikie's part not to have recognized Lapworth's stature as a geologist sufficiently from the beginning.

By 1884 or 1885 it was becoming more widely apparent that the main problem in the northwest was now the petrology of the metamorphic rocks, rather than the sequence and structure of the Assynt Series. Heim's work gave Lapworth all he needed in order to offer a convincing account of the geometrical structures of the sedimentary rocks. But it was otherwise with the metamorphics. In his correspondence, and in one of his publications (1886, 1026), Lapworth referred to the significance for his investigations on the Scottish metamorphic rocks of the work of the German geologist Johannes Georg Lehmann (1851–1925), lecturer in mineralogy and geology at Bonn, and formerly of the Geological Survey of Saxony. The work of Lehmann (1884) was concerned with the metamorphic rocks of Saxony (to which reference was made in chapter 6). However, Lapworth maintained privately that he had arrived at ideas similar to those of Lehmann quite independently, which is not unlikely seeing that Lapworth's fieldwork in the north was finished before Lehmann's book was published.[81]

In his monograph of 1884, Lehmann discussed the mining area of Saxony where metamorphic rocks called "granulites" (in the German sense) were to be found. As previously mentioned, Saxony was the type area for such rocks. There is a large ellipse-shaped metamorphic complex, about forty miles long, to the north of Chemnitz, which had long been of interest to miners and geologists. Carl Friedrich Naumann (1797–1873), of the University of Leipzig, had mapped the area, representing the center of the ellipse as granite, and the main body as granulite, round which there was a margin of schists and clay slates.[82] He regarded the granulites (along with the granite) as eruptive and responsible for producing the schists (Naumann 1856), which was in keeping with the general theory of the origin of foliated rocks that he espoused (Naumann 1847, 1848). For this work, he was able to draw on the authority of a considerable number of previous authors, including Darwin, for the view that foliations might be found in eruptive rocks. Naumann made no clear distinction between granites and gneisses, regarding them as "brothers" (Naumann 1847, 302).[83]

But Lehmann (who dedicated his work to Naumann) disagreed with the earlier interpretations. Lehmann examined the many different rock types in the elliptical complex in detail, both in hand specimens, and with the aid of the microscope. The thin sections strongly suggested that the materials of the

81. Letter A29, Lapworth to Bonney, Dec. 28, 1884.
82. On Naumann, see Kobell (1874).
83. On page 424, we find Darwin introducing the distinction between cleavage and foliation for the first time. (Darwin's *Geological Observations on South America* was first published in 1846.)

granulites were produced by mechanical deformation and chemical recrystallization, giving minerals such as mica and pyroxenes. The material of the granulite appeared to have yielded under pressure. But there also appeared to be stretching of the rocks and a fragmentation or brecciation of the material, and the appearance of "slide-surfaces" (*Gleitflächen*). Thus the processes did not seem to be due to heat in any special way. Lehmann (1884, 248) wrote of the formation of a rubblelike product, or a groundmass consisting of fragments and mixed parts in which larger bits of the original rock were embedded. Recrystallization to mica would give a schistose character to the rock, which, of course, had nothing to do with any original sedimentation. All this supposedly took place in a structure quite different from that under debate in the Northwest Highlands. Nevertheless, it may be seen that there were features of Lehmann's thinking that would have been useful to those seeking to uncover the "secret of the Highlands."[84]

The parallel features of Lehmann's and Lapworth's thinking had to do with the processes that Lapworth compared with the action of a mill—the rolling out of rocks along the lines of thrusting to form "mylonites," of composition dependent on that of the rock being deformed; also, the formation of schists by the stretching, folding, and shearing of the rocks under the influence of the processes of mountain building, so that the laminations arose from the action of mechanical forces and had nothing to do with planes of deposition. Thus foliations or planes of schistosity could not have any chronological significance.[85] Thinking in this way, Lapworth regarded the eastern schists (Moines) of the Eriboll region as metamorphosed rocks of mixed origin. (For Lapworth the analogy was between the Moine Schists and the Saxon granulites, not—as Nicol perhaps envisaged—between the "eastern gneiss" and the Saxon rocks.) Geikie was well aware of Lehmann's work, and gave a sympathetic and lucid review of it in *Nature* (Geikie 1884b). However, while he appreciated the idea of the schists of the Saxon granulite complex as being metamorphic in origin, and not Archaean, he did not apply this notion to the Northwest Highlands, whereas Lapworth's construal of the Moine Schists was one of his main contributions to the theory of the Highlands.

It should be noted that the major point of difference between Lapworth and Callaway had to do with the interpretation of the eastern metamorphic rocks. Callaway had distinguished Hebridean and Caledonian units, the latter being divided into a Hope Series of schists and the Arnaboll Series of gneisses. Lapworth also had an Arnaboll Gneiss, and a set of Sutherland Schists; but the former were, for Lapworth, Archaean, while the latter were thought to have been produced some time later than the Silurian period. Moreover, for Lapworth, the Sutherland Schists might be formed from the Archaean rocks

84. For further aspects of the history of metamorphic theory in the nineteenth century, see: Vallance (1984).

85. For a comparison of the views of Lehmann and Lapworth, see the statement drawn up by Teall, intended to serve as an introduction to the printed version of Lapworth's July 4 (1884) paper (Lapworth Archive, H95). Teall's statement was never published, however.

themselves, from the sediments of the Assynt Series (which he regarded as Ordovician, equivalent to the Survey's Lower Silurian), or from igneous rocks that might have been intruded in the region. In some places, one might be able to recognize what the rocks had been originally; but as one moved further east, the distinctions became wholly obliterated (Lapworth 1883–84, 439–40). All this was essentially the kind of argument that might be found in the work of Lehmann, with the foliations of schists and gneisses (or granulites in the German sense) being due to pressure and the consequent restructuring of rocks.

Following the presentation of Lapworth's ideas to the public (though not with the magnificent care and detailed evidence that had characterized his Moffat and Girvan papers), and the preliminary announcement of the views of the Survey in 1884, the Highlands controversy was moving toward its conclusion, so far as the "internal" history of geology was concerned. But its social ramifications were only just beginning. Reviews of the history of the controversy were produced by Bonney, in his presidential address to the Geological Society in 1885 (Bonney 1885, 50–57), by Hicks in a paper to the Geologists' Association the same year (Hicks 1885–86), by Judd in his presidential address to Section C of the British Association when it met in Aberdeen in September 1885 (Judd 1886), and by Lapworth himself in a paper published in the *Geological Magazine* (Lapworth 1885).

Hicks's paper was marked by a somewhat transparent attempt to show that his work of 1878 gave him priority for all that followed. Judd's address was more interesting in that it marked the final reinstatement of Nicol as a reputable geologist, after years of neglect, even vilification, on the part of Murchison. Judd asserted that Nicol had mastered the great Highland problem in "all its essential details" in his paper at the Aberdeen meeting of the British Association in 1859. It was fitting that Nicol should thus be posthumously vindicated in his home city, where he had previously been vanquished by the all-powerful Murchison. Even if Nicol did not live to see his reputation restored in so public a manner, it must have been gratifying to his widow, who was recorded by Woodward (1908, 197) as saying, when told of Judd's kindly remarks concerning her deceased husband: "I knew well James would be right." Why Judd may have been so anxious to record the victory of the amateurs will be examined in chapter 10, and there is some reason to regard his assessment as something of an exaggeration. Nevertheless, he paid tribute to Geikie for "his loyalty in accepting results so entirely opposed to his published opinions, and to his promptitude in making his fellow-workers in geology acquainted with these important discoveries." We may doubt, however, that Geikie took pleasure in Judd's further remarks:

> Unfortunately called upon while still young, and with but little of that
> ripe experience which he has since gained, to grapple with the most

intricate of problems . . . [Geikie's] own judgement yielded, though not
without serious misgivings, when opposed to the ardent confidence of
a companion and friend, whose reputation in the scientific world com-
manded his respect, and whose previous achievements had won his
complete reliance. (Judd 1886, 1000)

Thus responsibility for the debacle was tactfully but conveniently focused on
the deceased Murchison, which was, no doubt, where it chiefly belonged.

Lapworth's 1885 paper, entitled "On the Close of the Highlands Controversy," did not bring forward any new information, but it made an interesting
public statement of what he desired for the relationship between the amateurs
and the surveyors: "Every investigator has a right to address himself to any
part of the work [of geological research] that he pleases, and the right, if he
deems it fitting to exercise it, to demand a full recognition of the importance
of his own contribution to the common stock of discovery" (1885, 102). This
point seems to have been well and truly won by Lapworth, as a result of the
efforts of workers such as himself, Callaway, and Hicks. But we may wonder
why it was that he felt so strongly about this question. The other participants
in the debate did not seem to regard it as a matter of such supreme consequence. I can only assume that Lapworth had suffered from Geikie's tongue
and had been dismissed as a mere amateur at some geological meeting or
meetings where he had expounded his views—very likely on the Moffat succession. But of such verbal remarks, I have found no trace on the historical
record, and they can only be hypothesized on the basis of Lapworth's preoccupation with his supposed "amateur" status—even after he had been appointed to his Birmingham chair.

The preoccupation with the role and status of the amateur geologist was
not, however, restricted solely to Lapworth. In his presidential address to the
Geological Society, Bonney (1885, 57) spoke on the same topic. He warned
that the laudable esprit de corps of the Survey might lead its officers to "speak
with some contempt of 'amateurs.'" Using a theological analogy (which may
have appealed to Bonney with his clerical background), he warned against a
division of the geological community into "establishment" and "nonconfor-
mist" camps. "Science," he said, "needs no infallible church and admits of
no pope."

While such sentiments no doubt attracted applause, it is doubtful that they
did much to alter Lapworth's views of Geikie. There is a long-standing tradition in the Survey that Geikie was strongly disliked because of his arrogant
and autocratic behavior (Wilson, 1985, 15).[86] It is difficult to document such a
claim, for thoughts like this are not usually expressed in writing, and if they
are, they may not find their way into archives. However, it is interesting to
see what the outsider Lapworth had to say about Geikie. In a letter to H. A.
Nicholson—Nicol's one-time colleague in Aberdeen—Lapworth stated that
Geikie would "never shew the slightest gratitude for anything on earth."[87] He

86. Wilson is a present member of the Survey staff.
87. Lapworth Archive, letter A30, Lapworth to Nicholson, Jan. 9, 1883.

"*tacks* so swiftly that he is rather difficult to get at." To Teall, Geikie was described as "a man who never forgives or forgets."[88] He was "a little man," and,
advised Lapworth, "the less you are beholden to him the better."[89] This certainly evidences some degree of personal animosity between Lapworth and
Geikie.

The context of Lapworth's advice to Teall to avoid being beholden to Geikie
is worth pursuing a little. As we have seen, Teall acted as a source of moral
support during Lapworth's crisis years of 1883 and 1884. Teall advised him,
presented his ideas to the geological community, and tried to speak up for his
priority claims, since Lapworth himself was not (it seems) much inclined to
push his own interests, adopting, rather, a "science for science's sake" attitude.
But Teall—though an "amateur"—was recognized by Geikie as being perhaps
the best petrographer in the country. Moreover, Geikie realized that the
Survey was not very strong in that department. If the organization was to do a
proper stratigraphical job in the largely unfossiliferous Highlands, bringing
its research standards up to the international level that was being set in Germany and elsewhere on the Continent, it would have to enlist the services of
someone like Teall.[90]

It is not altogether surprising, then, that as early as 1885 we find evidence
of Geikie's overtures to Teall to take a position as petrologist at the Survey.[91]
This despite the fact that Teall was well known to be a friend and ally of Lapworth. Lapworth urged his friend to think of his own career and act in whatever way was best suited to his interests.[92] But in point of fact, Teall did not
actually join the Survey staff until 1888, by which time much of the heat had
gone out of the Highlands controversy; his treatise, *British Petrography,* was
completed; and the Survey officers were well on the way to sorting out the
structure of the Durness-Eriboll and Assynt regions and preparing their detailed maps. Thereafter, Teall provided essential service to the Highlands work
and contributed very significantly to the final publication of the great memoir
of 1907. In the 1880s, before he joined the Survey, he made one major contribution to the Highlands work, in the form of a paper'that showed that one of
the main claims made by Lapworth was entirely plausible: namely, that all
sorts of different rocks, including igneous rocks, might be altered by pressure
(perhaps assisted by heat) into metamorphic rocks. This was one of the claims
to be found in Lehmann's treatise, and it was of the highest importance for
Lapworth's theory as to the formation of the "Sutherland (Moine) Schists."

The work in which Teall proved the transformation of igneous into metamorphic material (Teall 1885) was carried out on two dykes that he examined
with his friend Blake in 1883. These were found near the little fishing village
of Scourie, about forty miles north of Ullapool, in the region of the "western"

88. Letter H84, Lapworth to Teall, Feb. 6, 1886.
89. Letter H84, Lapworth to Teall, Feb. 6, 1886.
90. The Survey's mainstay in the field of mineralogy and petrology, Frank Rutley (1842–1904),
had resigned in 1882 to take up a lectureship in mineralogy at the Royal College of Science.
91. The matter is mentioned in a letter from Lapworth to Teall, dated Jan. 4, 1885 (letter H82).
92. Lapworth to Teall, letter H83, Feb. 3, 1885.

gneiss (Archaean / Lewisian / Hebridean / Laurentian).[93] The dykes, which ran inland in a southeasterly direction from the coast, clearly cut across the strike of the main mass of gneiss in the district. In part, they appeared igneous in character (dolerite), but in part they displayed the characteristic foliation of a metamorphic rock.[94]

Teall examined the rocks in the dykes with care, both for their mineral contents and their chemical composition. It was apparent that both the foliated and the unfoliated portions were much the same in chemical composition. Moreover, examination of specimens in thin section revealed a gradual transition from dolerite to gneiss, and in particular mineralogical transitions from crystals of augite to hornblende.[95] Within individual hand specimens, transitions could be perceived between the granular dolerite and the foliated gneiss. It appeared, therefore, that the gneiss was formed *from* the dolerite, the alteration only being partial, so that the transitional stages might be studied. The evidence indicated that the transition took place under high pressure, the rocks becoming plastic and acquiring a foliated character without melting. Moreover, Teall cited evidence that augite was the stable form at high temperatures, while hornblende was the stable form at lower temperatures. It thus followed that the transition (involving formation of hornblende) had not occurred at high temperatures.

All this was very satisfactory from the point of view of the theory that Lapworth was advancing. It provided a Scottish example of the processes that Lehmann had been suggesting for the rocks of Saxony. It showed, moreover, that igneous rocks could be converted directly to metamorphic rocks by the action of pressure and without undue heat. Hence it made plausible Lapworth's suggestions for the formation of the "Sutherland Schists" (Moine Schists) from a complex mixture of sedimentary, igneous, and metamorphic rocks by the action of intense lateral pressures.

It may be pointed out, however, that when Teall met Lapworth in the Highlands in 1883, he could not, according to his own admission, immediately understand the suggestions that Lapworth made to him concerning the processes of dynamic metamorphism (Teall 1885, 143).[96] At that time, Teall col-

93. The "Scourie Dykes" have subsequently become a major topic of geological research in the twentieth century, and have been a central focus of debate in studies of the Lewisian Gneiss and its history. See Barber et al. (1978, 8 and passim).
94. Dolerites are a family of igneous rocks, dark and medium-grained, with minerals such as calcic plagioclase, augite, and olivine, and in many cases with a characteristic "ophitic" texture—in which the augite crystals "wrap round" those of the feldspar. The name was suggested by the French mineralogist R.-J. Haüy, since the dolerites were so difficult to characterize petrologically.
95. Augite is an important member of the family of minerals known as pyroxenes, with green, brown, or black monoclinic crystals. A complex alumino-silicate of calcium, magnesium, iron, and aluminium, it is one of the chief dark constituents of basic igneous rocks.

Hornblende is a common dark mineral with monoclinic crystals, belonging to the group known as the amphiboles. It is found typically in igneous rocks, and in metamorphic rocks derived from igneous rocks. Chemically, hornblende consists of a complex alumino-silicate of calcium, magnesium, iron, sodium, and aluminium. The composition is variable.
96. See also Lapworth Archive, document H95.

lected specimens under Lapworth's guidance, but it was not until he had the opportunity to study their thin sections under the microscope, and had read Lehmann's 1884 treatise, that he felt he properly understood the arguments that Lapworth was advancing. Thereafter, Teall's skills as a practical petrographer were at Lapworth's disposal in the furtherance of his views. Thus the Scourie Dykes helped to promote the new ideas concerning the structure and history of the Northwest Highlands that were now emerging. But it seems that Teall, like Blake and everyone else, needed the assistance of Lapworth's eyes to make sense of the rocks of the Northwest.

This brings to a close what needs to be said about the "amateur" contributions to the understanding of the geology of northwest Scotland. It was now the turn of the "professionals" of the Survey, who, it must be said, soon carried matters far beyond what the lone geologists such as Lapworth had been able to accomplish. Chiefly through the remarkable efforts of Benjamin Peach and John Horne, the Survey was able to transform a less than distinguished performance into a major triumph. Their mapping would prove one of the outstanding pieces of geological work accomplished in the nineteenth century. In the next chapter, then, I show how the Survey "snatched victory from the jaws of defeat," and in chapter 10 we shall examine the consequences of this "victory" for the subsequent history of the Survey.

9

The Professionals Vindicated: The Work of the Surveyors in the Northwest Highlands

Following the strong challenge by the geologists of the amateur party—Hicks, Bonney, Judd, Callaway, Lapworth, and so forth—to the doctrines that had been long propounded by the Survey in the Highlands, one of Geikie's first tasks as director general was to direct a strong detachment of his best surveyors to look into the problem and to settle it once and for all, one way or another. By 1882, it was in any case appropriate to do this. The preliminary surveys both of Ireland and of England and Wales were nearing completion, but there was still a great deal to do in Scotland. It was recognized that a satisfactory survey of Scotland depended on having a firm stratigraphical base for the whole geological sequence. Clearly, this was to be found in the Northwest Highlands and the Hebridean islands, as Murchison had realized many years before.

On taking up the reins in London, Geikie did not immediately relinquish control in Scotland, and it may be that one of the reasons that prompted him to do this was his concern to keep a close watch on what was happening in the northwest, and to direct matters in such a way that his interests there were secured. Thus it was that although Henry Howell (who had first initiated Geikie into the art of field mapping when he joined the Survey in 1855) was appointed director for Scotland in 1882, he continued for some time to finish off work that he was doing in the north of England. According to Geikie's autobiography (1924, 203–4), this decision was reached by "the authorities"; but in fact it was a determination that he made himself—and it was thought to typify the arrogance of the man (Wilson 1985, 15).

Yet one cannot but admire the energy and enthusiasm with which Geikie entered into his new field of endeavor, for which he had been preparing himself during the whole of his career. As we have seen, he got out his massive *Text-book of Geology* in 1882, and this major task was brought to a successful conclusion only a few months after he assumed his new office. The same year, he also produced an anthology of some of his more popular writings: *Geological Sketches at Home and Abroad*. There were staff troubles among the Survey officers in Ireland, which he had to try to smooth over. He also had to try to meet all the staff in the field. The government was calling for the rapid completion of the basic field mapping, and in order to do this it was determined that the information should be entered on one-inch maps, even though this was an almost impossible requirement where complex areas such as those encountered in Scotland were concerned. On top of all this, Geikie had to tidy up his loose ends in Scotland, including making his farewells to the University of Edinburgh, and move himself and his family (with four children) down to London, where they took up residence in Harrow.

There had been serious questions about the workings of the Survey toward the end of the period of Ramsay's directorate. The matter was raised in parliament in both the Commons and the Lords, as a result of a highly critical letter appearing in the *Times* in June 1880.[1] The letter, signed "Observer," made the interesting point that the Survey was, by its very nature, a temporary organization, and therefore it was in the interests of the officers to drag their work out as long as possible. It was also pointed out that since it was becoming the practice for the surveyors to settle down somewhere with their families, rather than being constantly on the move, there was a further tendency for work to proceed slowly. The issues were also discussed in the pages of the *Geological Magazine,* but not in a manner specifically critical to the activities of the Survey; rather, consideration was given to the sorts of things that a survey could and should be able to do.[2]

The government did not allow such criticism to occur without taking any action, and the Department of Science and Art, which was responsible for the administration of the Survey, began to make some searching inquiries. Ramsay endeavored to defend his organization as best he could, but he was in seriously declining health and found it impossible to protect himself; and the government determined to bring about sweeping changes.[3] These included Ramsay's early retirement, a weeding of staff, completion of the survey work on the one-inch scale with minimum delay, discontinuation of the publication of the six-inch maps, and an examination of how long it might take to complete the primary survey work.[4] The estimate for the completion dates was subsequently given by Ramsay: England and Wales, two and one-half years; Ireland, seven and one-half years; and Scotland twenty-two years.[5]

These, then, were some of the constraints under which Geikie was required to function as he assumed his directorship. It should not be supposed, therefore, that all the courses of action that he undertook—which might have seemed somewhat strange to the outside observer, or indeed to the historian of geology—were ones that he would have preferred to adopt if he had had a freer hand. But we may imagine that someone like Lapworth, who had exam-

1. "Observer," "The Geological Survey," *Times* (London), June 22, 1880, 10, col. 6. On the discussions in Parliament, see *Hansard,* ser. 3, 254: 255 (July 12, 1880), 1194–95 (July 23, 1880).
2. See "The Geological Survey," *Geological Magazine* 8 (1881): 39–43.
3. The committee that looked into the matter consisted of Earl Spencer, W. Mundella, Sir F. Sandford, and Colonel J. F. D. Donnelly (Public Records Office, DSIR/9, file 10, 5/112). None of these men were scientists, but as we shall see Donnelly in particular was one of the most distinguished civil servants in the nineteenth century who were concerned with scientific matters.
4. Public Records Office, DSIR/9, file 10, 5/112 (decision taken Feb. 21, 1881).
5. Ibid., 8/112 (Ramsay's letter, dated July 20, 1881).

ined the graptolite zones inch by inch at Dobb's Linn and elsewhere, would have thought it quite absurd to attempt to map the Scottish Highlands using one-inch field maps.

The two surveyors who were chosen to initiate the attack on the Northwest Highlands, beginning in 1883, were Benjamin Neeve Peach (1842–1926) and John Horne (1848–1928), both of whom have been mentioned previously, but who should now be introduced more fully.[6] As we saw in chapter 3, Peach's father was one of the early contributors to the study of the geology of the Northwest Highlands. It was partly as a gesture of thanks to Charles Peach for finding the celebrated Durness fossils that Murchison generously arranged for his son, then aged seventeen, to study science at the Royal School of Mines in London, under such men as the geologist Ramsay, the naturalist and Darwinian Thomas Henry Huxley, and the notable German teacher of chemistry A. W. Hofmann.[7] Peach arrived in London in 1859. Then, with Murchison's recommendation, he joined the Survey staff in 1862, transferring to the Scottish branch five years later, at the time when Geikie assumed responsibility for the northern office (though Peach had been working in Scotland well before that date). He received valuable training in practical palaeontology in his early days from the Survey palaeontologist, Salter.

Peach did substantial work in the mapping of the Scottish coalfields, and in large areas of the Old Red Sandstone, with its associated volcanic rocks. By 1879, he was acting palaeontologist for the Scottish branch of the Survey and he was promoted to the rank of district surveyor in 1882. However, although he did so much important palaeontological work, his main and lasting reputation rested on his unraveling of the structures of the Southern Uplands and the Northwest Highlands, which culminated in the two great memoirs on these topics (Peach and Horne 1899; Peach et al. 1907). For this work, Peach's special skill in discerning (or perhaps intuiting) the internal structure of a mountain or a region on the basis of its general appearance was given its greatest opportunity, and it was here that he outshone all other geologists in Britain in the nineteenth century, except perhaps Charles Lapworth.

Peach's skills in the field have been admirably described by his colleague and great admirer Edward Greenly (1861–1951):

> Those who knew the real Peach, the master in tectonics, were those
> who have sat with him facing some mountain full of complicated struc-

---

6. In his autobiography, Geikie (1924, 214) gave the starting date as "the spring of 1884," but this was clearly a mistake—just possibly a deliberate one. On Peach see Macnair and Mort (1908, 259–64); Horne (1926a); Bailey (1926); Greenly (1928–32); (1938, 2: 515–21); Anderson (1980). The volume of Macnair and Mort contains some brief autobiographical notes of Peach.

On Horne, see Flett (1928–29); MacGregor (1928); Gregory (1928, 1929); Greenly (1928, 2: 506–14).

7. In his obituary of Peach, Horne stated that Murchison "completed the education of young Peach." This leaves it unclear whether Murchison assisted financially. But it is a reasonable assumption that he did do so.

tures. Peach's eye discerned every feature on its flanks; here an ill-defined space of colour, there a gently curving line, stopping perhaps against some other curving line. In a few moments, he had perceived what all this meant, and on some scrap of paper he would rapidly draw a section through the mountain's very heart, revealing to his astonished and delighted friend fold upon fold, thrust upon thrust, sometimes with the added perplexity of an uncomfortable overstep brought forward on a major thrust-plane from a buried region far away to the eastward. (Greenly 1928–32, 6)

I have had the privilege of accompanying Peach when he was mapping a hillside where numerous thrusts emerged from beneath a major thrust-plane, and was amazed at the rapidity with which he perceived outcrop after outcrop, and the decision with which he swept in their lines upon his maps. . . . Oh for an hour of Benjamin Peach! With him, what a delight it was to take a map along a mountainside. (ibid., 7)

Greenly's essay had many other pleasing anecdotes concerning Peach—but regretfully they cannot all be accommodated here.[8] For Sir Edward Bailey and others, Peach was regarded, in character and physical appearance, as akin to a lion or eagle (Bailey 1926, 189).

Besides his geological skills, Peach also had considerable artistic talent, as

8. We may, however, record a song jotted into a notebook of the young Surveyman, Henry Cadell (see below), who worked with Peach in the heroic mapping of the Northwest Highlands in the 1880s. It gives a further insight into Peach's interests and character, and shows the esteem in which he was held by his colleagues.

> D'ye ken Ben Peach with his shoulders broad
> His dimpled cheeks and smiling nod?
> D'ye ken Ben Peach with his reel and rod
> As he starts for the loch in the morning?
>
> Chorus
>
> Yes I ken Ben Peach and Jock Scott too
> The mallard wing and the black Tulu(?)
> You should see him at sport in Kylesku
> With a whale on his line in the morning.
>
> Chorus
>
> He lived at Durness for many a day
> With old Robert Sutherland at Sango Bay
> And was once nearly killed in a furious fray
> With a Frenchman at 1 in the morning.
>
> Chorus
>
> He tried camp life near Achamore
> But the rain came down in a steady pour.
> So all he could do was to lie and snore
> Till seventeen o'clock in the morning   (continues)

may be seen by the booklet reproducing a number of his drawings and paintings that was issued by the Institute of Geological Sciences in 1980, to accompany an exhibition of his artistic work (Anderson 1980).[9] Doubtless, the artistic facility greatly assisted Peach in his cartographic work. However, there was another side to Peach's character, in the view of some of those who had known him personally. According to Geikie, while Peach was the ablest field geologist in Britain in his day, he had a great "repugnance to the use of the pen," and "he would rather walk a mile than write a letter" (Geikie 1924, 244). Also, according to Bailey (1952, 91), Peach did not read widely in the geological literature, but preferred to keep himself informed by listening to the conversation of his colleagues. However, my examination of the archives pertaining to the Highlands Controversy suggests to me that Peach was quite a prolific letter writer, and that his correspondence was agreeably warm. After his retirement in 1905, he published a substantial volume on fossil crustaceans of the Carboniferous period (Peach 1908). Even so, it is true to say that the bulk of the writing for the great memoirs on the Southern Uplands and the Northwest Highlands was done by John Horne rather than Peach. Peach was the great conceptualizer—the man whose capacious mind could swiftly grasp a complex structure whole and reduce it to an intelligible section or diagram. This, of course, was the skill on which Murchison had so prided himself. But he was *too* hasty in his judgments, and could make egregious mistakes, with what results the present study has been at pains to demonstrate.

Horne was born at a village in Stirlingshire not far from Glasgow, and was educated in science at Glasgow University, though he did not complete his degree there. Instead, he applied for a position in the Scottish branch of the Survey and joined the staff as a young recruit in 1867, the year that Geikie took over the Scottish directorate. Almost from the beginning, Horne began work with Peach, to whom he was initially a kind of apprentice, so to speak; and

---

Chorus

Then here's to Ben Peach and his sunny smile
As he comes home at night from a day on the Kyle.
May the fish aye rise in a splendid style
To the Peach-flavoured flies in the morning.

Chorus

For the ricket of his reel brought the fish from their bed
And the swish of his line high over his head.
They hurried up in shoals to be all struck dead
By a wave of his wand in the morning.

(H. M. Cadell, "Notes on Geology of Sutherland. Mapping begun at Durness by Messrs. Peach Horne & Hinxman in summer of 1883. Afterwards continued southward by Messrs. Peach, Horne, Hinxman, Clough & Cadell in 1884.1885," British Geological Survey, Land Survey Archives, Edinburgh, 95.1, p. 176.)

9. In my judgment, however, Peach's talent in this line was not as great as that of Geikie, displayed in the beautiful collection of Geikie watercolors held at the Haslemere Educational Museum.

thus began their lifelong collaboration and friendship, which has become legendary in the annals of Scottish geology. Horne's early surveying was done in the south of Scotland, and he was inevitably associated with the debacle over the Southern Uplands. But it was Horne who was the first Survey officer to "see the light" (Bailey 1952, 93) and recognize the merit of the case that Lapworth was developing.

Peach and Horne were dissatisfied with the way in which the Survey prevented them from putting forward their own ideas as independent scientists on areas that were being examined by the government officers. And so they used to spend their vacations in the far north of Scotland (the Orkneys, Shetlands, and Caithness), away from the scene of official action, to collect material for papers that they might publish independently. Thus it came about that they did some important joint work on glacial geology (Peach and Horne 1879, 1880, 1881), and in the Shetland Islands they were the first to describe in detail the volcanic rocks there in the Old Red Sandstone.

In the late 1870s and early 1880s, Horne did much mapping in the region round the Moray Firth (Nairn and Inverness). But then he was called by Geikie, with Peach, to try to unravel the intricacies of the Northwest Highlands; and it is this work with which we shall be chiefly concerned in the present chapter. Then, with Peach, Horne gradually returned to take up the problems of the Southern Uplands, which fully vindicated Lapworth's work and led to the publication of the great Silurian memoir (Peach and Horne 1899). Horne was a very clear thinker and lucid writer, and an excellent organizer—indeed the perfect foil for the brilliant but less systematic Benjamin Peach. Thus on Geikie's retirement and the associated reorganization of the Survey, Horne took over the running of the Scottish branch of the Survey, a position he held from 1901 until his retirement in 1911. Horne did much of the writing for the massive Northwest Highlands memoir (Peach et al. 1907), though Geikie did the final editing. In his later years, Horne did a good deal of editorial work and was active in several scientific societies in Edinburgh, Glasgow, and elsewhere. He and Peach were honored by the award of the prestigious Wollaston Medal of the Geological Society in 1921, and Horne received honorary doctorates from St. Andrews, Aberdeen, and Edinburgh. In their old age, the two friends were preparing a last monograph on Scottish geology, but with Peach's death in 1926 Horne had to carry on alone. Tragically he too never saw the final outcome of his life's work, though the volume was issued posthumously (Peach and Horne 1930). One of Horne's last literary acts—the composition of his obituary of Peach—must have been a truly heartbreaking ordeal for him.

Though Horne married and was survived by two sons and a daughter, one cannot help but think that it was Peach who was his real life-companion. In the words of J. S. Flett, a later director and historian of the Survey: "Their work became so perfectly blended that no one could tell the part contributed by each. Each of them knew the other's mind so thoroughly that he knew what

Fig. 9.1. Benjamin Peach and John Horne outside Inchnadamph Hotel in 1912
(Bailey 1926, plate 18, facing p. 188).

he would say before it was said" (Flett 1928–29, viii). Their warm and whole-
hearted relationship is well displayed in the photograph of them taken outside the Inchnadamph Hotel in 1912, on the occasion of a British Association
excursion that they led that year from Dundee to the Northwest Highlands.
This picture is reproduced in figure 9.1.[10]

In giving an account of the Survey's work in the Northwest Highlands, then,
it is scarcely possible to distinguish exactly the separate contributions of Peach
and Horne, or for that matter, of the other surveyors, Charles T. Clough
(1852–1916), William Gunn (1837–1902), Lionel W. Hinxman (1855–1936),
and Henry M. Cadell (1860–1934); and I shall, to a large degree, treat them
as a homogeneous group. It may be said, however, that Peach and Horne began the fieldwork in 1883, and they were later joined by the other workers.
Prior to the appearance of the first major publication on the surveyors' work
(Peach et al. 1888), Peach and Horne were chiefly responsible for the north-

10. This photograph (taken by Professor S. H. Reynolds of the University of Bristol) was first
published in the *Geological Magazine* in 1926. A copy is to be seen to this day in the foyer of the
Inchnadamph Hotel.

ern work at Durness and Eriboll, and at Assynt; Cadell mapped the line between Eriboll and Loch More; Clough (who was revered in the Survey for his artistic skill as a mapworker) worked from Loch More to Glas Bheinn (i.e., the Loch Glencoul area between Loch More and Assynt); Hinxman mapped from Elphin to Strath Kanaird (taking in the famous Knockan Cliff section); and Gunn mapped from there south to Little Loch Broom (thus including the Ullapool region). From 1889 until 1895, Greenly was also a member of the field party.

As we have seen, Peach, Horne, Teall, Blake, and Lapworth met at Durness early in the season in 1883 (Teall and Watts 1921, xxxv). We know that by that date Lapworth had largely reached an understanding of the general structure of the northwest. We know too that he did not display all his understanding to Blake on that occasion. It is also well established that all parties concerned maintained after the event that the understandings reached by Lapworth on the one hand, and the government surveyors on the other, were entirely independent. Yet the matter is not entirely clear-cut, as the following details reveal.

Years later, after Lapworth's death, his biographer and former colleague at Birmingham, William Watts, wrote to Teall seeking some first-hand information to put into his obituary essay on Lapworth's life and work (Watts 1921b). In commenting on the draft of Watts's essay, Teall wrote: "When [Blake and I] met Peach & Horne at Durness [Lapworth] told them that the solution of the problem could not be found there & that they must look for it in Eriboll as he had done the previous year (1882). They followed his advice in 1883, found the clue, and at once rushed into print."[11] In another letter, Teall wrote: "[Lapworth] insists that the Survey work was done independently. Was it? If Geikie did not know Peach & Horne are much to blame or I am much mistaken. I hope I am."[12]

These statements are somewhat enigmatic, as no doubt they were intended to be, considering that Peach, Horne, and Geikie were all still alive in 1920. At least they show that Lapworth told Peach and Horne that he had solved the fundamental problem of the "secret of the Highlands"—which is scarcely surprising since Lapworth was already beginning to publish his results quite early in 1883. They also tell us that he advised the surveyors that they would do better to look carefully at Eriboll, rather than Durness. But whether he told them more than that remains a mystery. On the one hand, it might seem strange that Lapworth would *not* tell the surveyors all he knew, for he was always on excellent terms with Peach and Horne. He must have had his field maps with him, which one would have expected him to show to them. On the other hand, we know that he didn't tell Blake everything at that time—although it appears that he did tell Teall (and presumably Blake also) quite a lot at Eriboll—when, for example, the equivalence of Murchison's upper and lower quartzites was so clearly demonstrated. But from another point of view,

11. Lapworth Archive, University of Birmingham, letter H99, Teall to Watts, Dec. 10, 1920.
12. Letter H97, Teall to Watts, April 5, 1920.

it would have been surprising if Lapworth *had* told Peach and Horne every-thing they needed to know in order to unravel the "secret," for this would surely have played into Geikie's hands; and Lapworth would understandably have been reluctant to assist any organization that had Geikie as its director.

Fortunately, however, we do have another piece of information on this head: the account by the former surveyor Edward Greenly, who maintained that he had discussed the matter directly with Peach and Horne. According to Greenly (1938, 2: 490), Peach informed him that they were told nothing by Lapworth, and he did not show them his maps. It might be thought that Greenly, a former surveyor and a devoted admirer of Peach and Horne, could possibly have been partial in his account. But there seems to be no very strong reason to doubt his statement.

Let us therefore leave the matter where it has long stood—with the state-ment that Lapworth and the surveyors worked out the "secret" independ-ently, with Lapworth at least giving them the clue that Eriboll was the best place to look for a solution to the problem. One thing was certain, namely, that outside the Survey (and perhaps within it to some degree) it was believed that it was in poor taste for Geikie, Peach, and Horne to publish their prelimi-nary account of the northwest in November 1884 (Geikie 1884b; Peach and Horne 1884), without mention of the discoveries of Lapworth and Callaway, especially when it was common knowledge that Lapworth had been forced by his illness to retire from the field. However, criticism was always directed against Geikie, rather than against Peach and Horne, for the publication was made at Geikie's behest. Yet in defence of Geikie it must be said that by that time Lapworth had already published three parts of "The Secret of the High-lands" and his letter of July 4 to Teall, which gave many of his main results. Callaway had also had his say in print.

In Teall's letter to Watts, quoted above, there is an indication that Peach and Horne rushed off to Eriboll from Durness as soon as they received the suggestion to do so from Lapworth. But there is reason to suppose that they did not go there until 1884. This appears from another letter from Teall to Watts, which actually says (in contradiction to his other statement): "I don't think the Survey men went to Erribol [*sic*] till 1884."[13] In addition, a hasty de-parture to Eriboll would have been contrary to the systematic character of official Survey work. If they had begun to prepare a map of the Durness re-gion, they would not, without instructions from Geikie, have moved on before finishing at Durness. And from some surviving correspondence, it seems un-likely that he would have issued such instructions early in the "campaign." In-deed, at first it seemed as if the results were turning out in favor of the Murchison-Geikie theory. Geikie wrote to Peach on May 21, 1883:

> I am delighted to hear that you and the rest of the party are so well established at Durness. I knew you would be astonished at the model-like simplicity of the Country. How one can talk about faults and inver-

13. Letter H100, Teall to Watts, Dec. 23, 1920.

sions is to me inconceivable, except on the view that he has a foregone
conclusion to support. But you and Horne are perfectly free to find
out the exact truth. If Murchison was wrong by all means let this be
clearly shewn. There will be not the least "Official Pressure" put on
you. I am as much concerned as anybody can be in knowing what the
true structure of the Sutherland ground is.[14]

Geikie went on to tell Peach of what he had heard of the meeting at the Geological Society (May 9, 1883), at which Callaway had presented his major
statement of his views (see chap. 7). It was, perhaps, unfortunate that Geikie
had not been present to hear Callaway's persuasive arguments in person. As it
was, he referred contemptuously to "Callaway's lucubration," and remained
unchanged in his opinions.

Geikie visited the surveyors at Durness in June 1883, but he apparently
found "the results entirely confirming Murchison's ideas and demonstrating
the ingenious perversities of Lapworth, Callaway and Hicks to be worthy of
their authors."[15] Unfortunately, we have no field notebook corresponding to
this visit. But in his annual report on the progress of the Survey for 1883,
Geikie stated that having visited Peach, Horne, and Hinxman in the field at
Durness he "took the officers over a series of important sections further South,
where I had in previous years established the order of succession, and which
will, I trust, expedite progress of the work this year" (Geikie 1884a, 284).

We have some further valuable information about the events of 1883 from
the autobiography of Edward Greenly (1938, 2: 487–96). At Sangomore on
the coast a little to the east of Durness, Peach and Horne observed Moine
Schist lying on the Durness Limestone, apparently conformably, and Peach
took it as confirmation of Murchison's regular upward succession. But Horne
was more skeptical and thought it might actually be a pseudoconformity corresponding to a plane of movement. His doubts were strengthened when they
moved on to Eriboll, where they found definite evidence of inversion of the
strata, and no limestone underlying the Moine Schist.

According to Greenly's account, when Geikie visited the surveyors he was
so impressed by the evidence at Durness and Sangomore that he was all ready
to dash off a paper to *Nature*, announcing that there was now definite evidence
for the truth of the Murchisonian doctrine. Yet Horne cautioned strongly
against this and succeeded in dissuading his chief from overhasty action:

Geikie was impressed by this. Masterful spirit though he was, he was
by no means lacking in prudence: such earnestness as this of Horne's
could not be without good reason. Accordingly he gave way. (ibid., 491)

14. British Geological Survey Archives, Keyworth, GSM 1/380, p. 163, Geikie to Peach, May 21,
1883.
15. Ibid., GSM 1/320, Geikie to F. W. Rudler, June 29, 1883. (Frederick William Rudler, 1840–
1915, was curator of the Museum of Practical Geology and librarian to the Survey. He was an
indoor geologist, who handled much of the Survey's routine administration at that period. He was
president of the Geologists' Association from 1887 to 1889.)

Even so, Geikie insisted on taking Peach and Horne south to look at the well-known sections at Knockan Rock, Ullapool, and so on.

But it seems that Horne was by now highly skeptical; and on discussing the matter back at the Survey headquarters in Edinburgh during the winter months of 1883–84, he came to think that the Murchisonian theory was untenable, with the Moines lying on quite different strata at different localities. This indicated either unconformity or movement: and the evidence at Eriboll clearly favored the latter hypothesis. Eventually, Peach too was convinced that the Murchisonian theory had to be abandoned when, in the season of 1884, he was mapping imbricate structures on the eastern shore of Loch Eriboll to the north of Heilam. The surveyors saw the Moine Thrust in section from a boat off Whiten Head; and the enigma of the schist at Far-out Head was finally dispelled by interpreting it as a klippe—a fragment of the Moine Schists that had at one time formed a cover right over the Durness area and now existed only as an erosional remnant at Far-out Head.

For Geikie to have tried to hold the line much further would have been hopeless. We learn from a statement made by Howell in 1899—at the time of the government inquiry into the affairs of the Survey that eventually terminated Geikie's period in office (see chap. 10)—that Geikie received a report from Peach in the field in July 1884, to the effect that the old Murchison-Geikie theory was "absolutely untenable."[16] Peach apparently wrote informing Geikie of the surveyors' developing views, on August 10, 1884, when he was on holiday with his family in Cornwall. Geikie replied:

> I have been reflecting on all that you have recently written to me and I think I see the solution of the problem & that your work will not only be of the utmost importance as regards the structure of the Highlands but will give us a new starting point in metamorphic studies. I have been working lately at metamorphic studies and have much to discuss with you.[17]

Again, we have a somewhat enigmatic statement. But it may be noted that on June 5 Geikie had published a favorable review in *Nature* of Lehmann's treatise on metamorphism (Geikie 1884b), and it is clear that he was bringing himself up-to-date with the recent developments in German petrology.[18] I suggest that he was hoping that with this new theoretical knowledge he might somehow be able to effect a reconciliation of the old ideas and the new. It was not to be, however. According to Geikie's autobiography, he visited his officers

16. Ibid., BGS 2/211, Statement of H. H. Howell to Wharton Committee, May 1899. (Unfortunately, we do not have Peach's actual report.)
17. Ibid., GSM 1/380, p. 176, Geikie to Peach, Aug. 15, 1884.
18. There is also evidence that Horne was familiarizing himself with German ideas on metamorphism at about the same time. In a paper presented to the Mineralogical Society on June 24, 1884 (Horne 1886), he reported work on metamorphic rocks in Aberdeenshire and Galloway, and contended that the observations were better construed in terms of German theory than according to the ideas of Sterry Hunt.

in the field in October 1884, in such bad weather that he was blown over at times. Nevertheless, he admitted: "I was completely convinced by the evidence so fully worked out by my two colleagues, that the Murchisonian view of the order of sequence in the north-west of Scotland [had to] be abandoned" (Geikie 1924, 214).

This sounds well, suggesting an honest acknowledgment of a change of heart when presented with new empirical evidence. And indeed the evidence was new to Geikie—very new—for so far as I have been able to determine, in all those years in which he had held to the Murchisonian doctrine of a conformable upward succession he had never been to Loch Eriboll, where the evidence that persuaded Lapworth, Callaway, Teall, Blake, and then Peach and Horne, was to be found. This was an extraordinary state of affairs indeed, and tells us something about the nature of scientific knowledge, which I shall discuss in chapter 11. But was Geikie a willing receiver of the information that was made known to him by Peach and Horne, as his empiricism would require? I think not. There is a letter from Lapworth to Teall, written shortly after Lapworth's visit to the Edinburgh office of the Survey in January 1885. It said:

> The story of the enforced conversion of G[eikie] in the NW was intensely amusing. They were almost afraid he'd never yield: and his aspect after yielding seems to have been pitiable.[19]

There could, of course, be a suspicion that while they were in the field Peach and Horne heard about Lapworth's letter of July 4, 1884, read at the meeting of the Geologists' Association that day. And thus they received the clue needed to unravel the succession at Eriboll. But this suggestion is refuted by a letter from Horne to Lapworth, dated March 19, 1885, written in response to a congratulatory letter from Lapworth to Peach concerning the surveyors' success in elucidating the Northwest Highlands structure, and publishing the preliminary results in *Nature*.[20] Horne's letter (which reads in some measure as if he was trying to "justify" the actions of the Survey in Lapworth's eyes) stated that Horne had read the July 4 letter in January 1885, "with no little excitement." This seems conclusive demonstration that the announcement to the Geologists' Association did not provide the signpost toward Peach and Horne's comprehension of the Eriboll region.

Horne's letter also confirms the view that he and Peach mapped the Eriboll area without having any definite expectation of finding what they did: "It came upon us with great surprise. I will never forget our feelings of wonder when we began to realize the full meaning of the *extraordinary* mechanical movements and the startling array of new facts bearing on the metamorphism." Apparently, having eventually convinced Geikie in the field (he was "profoundly impressed with the work"), Peach and Horne accompanied him south into Ross-shire, to revisit the old sections (probably at Ullapool and

19. Lapworth Archives, Letter H79, Lapworth to Teall, Jan. 13, 1885.
20. Letter A35, Horne to Lapworth, March 19, 1885.

Loch Maree), and to see whether they would bear the new interpretation suggested by the Durness-Eriboll area. At one spot, they found some Torridon Sandstone converted into schist, and this seemed to be regarded as "crucial" evidence by Geikie.[21] Convinced at last, Geikie returned to London, and the Peach and Horne report, with Geikie's prefatory recantation, was published only a few days later (on November 13). Here Geikie introduced the term "thrust plane" into the literature for the first time.

Horne was anxious to point out to Lapworth that when the full accounts were published there would be a proper statement of the results of all previous workers in the northwest. Indeed, this promise was fully kept, though the first major paper on the subject did not appear until 1888 (Peach et al. 1888), and the full memoir only in 1907 (Peach et al. 1907).

Horne emphasized to Lapworth that the new discoveries in the northwest were "achieved without the slightest friction between [Geikie] and ourselves." But this does not mesh entirely with what Lapworth recounted to Teall about his visit to Edinburgh early in 1885 (see p. 258). Later that year, at the Aberdeen meeting of the British Association, when Judd gave his partisan history of the Highlands controversy and his vindication of Nicol, it appears that Judd and Peach did not get on too well together. This was reported by Teall to Lapworth, who replied with distinct heat:

> You must recollect the atmosphere of the Survey office in Edinburgh for the last 10 years has not been a well ventilated one. Everything has been worked on the principle of a brilliant genius—the isolated leader of the only geologists in the country—to whom all matters in geology of necessity belong, being badgered and vilified by a set of ? ignorant curs, outside society, outside decency, to whom Judd, Bonney, Nicholson, Page[22] & a few others who ought to know better have slyly given aid & encouragement & against whom the feelings of the Lord's anointed and some of his more faithful followers is naturally very sour and bitter.[23]

21. The exact locality was not stated by Horne, but he mentioned that the metamorphosed Torridon Sandstone was to be found *east* of the thrust plane. Such a state of affairs is to be found, for example, in the exposures at Loch Coulin (near Loch Clare) to the south of Kinlochewe (Loch Maree), where Torridonian lies caught between the Moine and Kinlochewe thrust planes. In the 1907 memoir, it is mentioned (p. 289) that Horne collected some metamorphosed Torridonian in this area, but the date at which he did this was not given. However, I suggest that it was in this area of Ross-shire, to the south of Kinlochewe, that Geikie finally yielded all hope of saving his well-loved theory. For according to the Murchison paradigm, there was no reason at all for altered Torridon Sandstone ("Cambrian") to be lying over the "Silurian" sediments. Yet it clearly did so at Coulin. (Possibly also the geologists finally climbed the hill between Glen Logan and Glen Bianasdail/Haasach, where much the same thing would have been found.)
22. David Page (1814–79) was noted for his many excellent texts on geological subjects that he published for Messrs. Chambers of Edinburgh. In 1871 he was appointed professor of geology at Newcastle. Page published a couple of minor papers on the Northwest Highlands, but he did not significantly influence the course of the controversy.
23. Lapworth Archive, Lapworth to Teall, letter H85, Sept. 21, 1885.

The old animosities had by no means been extinguished. But the letter also suggests that there was indeed a group of "insiders" and "outsiders" in the geological community. So while relationships could remain quite cordial between Geikie and his staff, external relationships were still strained, despite the best efforts of Horne to put matters in the most favorable light.

In writing his recollections of the affair many years later, Greenly emphasized the vitally important role played by Horne. As in the Southern Uplands, he was the first to come round to the view that the official doctrine was mistaken; and it was he who dissuaded Geikie from getting himself deeper into the mire with a paper to *Nature* in 1883 describing the Sangomore sections. We may gather that it took considerable strength of character to oppose Geikie:

> I served under Geikie, and I know that there was about him something which, to say the very least, did not invite opposition from the members of the staff. I have also been told, and can well believe, that before 1884 this was much more pronounced than when I served under him [in the 1890s]. . . . Yet [Horne] boldly confronted his formidable Chief. Without Horne's discernment and courage, there would certainly have been a dire disaster to the Survey. (Greenly 1938, 2:495)

This gives a clear indication of the characters of Geikie and Horne; and also of the psychological effect that the events of 1883–84 had on Geikie. He must have been severely shaken by what happened, in relation to his reputation and to that of the Survey as a whole.

Let us turn now to consider what the surveyors accomplished during their many years of labor in the northwest. By the time of the preliminary announcement of their results at the end of 1884, we can be confident that they had made a close examination of the terrain round Durness and Eriboll, mapping it with care as they went. We do not have their draft field maps, and the final result of their efforts at Eriboll was only deposited in the Survey office in Edinburgh as late as 1913.[24] But the hand-drawn six-inch map shows the main lines of survey that were established in 1883–84.

In their paper of 1884, Peach and Horne proposed a stratigraphical division of the rocks at Durness and Eriboll that was rather different from that suggested by Lapworth. Peach and Horne had more divisions of the Durness Limestone, while Lapworth had more for the quartzite. The surveyors' proposed classification is outlined below (Peach and Horne 1884, 32–33):

Newer Gneiss
10. Gneissose flagstones
 9. Siliceous schist
 8. Frilled schist
    Green schist

---

24. British Geological Survey Map Library, Edinburgh, Sutherland O.S. 1:10, 560. Sheet #5 (1913), reg. no. 044149.

Lower Silurian Series[25]
 7. C. Upper or Calcareous Series ("Durness-Eriboll Limestone")
          VII. Durine Group
          VI. Croisaphull Group   [Various
          V. Balnakiel Group   subdivisions
          IV. Sangomore Group   of the
          III. Sailmhor Group   Durness
          II. Eilean Dubh Group   Limestone]
          I. Ghrudaidh Group

      B. Middle Series (partly calcareous, partly arenaceous)
 6.   Upper Zone Serpulite grit
 5.   Middle Zone Brown, calcareous grits,
                      quartzites, and cleaved   [Equivalent
                      shales   to former
          Lower Zone Calcareous mudstones and   fucoid beds]
                      dolomitic bands,
                      with worm-casts

      A. Lower or Arenaceous Series [Equivalent to former quartzite or
               "quartz rock"]
 4.   Upper Zone Piped quartzite [Equivalent to
                      former "pipe rock"]
 3.   Lower Zone False-bedded quartzite
                      Basal breccia

"Cambrian"
 2. Sandstones and conglomerates ["Torridonian"]

Archaean Gneiss
 1. Gneiss, granite, pegmatite, etc.

As may be seen, this sequence (which the reader may wish to compare with that of Lapworth shown on page 243 and also with figure 9.2) was not fundamentally new, though there were changes in emphasis and points of detail compared with the earlier investigations of Callaway and Lapworth. We notice particularly the increase in the number of subdivisions of the Durness Limestone, and one may wonder how this was accomplished. The limestone subdivisions (which have been retained into modern publications) must undoubtedly have been established at Durness, for not all of them appear at Eriboll. The surveyors used partly lithological subdivisions, and partly ones based on fossils. The main criteria were in fact color and texture, with some attention also given to chemical composition.[26] Assuming a generally eastern dip across the Durness peninsula, the several layers could have been recog-

25. Today the upper portion of this is designated as Ordovician, the lower as Cambrian.
26. That is, whether or not the rock was dolomitic (magnesium bearing).

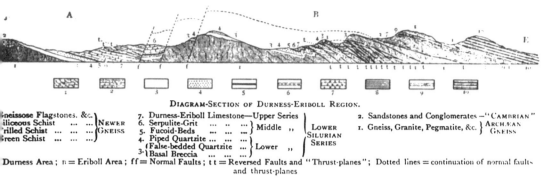

Fig. 9.2. Section of strata in Durness-Eriboll region, according to Peach and Horne (1884, 33).

nized one on top of the other, even though disturbed by (normal) faulting. The pattern is complex, but not impossible to make out with careful mapping; and Peach and Horne would have been able to go over the ground carefully, fairly close to civilization, with a shore line offering plenty of exposures.

Then coming over to Eriboll, they would have had to map their ground with extreme care, but as this was done the pattern might gradually unfold. Two major thrust planes were discerned, which they called the Arnaboll Thrust and the Moine Thrust. But below the more westerly Arnaboll Thrust, which was recognized as bringing forward the ancient Archaean gneiss over the "Lower Silurian" sedimentary rocks, there was an immensely complicated "imbricate" structure, consisting of chopped-up portions of Serpulite grit, fucoid beds, etc., giving an appearance on a map rather like the coat of a zebra. Each layer was recognized as being formed by a minor thrust fault, usually of higher angle than the main Arnaboll and Moine thrusts. The Moine Schists, brought over from the east by thrust faulting, were the highest rocks in the series, geometrically speaking, but what their actual age and provenance might be could not be stated with certainty.

After two years of fieldwork, with a superb performance in geological cartography, the surveyors were able to offer the section shown in figure 9.2 to the public (Peach and Horne 1884, 33). The section was later elaborated considerably, but in this figure the basic understanding was established for the geometrical structure of the Durness-Eriboll area as it is accepted still today. The reader may wish to interpret the figure in terms of the thrusting of material from the east. But the eastern materials (e.g., 10, the Moine Schists) may, according to the Peach-Horne (and Lapworth) view, be thought of as having been formed during the process of their emplacement, rather than having been deposited somehow on the underlying rocks or (as Murchison supposed) having somehow been metamorphosed *in situ*. It would also appear that there were at least *two* phases of faulting, in that the trough of the Durness region has been dropped down by normal faults. Were this not so, the Durness Limestone, et cetera, might have been many hundreds of feet

Fig. 9.3. Section of strata through Loch Assynt, Cnoc an Droighinn, Ben an Fhurain and Ben More Assynt (Institute of Geological Sciences, 1965), based on original survey by Peach, Horne, Clough, Hinxman, and Cadell, 1892. By permission of the Director, British Geological Survey: British Crown copyright reserved.

higher in the Durness area, and thus would long since have been worn away by erosion.

In the next few years, the surveyors gradually worked their way southward, until eventually the whole "zone of complication" from Durness and Eriboll to Skye was mapped. Naturally, much attention was given to the Assynt region, where the usual cover of Moine Schist has been stripped back by erosion, so that the immensely complicated underlying structure of multiple faults and folds is revealed. To give an idea of the magnitude of the cartographic task facing the surveyors, and the skill and elegance with which they performed this task, I reproduce in plate 8 a small portion of the six-inch map for the Assynt region, showing specifically the area round the Inchnadamph Hotel at the head of Loch Assynt. Murchison's famous Cnoc an Droighinn section ran from the lakeside at Inchnadamph in a northeasterly direction. The colors were keyed as follows:

Fawn	Alluvium	
Dark blue a^{III}	Sailmhor Group	⎱
Light blue a^{II}	Eilean Dubh Group	⎰ Durness Limestone
Blue a^{I}	Grudaidh Group	
Yellow a^4	Serpulite Grit	
Dark olive green a^3	"Fucoid Beds"	
Light mauve a^2	"Pipe Rock"	⎱ Quartzite
Darker mauve a^1	Basal quartzite	⎰
Plum color	Torridonian (not appearing on illustrated portion of map)	
Red A	Gneiss	
Dark red D_i	Diorite	
Orange F	Felstone and porphyry	
Crimson/maroon B_g	Dolerite-epidiorite and Hornblende-schist dykes	
Purple	Ultrabasic dykes	

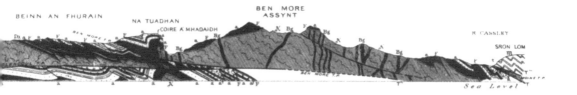

It is instructive to compare this map with that produced by Heddle in 1881 (plate 7). One may, perhaps, see in this why some officers of the Survey felt that their work was so markedly superior to that of the amateur party that it bore no comparison, and why all authority in British geology should be vested in the judgments of the official government organization.

In their publication of 1888, the surveyors provided a considerable number of additional sections of various parts of the "zone of complication," but instead of reproducing any of these I offer in figure 9.3 a section through Inchnadamph and Ben More Assynt taken from the Survey map that was eventually published in 1923 (Institute of Geological Sciences 1965, second accompanying section).[27] This is based on Sheets 101 and 107, issued in 1892, according to the original surveys of Peach, Horne, Clough, Hinxman, and Cadell. I use this later map because the structure that the surveyors established in the nineteenth century has substantially stood the test of time in the twentieth century; and in these sections we may see displayed the final outcome of the Highlands controversy, as it concerned the understanding of the three-dimensional geometrical structure of the Northwest Highlands. The section shows two major thrust planes: the Glencoul Thrust and Moine Thrust. If one "stretches out" the section mentally, the rocks of the Cnoc an Droighinn hill would then be well over to the east, where it might have formed an uninterrupted set of layers connecting with the quartzite (*a*) to the west of Loch Assynt, and up the slope of Ben Gharb. But by being thrust to the west along the plane of the Glencoul Thrust, for example, the quartzites (etc.) became crumpled and faulted, and in some places (as at Ben an Fhurain), the underlying "archaean" gneiss (*A″*) was forced up to the surface. This was Nicol's "igneous" rock, or Heddle's "Logan Rock." Examination of the section across Loch Assynt to Cnoc an Droighinn will show how Murchison was deceived into thinking that there were two quartzites—to the west and the east of the loch.

At the right (east) of the section will be seen the Moine Schists (*m*), also

27. This map was reprinted in 1947, and in 1965 with minor emendations. It was reissued as a third impression in 1969.

thrust in from the east, along the Moine Thrust plane. According to the theory developed by Lapworth, and by Peach and Horne, this schist was *produced* as a result of the thrusting process, whereas the gneiss (A''), though altered by the tectonic activity to some degree, was chiefly *moved* by the thrusting, rather than generated by it.

At Assynt, the Moines are well to the east, presumably as a result of erosion, though they may never have extended so far to the west at that locality. But they curve round to the south, extending much further west so as to be brought directly over the Durness Limestone (a^1 and a^{11}). Thus we have the famous section at Knockan Cliff. If the thrust plane there is construed as a bedding plane—which is a likely construal if the notion of a low-angle reverse fault is lacking—then we can easily be convinced of the Murchisonian hypothesis of a regular conformable ascending stratigraphical sequence (though without an "upper" quartzite at that point). It should not be thought surprising, therefore, that Murchison construed things the way he did. Hudleston, it will be recalled, regarded this section as "unassailable." It must be emphasized, however, that Murchison only seems to have driven along the road at the foot of Knockan Cliff, examined that section, and walked about a bit. He did not do anything like enough fieldwork in the district in order to form a hard-and-fast judgment on the structure of the region. In particular he did not make himself a detailed map of the area. The same must be said of Geikie.

From this point in my narrative, I shall regard the problem of the geometrical structure of the Northwest Highlands as essentially solved, at least so far as nineteenth- and early twentieth-century geology was concerned. Geikie was so satisfied with the results that in the summer of 1885 all the officers of the Survey engaged in Highlands work were sent on a visit to Sutherland to study the sections under the guidance of Peach and Horne, so that they might understand at first hand the new basis that the Survey was adopting for the investigations of the geology of the Highlands (Geikie 1886, 327). Needless to say, however, there were numerous further problems to be handled concerning the ages of the various rocks, the tectonic history, and the processes of metamorphism. These matters depended particularly on the results of petrological investigations. There were also interesting attempts made to simulate the tectonic processes by means of laboratory investigations. For the remainder of this chapter, then, we shall concentrate our attention on these matters, before taking up further social issues that flowed from the controversy in the following chapter.

In their preliminary announcement of their findings at Durness and Eriboll, in November 1884, Peach and Horne offered much the same view as Lapworth concerning the eastern schists and gneisses. That is, between the two great thrust planes a rock (Logan Rock) could be discerned that was still recognizably Archaean, though altered a good deal. But the uppermost layer of "Newer Gneiss"—above the upper thrust plane—was apparently a rock of multiple origin.[28] Careful examination revealed several stages of alteration with rocks metamorphosed into new materials, and new planes of foliation could be seen superimposed on old ones. For example, the stages of alteration of an igneous rock into a green schist and hornblende gneiss into hornblende schist could be seen; and quartzite could be seen with the quartz particles being elongated in a common direction.[29] Further east, all original differences seemed to have been obliterated, and everywhere was schist, with some new minerals like garnet and actinolite appearing.[30] All this accorded with the kinds of arguments that had been recently advanced by Lehmann in Germany.

The increasing emphasis on problems of petrology and metamorphic processes, as opposed simply to the structures that might be revealed by mapping, was apparent in the paper of 1888—the first full-scale account of the Survey's new discoveries in the northwest. By 1888, the surveyors had had the opportunity to give close examination to the rocks of the western gneiss (Archaean), as well as those of the zone of complication running southwest from Eriboll. Within the western gneiss, certain patches of unfoliated basic igneous rocks (gabbros, peridotites, etc.) were found, running in belts parallel with the general foliation.[31] These could be found gradually merging into poorly foliated basic gneisses, and then into ones with more perfect foliation. The conclusion, then, was that the western gneisses had originally been basic igneous rocks and that these had been converted into basic gneisses by some kind of metamorphic process or processes. There were structures in the rocks also that suggested that they had been formed—at a period long prior to the post–"Lower Silurian" movements with which we have hitherto been chiefly concerned—by the action of great pressure, with concomitant thrust faulting.

There were also numerous dykes within the gneiss, a pair of which had already been studied in detail by Teall, as we have seen in chapter 8. Many of these dykes showed clear evidence of their having been subject to deformation subsequent to their emplacement. As Teall had shown, the movements had in some cases produced alteration of the material of the dykes into metamorphic rock. But in addition, there was evidence for three distinct lines of movement having affected the dykes. For example, by consideration of the deformation produced in the dykes and the associated mineralogical changes occurring in the dyke rocks, it was found possible to determine the direction of the forces that had produced the changes. Figure 9.4 shows clearly how the direction of movement might be inferred from the present appearance of the dyke (Peach et al. 1888, 394).

28. "Moine Schist" in modern parlance.
29. The color of green schists is due to the mineral chlorite.
30. Garnets form a family of minerals, some of which are used as semiprecious stones or as abrasives, but commonly found as dark-red, poorly shaped crystals in schists characteristic of a particular intensity of metamorphism. The type here would have been almandine, an iron aluminium silicate. Actinolite, a calcium magnesium iron silicate, is a common metamorphic mineral of the amphibole family.
31. A gabbro is a dark, coarse-grained basic igneous rock formed by deep-seated crystallization of magma; peridotites are even more basic and lack the feldspars found in gabbros.

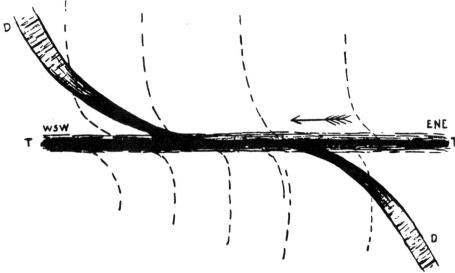

T. Thrust-plane.
D. Dyke, becoming attenuated and deflected and increasingly schistose as it
approaches the plane, and being reduced to a " crush-rock " in immediate vicinity
of the thrust. Displacement about ¼ mile.
Dotted lines indicate the strike of the gneiss, the planes being deflected as
they approach the thrust.
The arrow shows direction of movement.
The parallel lines indicate the newer schistosity produced in the gneiss within
the influence of the thrust.

Fig. 9.4. Figure showing effect of thrust plane on a dyke, according to Peach et al.
(1888, 394).

Having identified the lines of disruption in the gneiss by examining the appearances of the dykes, as shown in figure 9.4, the surveyors were then able to
gauge the effect of these lines of dislocation on the gneiss rocks themselves.
They could be seen to be thrown into a series of sharp folds running parallel
to the line of dislocation, and a new set of foliations (and associated mineralogical changes), superimposed on the original foliations, might be discerned. But
all these changes were not found to affect the overlying Torridonian rocks at
all. The changes were therefore manifestly pre-Cambrian in age and marked
huge episodes of ancient tectonic activity.

In their 1888 paper, the surveyors provided no less than fifteen sections of
the main zone of complication in illustration of their work, at various localities
as far south as Ullapool and Loch Broom; but essentially they told the same
kind of story, and so need not be reproduced here. Some modification of the
ideas expressed in 1884 was offered in that it was now thought that there
could be thrust faulting without there necessarily being a preceding folding.
Also, more than two major lines of thrust were now envisaged. Nevertheless,
the story proposed was essentially the same as that offered four years earlier.

However, a good deal of new information was supplied respecting the conversions of the various types of rocks into metamorphic materials. Between Eriboll and Assynt, the Archaean Gneiss could be found rolled out into mylonite with "thin folia like leaves of paper" and dark hornblendic gneiss could be seen passing into "frilled schist." The "Cambrian" (Torridon Sandstone) could be seen to be altered in such a way that the "original gritty matrix ha[d] been converted into a fine micaceous or green chloritic schist, showing exquisite 'flow-structure,' winding round the elongated pebbles [in the original sandstone] in wavy lines." From the "Silurian" sediments, the quartzites became sericite schists.[32] The "pipes" of the "pipe rock" became "flattened like strips of paper"; the "fucoids" disappeared from the fucoid beds, which, however, acquired new division planes coated with mica; the Serpulite grit became a quartz schist; and the limestones became crystalline. Thus the changes induced by pressure and movement became imprinted on the various kinds of rocks.

Within the Moine Schist could be seen a zone of hornblendic and micaceous schist, studded with garnets, which seemed to mark a former extensive sheet of igneous rock, since some trace of the original igneous matter could apparently be discerned.[33] The planes of foliation in the Moines were inclined to the thrust plane beneath them. The movements seemed to have been intermittent, since some division planes could be seen to be truncated by subsequent thrusts. Unlike the mylonite beneath the thrust plane, which contained distorted and broken grains, the rocks of the Moines themselves all had complete grains, perhaps suggesting a higher temperature during their formation, which would have made recrystallization possible as metamorphism occurred. But what appeared to be occasional residues of resistant material ("eyes," or *Augen*) from the original rock could still be discerned.

By 1888, Peach thought it possible to attempt a reconstruction of the physical conditions that might have obtained during the time of the deposition of the "Lower Silurian" sediments, and he offered the following account (Peach et al. 1888, 405–8).[34] The basal unit of the quartzite represented rapidly accumulated sand beds on a new sea floor, with insufficient food supplies to support animal life. But with the slowing of the accumulation of the sediment, conditions became more propitious for organic life and worms could make their burrows in the sand, giving rise much later to the well-known "pipe rock." New kinds of worms appeared in the sediments known as the fucoid beds, the supposed fucoids ("sea-weeds") actually being the "matted network of their flattened excrements." The Serpulite grit represented a shallowing of the sea, with the deposition of coarser sediments, but during the subsequent deposition of the limestone there was virtually no "clastic" material laid down,

32. Sericite is a form of mica, produced by alteration from feldspars.
33. Today it might be construed as a zone corresponding to a particular intensity of metamorphism, with the garnets as indicators of that degree of metamorphism—though the generation of garnets requires an appropriate bulk composition.
34. The paper made it clear that the historical reconstruction suggested was specifically due to Peach, not the other surveyors.

merely the remains of minute marine organisms.[35] The presence of worm casts and the absence of corals indicated that the limestones were the detritus of calcareous or siliceous pelagic (open sea) organisms—not coral reefs. Although there were shells to be found, these did not suggest the presence of shell banks, for most of the forms were chambered and might have been free-swimming, like the modern *Nautilus*. The fossils had North American affinities, rather than being similar to the organisms found in the Cambrian or Lower Silurian in Wales or England. This suggested to Peach that there must have been a shallow sea stretching across the region that is now occupied by the Atlantic Ocean, and that there was some kind of barrier that cut off the northwest of Scotland from the rest of Europe.

So far as I am aware, this was the first significant attempt to reconstruct the sequence of the conditions of deposition of the rocks in the Northwest Highlands; and it was an auspicious beginning. It may be noted, however, that it was a well-established interest of the British Survey to attempt to reconstruct the conditions of deposition for the different strata, and more broadly palaeo-ecologies or the conditions of ancient environments. Indeed, according to Secord (1986b), it was a characteristic feature of the Survey, considered as a "research school," going back to the days of Sir Henry De la Beche. Thus Peach's thinking was consonant with a long-standing tradition in the Survey.

Thinking of a very general problem in historical geology that had been raised on numerous occasions in the nineteenth century, the authors concluded their paper by maintaining that they had shown that metamorphism could occur in more than one geological period. By attention to the Scourie Dykes, they had proved its occurrence in pre-Cambrian times and also at some time subsequent to the deposition of the Durness Limestone.

The paper of Peach and Horne was received with considerable enthusiasm, according to the comments recorded in the *Quarterly Journal of the Geological Society,* which included some appreciative remarks by Lapworth, Teall, and Judd. It may be worth noting that the paper was actually delivered on behalf of the survey by Geikie, in the presence of Peach and Horne, even though there is little reason to believe that the director general contributed very substantially to the results attained.

A long-standing tradition in Scottish geology, going back to the work of the friend of James Hutton, Sir James Hall (1761–1832), was the attempted laboratory confirmation of general theories propounded to explain field observations. It is interesting, then, that we find published in 1888 the results of work done by Henry M. Cadell (1860–1934), which sought to show that the thrusting processes envisaged by Lapworth, Peach and Horne, and various Continental geologists, could be simulated in the laboratory (Cadell 1888).[36] Cadell

35. "Clastic," that is, made of broken fragments of other rock—such as pebbles or sand grains.
36. On Cadel, see Jehu (1934), and the anonymous article in the *Scottish Geographical Magazine* 50 (1934): 169–71.

had studied under Geikie at the University of Edinburgh and also at the Clausthal Royal Mining Academy in Germany. As a young man, appointed to the Survey in 1883, he was one of the party that did the mapping in the Northwest Highlands in the 1880s. However, Cadell was fortunate to come from a wealthy background, and he resigned from the service when he inherited his family estates at Grange, Linlithgowshire, in 1888. But this did not bring his scientific career to an end. He traveled extensively overseas, continued to publish, and was a founder member of the Royal Scottish Geographical Society. He did important work on the Scottish oil shale deposits and published some valuable semipopular geological books, including his *Geology and Scenery of Sutherland,* which referred specifically to the region to which the Highlands controversy pertained.[37]

Besides the published account of Cadell's "experiments in mountain building" (Cadell 1888), there are two sets of notes preserved in the Edinburgh branch of the Survey's archives, one of which appears to describe the direct results of the experiments as they were performed at his home at Grange, while the other contains notes on the literature of "experimental mountain building" by various Continental authors, some theoretical notes, experimental observations, figures, and photographs of the work.[38] According to the first of these manuscript notes, the work was begun on January 25, 1887, and was completed on January 29. Cadell had a photographer on hand to record the results of the work, and he was assisted on three of the days by Robert Lunn.[39] On the last day, Peach came out to Grange to inspect the results of the experiments.

Cadell was certainly not the first to perform experiments on the effects of lateral pressure on materials that simulated the materials of the earth's crust. In both the published and the manuscript versions of his work, he referred to authors such as Alphonse Favre, F. A. Pfaff, Mellard Reade, and Gabriel Auguste Daubrée. Thus the general approach that might be followed was already mapped out for Cadell. He prepared for himself a rectangular box, about five feet long, with one movable end, which could be forced forward by the action of a screw running in bearings bolted to an extension of the base of the box. After the application of the pressure to the layers of material in the box, its side box could be stripped away and the results inspected and photographed. Then an "idealized" picture of the results was prepared, and the photographs and figures were published together.

37. See also Cadell (1895). In his later years, Cadell took part in the affairs of local government by his appointment as deputy lieutenant of the County of West Lothian. His field notebooks (see note 8) show that he was an excellent artist, like so many other members of the Geological Survey in the nineteenth century.
38. H. M. Cadell, "Experimental Researches in Mountain Structure. Experiments by H. M. Cadell at Grange 25[—29] Jan. 1887," in "Notes on the Geology of the N.W. Highlands. No. 2," British Geological Survey, Land Survey Archives, Edinburgh, LSA 96.5, pp. 26–41; LSA 96.6.
39. Lunn started at the Survey as a porter, but by 1881 he was a general assistant. Later he was promoted to the position of clerk and curator of maps. According to Flett, he had considerable geological knowledge and artistic flair, and he frequently performed valuable service as photographer for the Survey. He was a year younger than Cadell.

   The main object of the exercise was to demonstrate that thrusting could be
produced in response to lateral pressures, as also could the production of imbricate structures. The main problem was to find a suitable experimental
material. After a number of substances had been tried, Cadell settled on alternate layers of white plaster of Paris and damp sand, which had the advantage
of producing a brittle material and also made possible the easy perception of
the results. The experiments succeeded admirably, and fine simulations of
thrusting were achieved.
   A point that had been previously puzzling to the surveyors was the question
of what happened to the thrust in the direction from which the lateral pressure appeared to act. Cadell supposed that the thrust planes must eventually
originate (geometrically, not historically) in great deep-seated synclinal folds.
He placed a layer of wax cloth below the material in his box, then a layer of
plastic clay, and then more brittle material made up with the help of plaster of
Paris. When lateral pressure was applied, the wax cloth buckled; the clay
above it folded correspondingly; and the overlying plaster of Paris layers developed thrust faults. The experiment showed how it was possible for a thrust
fault to bend downward and pass into an ordinary fold below. With further
pressure, another system of folding and faulting was developed, showing how
a series of thrust faults might be formed one above the other, just as had been
observed in the field in the Highlands. Cadell noted that the folding and faulting could be observed to be taking place at one and the same time. Thus doubt
was thrown on Heim's suggestion that thrusting arose as an extreme outcome
of folding (which was the view that Lapworth had adopted).
   Finally, Cadell repeated an experiment that had originally been performed
by Favre. He used a piece of stretched rubber sheet as the base for the layers
of test material, and the rubber was then allowed to return to its usual dimension by release of the tension. The result was that when the upper layer of
folded clay was stripped off, the lower surface was found to display a "series of
minute corrugations transverse to the direction of movement and parallel to
the main ridges [formed] above" (Cadell 1888, 355). The experiment was, I
suggest, inspired by the more general theory of a cooling and contracting
earth, such as was generally espoused in the nineteenth century. But according to Cadell, it suggested how great areas of vertically cleaved rock might be
found at the roots of ancient mountain systems. In general, he found that the
more rigid the material used, the better were the phenomena of thrusting
exhibited.

All this was immensely satisfactory so far as concerned the general theory
that was being developed to account for the structures revealed by regional
mapping in northwest Scotland. As I have said, in regard to empirical information the Highlands controversy was now settled. There was, however, at
least one major development still to occur. This had to do with the geological
age of the whole Northwest Highland sequence—a matter of importance that
must necessarily attract our attention.

It will be recalled that the Durness fossils had long been regarded as Lower Silurian, and the unfossiliferous strata lying unconformably beneath the quartz rock had been assigned to the Cambrian by Murchison. But as events were now to show, this was another mistaken assumption, though I would hasten to say that at the time that Murchison made the suggestion it was a very reasonable one.

As far back as 1844, the American geologist Ebenezer Emmons (1798– 1863) had been promoting a system for American geology that he called the "Taconic" and that appeared to contain a "primordial" fauna, in rocks apparently below strata that were equivalent to Murchison's Silurian in Europe (Emmons 1844). Subsequently, a genus of trilobite, to which the name *Olenellus* was given, was found among the fossils of this system.[40] At that time, the Cambrian of Sedgwick was not well established, with its own characteristic fauna. Consequently, Emmons and a number of supporters favored the use of the Taconic system rather than the Cambrian. The battle on this issue raged hotly for many years, but the details need not concern us here.[41] The important point is that by the 1880s the *Olenellus* genus had come to be recognized as a characteristic indicator of Lower Cambrian rocks. *Olenellus* fossils were found in Norway by Lapworth's correspondent Linnarsson, the Swedish palaeontologist, in 1871, but they were at first thought to belong to a more recent genus, *Paradoxides*. A revised identification was given by Brøgger four years later. Then in 1882 Linnarsson found the *Olenellus* fauna at the base of the Cambrian in Sweden. The same fauna was found soon after in Russia.[42]

With Lower Cambrian rocks found in America to the west and in Scandinavia and Russia to the east, it was a challenge to British geologists to see if they too could locate the characteristic *Olenellus* fauna in their own country— the by-now accepted type region for Cambrian strata. As usual, it was Lapworth who led the way. In 1885, he found what appeared to be fragments of *Olenellus* on the flanks of the hill Caer Caradoc, in Shropshire, but they were not sufficiently well preserved to warrant scientific description. In the summers of 1887 and 1888, however, as a result of the efforts of a fossil collector employed by Lapworth—Mr. Henry Keeping of Cambridge—sufficient material had been obtained to make a firm identification possible. Lapworth (1888, 485) named the species *Olenellus callevei* in honor of his co-worker Charles Callaway, who had been the first to find fossiliferous Cambrian rocks in Shropshire.

This discovery of the *Olenellus* fauna in Shropshire prompted some interesting speculations in Lapworth's mind. The rocks in which the Salopian fossils were found overlay a unit called the Wrekin Quartzite. Below this lay the unit called the Longmynd, which was therefore very likely pre-Cambrian in

40. In the first instance, the genus was named *Olenus*.
41. See Hunt (1883); Merrill (1924, 594–614); Schneer (1969). The difficulties and controversies arose in part, as they did with respect to the Southern Uplands and the Northwest Highlands of Scotland, because of the highly folded and faulted terrain and the consequent failure to recognize the repetitions of strata.
42. These details are drawn from Lapworth (1888).

age. Lapworth suggested (ibid., 486) that the Scottish equivalent of the Long-myndian might be the Torridon Sandstone; and the Wrekin Quartzite might have its equivalent in the quartzites of the Northwest Highlands. It would therefore follow that the whole stratigraphical sequence in the northwest might be tied up, with convincing analogies to the English and Welsh strata, if some fossils of *Olenellus* type might be found there in the rocks above the quartzite. Thus a new challenge was presented to the surveyors.

Again Peach and Horne opened out for the Survey the bridgehead initially established by Lapworth, though in this case they did not themselves actually discover the *Olenellus* fauna in Scotland. This was done by their colleague Arthur Macconochie, who had been employed by the Survey as a fossil collector since 1869.[43] By 1891, the surveyors had moved further south to map the region known as Dundonnell Forest, which lies between Loch Broom (Ulla-pool) and Loch Maree (Kinlochewe). This was the region of Geikie's "long march" of 1860, when he had hurried by much that would have repaid closer study. Along the line of the track, which runs north-south near Loch Nid, there is a good scarp on the eastern side of the valley, which has Moine Schist thrust over Torridon Sandstone, which is in turn thrust over a clear sequence of quartzite, fucoid beds, and Serpulite grit. (The Durness Limestone is here cut out by the thrust fault.) To the west of the valley, the quartzite can be seen very clearly lapping up the hillside of Torridon Sandstone. All this, the surveyors, forewarned and forearmed with their extensive experiences on the more perplexing terrain further north, could read almost like an open book. Small streams had cut their way through the eastern scarp near a very remote homestead called Achneigie, and it was there that Macconochie found the specimens of *Olenellus* fauna, within a dark shale belonging to the fucoid beds. On further searching, more specimens were found at several localities, and they have since been found at various localities along the Eriboll-Skye line, including Loch Assynt, though they are nowhere common. In their paper describing the fossil finds, Peach and Horne (1892) described the fossil localities in detail, with sections, and they provided detailed palaeontological descriptions of the specimens, with figures.[44] In fact, the species discovered was a new one, and it was tactfully named *Olenellus lapworthi*. A couple of years later, Peach (1894) was able to report some further discoveries of *Olenellus*, with several new species. This time the fossils were discovered at the southeast end of Loch Maree, on the slopes of the mountain named Meall Ghiubhais. This was an interesting locality: a klippe with the rocks completely overturned by the folding and thrusting process, such that a patch of Torridon Sandstone *overlay* an accustomed sequence of Torridon Sandstone, quartzite, fucoid beds,

43. It will be recalled from chapter 8 that Macconochie participated in the debate over the Southern Uplands. He became an assistant in the Survey's fossil department in 1889. Toward the end of his career, he became a lecturer in geology at the Herriot-Watt College in Edinburgh (1906–20).
44. A preliminanry announcement of the discovery was made by Geikie at the 1891 meeting of the British Association (Geikie 1892a). He reported that the recognition of the likely fossil-bearing horizon was made by Horne. Macconochie discovered the fossil fragments; Peach thought that

and Serpulite grit. But by this time such peculiarities were easily construed by the surveyors. Again the fossil finder was Arthur Macconochie.

Thus Lapworth's earlier speculation was vindicated. The fucoid beds and the Serpulite Grit were definitely Lower Cambrian in age, with a quartzite forming a base of the system, as in Shropshire. And since this quartzite was separated from the underlying Torridon Sandstone by a marked unconformity—which as it happens is seen with particular clarity near Loch Nid—it followed that the unfossiliferous Torridon Sandstone was pre-Cambrian, not Cambrian, as had long been assumed.[45] The Torridonian was, as Lapworth had suggested, analogous to the pre-Cambrian Longmyndian of Shropshire. The whole argument hung together very nicely, and the sequence thus established has been retained into recent times, except that arguments have subsequently been advanced for the Ordovician age of some members of the Durness Limestone. Of Murchison's great tracts of Silurian territory in the northwest of Scotland, nothing now remains.

A further important investigation from this period, this time of a mineralogical and petrological nature rather than palaeontological, was accomplished by Teall, in association with Horne, through their investigations in an area not mentioned hitherto in our account—Loch Borolan. This is a fairly small loch on the side of the road running south from Loch Assynt—but the one turning off toward Bonar Bridge on the east coast, rather than southwestward toward Ullapool. To the north of Loch Borolan, there is a substantial mass of granite-like rock, forming the hill called Cnoc na Sròine. This had long been known to geologists, appearing even on the map of Cunningham (see plate 2)—though not on that of Macculloch (see plate 1)—and it was regularly mentioned by subsequent investigators.

The mass of igneous rock had obviously intruded into an area where limestone was present, and as a result the so-called Assynt Marbles had been produced, giving rise to a rock of minor economic significance that had attracted the attention of the very earliest geological visitors to the northwest, such as Williams and Macculloch, and that had been quarried on and off near Ledbeg for generations. The investigation of Horne and Teall (1895) displayed three

---

they were *Olenellus;* and the Survey palaeontologists in London, George Sharman and Edwin Newton, confirmed the identification.

45. This conclusion did not go wholly without challenge. In a paper to the British Association in 1892, Blake (1893) maintained that the Torridon Sandstone did not look like pre-Cambrian rock, but rather younger, and was very likely Cambrian. Further, some evidence had been adduced by the American geologist C. D. Walcott for the rocks of the Durness Limestone as being Ordovician; in which case the Torridon Sandstone could well be Cambrian. Blake's logic was sound, in the light of the information then available, but no one seemed to be much interested in what he had to say, for the geological community was enjoying the experience of seeing a new consensus emerge and did not wish to see this held up by objections from second-rank geologists. In fact, Blake's conclusion has not been sustained by subsequent work, and the Torridon Sandstone continues to be classified as pre-Cambrian.

important things. First, the intrusion of igneous material was manifestly later than the period of deposition of the Cambrian sediments such as the Durness Limestone. Second, it was earlier than the post-Cambrian earth movements, for the igneous rock could be seen to have been affected by, for example, the Ben More Thrust. This suggested that there could have been a substantial interval between the deposition of the Cambrian sediments and their extraordinary deformations in the subsequent earth movements.

But from the point of view of the petrographer, Teall, the igneous rock itself at Cnoc na Sròine was of special interest, for he showed it to have a mineral composition different from any previously known to science. Hence the term "borolanite" was coined to refer to a new family of rocks, which may be called melanite-nepheline-syenites.[46] In a later paper, Teall (1900) displayed the analogies of his borolanites to rocks that had previously been investigated by Brøgger in Christiania (Oslo). Thus the Scottish discoveries were integrated with work being done elsewhere in Europe. The details of the petrography of borolanite need not detain us here. But the point may be made that Teall's papers reveal the very considerable increase in sophistication of the Survey in mineralogy and petrography as a result of Geikie's judicious appointment of Teall to the staff.

As we shall see in the next chapter, there were a number of rumblings within the British geological community in the 1980s that eventually led to the establishment of a government committee of inquiry into the affairs of the Survey. To a large extent, the problems that gradually developed had their origin in the Highlands controversy. On the surface, however, the resolution of that controversy seemed a grand triumph for the Survey. Indeed, it might properly be said that the Northwest Highlands furnished the Survey's outstanding piece of geological research during the nineteenth century, and the officers' most accomplished map work. During the 1890s there was a steady flow of geological visitors to the northwest, particularly to Inchnadamph, and they marveled at what Peach and Horne, and their colleagues, had managed to achieve in sorting out the extraordinarily complex structures of the district, envisaging the three-dimensional structures as if they could see into the very hearts of the mountains. We have records of some of these visitors to the northwest, and some of them published papers as a result of their visits. There were, of course, many others who came to marvel, but who left no permanent record of their stay. The same is true to the present, where the bar of the Inchnadamph Hotel provides the appropriate venue for the informal exchange of news and views, and where generations of students have received instruction and have gained insights into geological theory and the social

46. That is, syenites containing the mineral melanite (a particular kind of black garnet) and nepheline (a type of sodium/potassium aluminium silicate belonging to the hexagonal system of crystals).

structure of the geological community. Thus the Assynt region rapidly became one of the best-known regions in British geology.

Among the overseas visitors in the 1890s, one may mention such names as Emmanuel de Margerie, Marcel Bertrand, Baron Ferdinand von Richthofen of Berlin, Albrecht von Penck (1858–1945) of Vienna, and the great systematizer Eduard Suess. The first two came in 1892, with an invitation from Geikie, who provided Peach as their official guide. Bertrand (1893) expressed his astonishment (even "terror") at the minute differences on which the Survey officers relied for their stratigraphical subdivisions, but which nevertheless proved reliable over large areas and were essential if the complex structures were to be elucidated.

In examining the rocks of the Assynt region, Bertrand recognized the analogies to the tectonic structures in Belgium and Switzerland with which he was familiar. However, it appeared to him that there were some important differences. The surveyors had found three main thrust planes at Assynt (the Glencoul, the Ben More, and the Moine, as well as the "sole" thrust), but the uppermost (the Moine Thrust) could be found to cut across the other two, in a manner that seemed different from what was found on the Continent. Thus it did not appear that one was simply dealing with three large folds that had passed over into, or had given rise to, three thrust planes. Also, in the Scottish case, the layers between the thrust planes had themselves been faulted, or, as it were, cut up into slices, which then allowed many repetitions; and the strike of the slices was generally oblique to the strike of the main thrust planes.[47] Thus the "imbricate" structure, which was so important to the structure of the Northwest Highlands was different from anything that Bertrand had seen on the Continent. In Scotland, it did not appear that the imbrication arose from a large number of relatively minor folds that had passed over, or had been converted into, a succession of faulted fragments. Bertrand suggested that the difference might be due to the comparative lack of plasticity of the Scottish rocks at the time of their subjection to lateral forces. The mechanism for their formation seemed to be supported by the experimental work of Cadell, who had deliberately chosen to use an experimental substance—plaster of Paris— that was quite rigid.

Von Penck's visit to the Northwest Highlands took place in 1897, as one of the official excursions following the International Geological Congress held in St. Petersburg that year. Not surprisingly, perhaps, he was the only congress visitor who managed to travel to that remote part of Britain, so far from Russia; and so it came about that he had Horne as his personal guide. As a result of his visit, von Penck published a substantial paper in German, giving readers information about the new discoveries in Scotland (von Penck 1897). His paper was also given in summary form in an English journal (von Penck 1898). He gave particular attention to the character of the Torridon Sand-

---

47. These features may be seen in the section of figure 9.3.

stone and proposed that it had been deposited under essentially desertlike
conditions, since it exhibited characteristics similar to those that might be
found today in wadis in the Sinai Peninsula. But as time went on, conditions in
the Torridonian appeared to have become more humid, until eventually the
region was inundated at the beginning of the Cambrian.[48]

Referring to the structures at Assynt and elsewhere, and to the experimental work of investigators such as Cadell, it seemed to von Penck that lateral forces might give rise to three different kinds of structures at different
depths. The "surface" structures were exhibited on the North German Plain;
the "intermediate" ones in places such as the Alps and the Appalachians; and
the "deep" ones in Northwest Scotland. Thus the Scottish work was being
fitted into a global pattern of tectonic structures.

Though the publication of *The Geological Structure of the North-West Highlands
of Scotland* in 1907 marked the culmination of the nineteenth-century geological work in the Highlands and was—with its six hundred and sixty-eight pages,
its forty-two chapters, its fifty-two photographic plates, its sixty-six figures (including several colored sections), its colored map, and its palaeontological,
chemical, and bibliographical appendices—a truly remarkable testimony to
the labors of Geikie and his co-workers in the field, laboratory, and study, it is
not necessary for us to subject the whole to close examination here. For we
have already seen the arguments contained in the book as they developed historically. Suffice it to say that the text guided readers in detail around all the
salient places along the zone of complication, for each of the principal systems: Lewisian, Torridonian, "Cambrian".[49] Then the post-"Cambrian" earth
movements were described, with descriptions also of Cadell's experiments. Finally a detailed description was given of the "eastern schists."

Following the publication of the great memoir of 1907, Peach and Horne
put the finishing touches to their work by supervising the construction of a
large three-dimensional model of the Assynt region, of which several copies
were made, for exhibition in various parts of Britain.[50] In conjunction with
this, they published a "guide," which gave a synoptic account of the geology of
the Assynt region, that would have been invaluable to amateur geologists visiting the Northwest Highlands (Peach and Horne 1914). It was undoubtedly
used by the twenty-eight participants in the field trip organized by the Geologists' Association in July 1914, which was conducted with Horne as leader
(Horne 1915).[51] In a sense, the description of the Assynt model provided the

48. Recent accounts of the palaeo-environment of the Torridonian rocks do not support von
Penck's view that some of the rocks at least represented desert conditions. See Johnson (1983,
49–51).
49. It may be mentioned that the upper part of the Durness Limestone is now regarded as Ordovician. The terms "Ordovician" and "Silurian" were *both* absent from the memoir's index.
50. The one most probably familiar to visitors is that presently exhibited in the Geological Museum in South Kensington, London. Another was located at the Royal Scottish Museum,
Edinburgh.
51. The guide appeared in the bibliography for this paper.

definitive account of the views of Peach and Horne on the geological structure of that part of Scotland.

There was one other major publication of Peach and Horne that treated the Northwest Highlands and that we should notice, even though it falls outside the main period of our interest. This was their *Chapters on the Geology of Scotland*. The book, which was published by the Oxford University Press in 1930, was, of course, issued posthumously, for Peach died in 1926 and Horne in 1928. It was a sad thing that neither of the two grand old men of Scottish geology, who had labored so long and hard, and so effectively, for the Survey, lived to see the final synthesis of their life work. But for our present purposes, the point to be made is perhaps more theoretical than personal, for the posthumous publication reveals that on one issue at least, Peach and Horne began to differ concerning the interpretation of the Northwest Highlands. The difference of opinion had to do with those difficult rocks, the Moine Schists. The question was, What were the Moines formed from? And when did their metamorphism take place?

When Lapworth first tackled these questions, he supposed that the Moines were *produced* from a number of different rocks, ranging from the Archaean, through Torridonian, to the Assynt Series ("Cambrian") (Lapworth 1884, 438). This opinion was adopted by the Survey also, in the first announcements of 1884, in the major paper of 1888, and in the memoir of 1907. Peach, however, having particular regard to evidence gathered near Loch Carron in 1892, came to the view that the Moines were chiefly metamorphosed Torridon Sandstone. Where patches of gneiss could be found within the general territory of the Moine Schists, he supposed that these represented an earlier metamorphic phase, corresponding to the west coast gneisses. Thus the eastern schists were thought to overlie portions of preformed *eastern* gneisses unconformably. Peach further thought that it was possible to correlate subdivisions of the Torridon Sandstone with specific subdivisions of the Moines (Peach and Horne 1930, 199). Thus all the regional metamorphism of the Moines was attributed to post-Cambrian earth movements.

But Horne began to have doubts about all this. In fact, by the end of his life (ibid., 200–201), he had come to the view that the rocks of which the Moines are made had been schistose since pre-Cambrian times, rather than acquiring their schistosity during some post-Cambrian episode or episodes of earth movement, as Peach envisaged. In fact, in Horne's view the Torridonian was not necessarily an original constituent of the Moines, since along the zone of complication for the region from Loch Glencoul northward up to Loch Eriboll, there was no evidence for there ever having been any Torridon Sandstone present; yet the Moines for that stretch were much the same as elsewhere (though mylonite is chiefly exposed at the western boundary of the Moine Schist for this part of the thrust plane). Horne further doubted the suggested correlation of subdivisions of the Torridonian with those of the Moines. He proposed, therefore (ibid., 201), that the supposed pre-Cambrian schists formed a land barrier separating the Cambrian sea in the Northwest Highlands from the sea in which the fossiliferous rocks of the Highland Border region were deposited. It may be mentioned that the long delay in the publication of *Chapters on the Geology of Scotland* arose in part from the fact that in the end Peach and Horne found that they could not reach agreement on the interpretation of the metamorphic rocks of the north of Scotland (Gregory 1928, 992).

Actually, Peach and Horne were not alone in their disagreements about the origins of the Moine Schists. Although the Lapworth view that the Moines were produced by earth movements and concomitant metamorphism was officially adopted by the Survey, and the theory was incorporated in the memoir of 1907, there was dissent to this view within the ranks of the Survey even in 1885, when Geikie sent all his Highlands staff up to the northwest to have Peach and Horne show them round the chief geological sites and expound the new views. Thus there is a report that one young surveyor, George Barrow (1853–1932), was not convinced of the Lapworth theory of the origin of the Moines (Green 1935, lxvi).[52] Barrow contended that metamorphism decreased as one approached the thrust belt, instead of increasing as might be expected according to Lapworth's ideas. Barrow also argued that the apparent bedding in the Moines was indeed true bedding, as Murchison and Geikie had long maintained. There followed a long drawn-out controversy about the origin and nature of the Moines, which warrants another book for its full examination and will not be undertaken here, though it may be mentioned that Barrow took the view that the Moine Schists were in fact Lewisian in age, and that their metamorphism also derived from that very early period of geological history.[53] The point to be emphasized is that although in one sense the Highlands controversy was solved by the work of Lapworth, Peach, and Horne, in another sense it was only a partial solution. The *geometrical* problem was solved

52. On Barrow, see Strahan (1913) and Thomas (1933). Barrow studied at King's College, London, and after graduating he became private secretary to Scrope, charged with reading aloud to the elderly geologist and retired banker, who had gone blind in his old age. It was on Scrope's recommendation to Ramsay that Barrow was appointed to the Survey in 1876. He was a skilled petrologist as well as field geologist, and performed major service in his surveying of the Central and Eastern Highlands, where he began to study the different grades of metamorphism in terms of the different minerals present. However, he was regarded as iconoclastic by his colleagues, and his views did not always agree with those of the rest of the Survey. He was eventually withdrawn from Scotland to do work on the younger rocks of England in 1900. His ideas never received their due recognition in the Survey until taken up and developed by Charles Tilley.
53. For further information, see, for example, Read (1934), Green (1935), McIntyre (1954), and Anderson (1979). Barrow's views were never published in a systematic way for the Northwest Highlands. For a statement of his views, see his comments (at page 111) on Tilley (1925) and Barrow (1895). Barrow's great contribution to the understanding of the metamorphism of the Highlands was to introduce the notion of "metamorphic zones," according to which areas of different intensity of metamorphism were characterized by the presence of different subsidiary minerals within the schists (Barrow 1893).

in essence, but the cause of the complicated folded and faulted structure was by no means known, and the history of when and how it all happened, and in what order, and what was metamorphosed into what, was certainly not fully elucidated in the 1880s. Indeed, it has provided geologists with sufficient problems to keep them occupied up to very recent times.[54]

This concludes my discussion of the main events leading to the understanding achieved for the geology of the Northwest Highlands in the period prior to the great memoir of 1907. That is, I have given some account of the main events that related to the Highlands controversy during the nineteenth century, or until the end of Geikie's directorship of the Geological Survey. It might be thought that that was the end of the matter so far as our historical account need be concerned. But this is far from the case. A considerable number of issues flowed from the controversy in the 1890s, which had profound implications for the history of the Survey, and which can tell us much about the way in which scientific debates may be conducted. We need, therefore, to examine some of the social issues over and above the more technical matters that have been considered hitherto. To these interesting questions we turn in chapter 10.

54. According to Johnson, "despite some eighty years' research neither the stratigraphical age of the Moines nor the order of succession of Moine lithostratigraphical units is known over the whole outcrop" (1983, 53).

10

The Impact of the Highlands Controversy on the Progress of the Geological Survey; The Wharton Committee's Inquiry

On January 5, 1893, *Nature* published an essay on the scientific work of Archibald Geikie, as a contribution to its series "Scientific Worthies." The author was the French geologist Albert Auguste de Lapparent (1839–1908), who held a chair at the Catholic University in Paris. Geikie and de Lapparent had a good deal in common in their scientific interests, particularly in the areas of stratigraphy, igneous activity, and the study of landforms. They were both very active in the affairs of their respective national scientific societies, the Royal Society and the Académie des Sciences, and de Lapparent, like Geikie, wrote a major textbook of geology (de Lapparent 1883).[1] Thus we find that the two geologists, who became acquainted at the International Geological Congress in Berlin in 1885, established a strong personal friendship, to which Geikie attested in his autobiography (Geikie 1924, 217, 224, 249). The link was no doubt strengthened by the fact that Geikie's wife was partly of French descent, so that the Frenchman felt at home among the Geikie *ménage* when he visited England.

In his eulogistic account of Geikie's work in the essay in *Nature*, de Lapparent launched into a discussion of the work in the Northwest Highlands, stepping boldly into territory where (British) angels might have taken care to tread more lightly. From what he said it might have appeared that the whole credit for the revision of the Murchisonian theory (with which, according to de Lapparent, Geikie was "never quite satisfied") was due to the Survey officers, working under their chief's able guidance and direction. Never a word was said about the contributions of Lapworth, Callaway, Hicks, and Bonney. No doubt such an omission did not seem so very serious to a French geologist, unacquainted with the tensions that had attended the work done in the northwest a decade before. But several members of the British geological community were evidently outraged, as is clear from a heated correspondence that immediately developed in the pages of the *Daily Chronicle*.

On January 14, the newspaper carried an unsigned column-length article discussing what de Lapparent's essay had had to say about the Northwest

1. De Lapparent was geological consultant for one of the early projects to build a cross-channel tunnel, thus furthering his links with Britain in a rather special way.

300

Highlands episode. The article, though obviously written with much spleen, is so engaging and is so pertinent to our present purpose, that I shall quote it virtually in full, for it gives vivid expression to many of the issues that underlay the Highlands controversy but that rarely surfaced in such direct manner.

There is in *Nature* an article by a French writer on Sir Archibald Geikie, Director-General of the Geological Survey, which is just now causing a good deal of talk among English men of science. Of course nobody is surprised at the fulsomeness of M. de Lapparent's eulogy. As *Nature* seems to exist for pushing the great official scientific syndicate of Huxley, Hooker, Geikie and Co., Limited—very strictly limited—which may be said to "run" science in England, M. de Lapparent would probably not have been permitted to write anything about a member of it unless it were fulsome. What has really amazed people is the audacity with which a famous historical bungle on the part of the Geological Survey is glossed over, and the Director-General not only credited with the work of those who exposed and corrected it, to his utter discomfiture, but actually covered with laurels for thus winning one of the most glorious scientific conquests of the century. The whole thing is delightfully characteristic of State-endowed science in England. If you are one of the official syndicate who "run" it, you may blunder with impunity and make your country ridiculous at the taxpayers' expense. Scientific men, who can correct you shrink from the task. They know that the syndicate can boycott them, and by intrigue keep them out of every post of honour and profit, and that the syndicate's satellites can write and shout down everywhere independent and nonofficial critics. They also know that if, perchance, some particularly intrepid person does succeed in exposing one of this syndicate, they can always, by the same means—after the public has forgotten the incident—suppress him, and boldly appropriate to themselves the credit of his work.

The geological secret of the Highlands, with the unlocking of which Sir Archibald Geikie is now credited, was really made a puzzle for more than half [*sic*] a century by the blundering of the Geological Survey and its director—General Sir Roderick Murchison—and famous courtier and "society" geologist of the last generation. In the Highlands he saw gneisses and ordinary crystalline schists resting on Silurian strata, and he foolishly held the sequence to be quite normal. The schists, he would have it, were not archaic formations, but only metamorphosed Silurian deposits. He also held that primitive gneiss was not part of the molten crust of the globe, but only sediments of sand and mud altered by intense pressure and heat. Murchison, not to put too fine a point on it, "bounced" everybody into accepting this absurd theory, and the whole forces of the Geological Survey, with its official and social influence, together with the unscrupulous power of the official syndicate which then, as now, jobbed science wherever it had a state endowment,

were spent in perpetuating the blunder and blasting the scientific reputation of whoever scoffed at it. But in the Natural History School of
Aberdeen University it *was* scoffed at. The late Dr. Nicol, Professor of
Natural History in Aberdeen, proved that Murchison and the Survey
were wholly wrong, his proof being as complete as the existing state of
science allowed. When he died, his successor, Dr. Alleyne Nicholson,
was too deeply absorbed in the work of his chair, and in those palaeontological researches which have made him famous, to engage in pro-
longed stratigraphical controversy with Murchison's successors and the
official syndicate; and so the question gradually dropped, the last word
being apparently with the Geological Survey. In shouting that last
word, no voice had been louder than Sir Archibald Geikie's. It is therefore diverting to find his official biographer stating in *Nature* that all
the time he was wrestling *in foro conscientiae* with doubts as to the
soundness of the official position, and that finally his "love of truth"
prompted him to order a resurvey of the whole Highland region. In
plain English, the taxpaper having had to pay for Murchison's bungling
survey, was because of his successor's "love of truth" to enjoy the luxury
of paying over again to correct it.

The real truth, however, is this:—When it was supposed that the
Aberdonians were finally crushed, there arose in England a young geologist called Lapworth, who had the courage to revive the whole
controversy and take sides with the Aberdeen school. As he developed
an extraordinary genius for stratigraphy he not only broke to pieces
the official work of the Geological Survey in the Highlands, but by revealing the true secret of the structure of that perplexing region, he
played havoc with the Murchisons and the Geikies and all their satellites, convicting them of bungling, and covering them with ridicule. Yet
not only is this achievement credited by *Nature* to Sir Archibald Geikie,
but not the dimmest allusion is made to Professor Lapworth's work in
connection with it, or to his epoch-marking paper in 1883 entitled
"The Secret of the Highlands." Geikie's surveyors, who were not sent
out till Lapworth's disclosures frightened that Geological Survey and its
official chief, are, however, credited with proving just what Lapworth
previously demonstrated—namely, that "Murchison had been deceived
by prodigious terrestrial disturbances, of which at the time nobody
could have formed an idea." Of course the Aberdeen school, who first
exposed the blundering Murchison and the Geological Survey count as
"nobody," . . .

. . . It was a terrible blunder, as the Aberdeen men persistently held,
and we do not wonder that Sir Archibald Geikie, who rose to place and
power by defending it, is anxious to have his connection with it veiled
by a friendly hand. But it is rather outrageous for this friendly hand to
give him the credit of conceding the very error which he defended to
the last, and deprive Professor Lapworth of the honour of having ban-

ished it from science. One of the most diverting things, however, . . . is that Sir Archibald Geikie is belauded because, when frightened by the stir Professor Lapworth's paper made in 1883, he was fain to send his surveyors to go over the Highlands again—he, as their official chief, ordered them "to divest themselves of any prepossessions in favour of published views, and to map out the actual facts." Old Colin Campbell, when he objected to the institution of the Victoria Cross, said it was as absurd to decorate a soldier for being brave as a woman for being virtuous. He did not foresee a still greater absurdity—that of eulogising a man of science because he instructed his assistants to tell the truth when conducting an investigation into his own blunders.[2]

It would be interesting to know who was the author of this little piece. It cannot, I think, have been any of the persons whom we have mentioned so far in our story, for there are certain misconceptions in the article that could scarcely have been found among the ranks of "core" geologists, or even the better informed "amateurs." Notably, there is the error of saying that the "secret of the Highlands" was locked up for half a century by the actions of the Survey.[3] The further objection may be made that the Survey had never formally examined the Northwest Highlands before 1883—though it was true enough that a succession of three directors had all held to essentially the same interpretation of the northwest stratigraphical sequence. We may note that the author of the article seemed as much concerned with the political aspects of the affair as the scientific. He had an objection to government-organized science, and the "jobbing" to which in his opinion it gave rise. Possibly he was someone who had unsuccessfully applied for some government position.

The point that the work done in the northwest by Murchison, Ramsay, and Geikie did not constitute an official survey and was not published by the Survey, but rather was by the way of "private" work, was promptly made by a correspondent who signed himself "One Who Knows."[4] Government science was also defended: the Survey officers were "a body of hardworking and underpaid servants of the State." The writer pointed to "the manner in which the Geological Survey is cramped and starved" and to the fact that "many of the field-geologists have been more than twenty years on the staff without promotion." One may conjecture that this contribution came from one of the Survey's own members.

The defence from "One Who Knows" did not, however, satisfy another correspondent, writing from the Savile Club, who signed himself "A Field Naturalist."[5] "Does 'One who Knows' mean," the writer asked rhetorically,

---

2. A clipping of this article, entitled "A Geological Blunder," together with the correspondence to which it gave rise, is held in the archives of the Geological Survey: 1/192.

3. It is possible, I suppose, that the author was thinking right back to the first paper of Sedgwick and Murchison in 1828. But that was before the Survey was even established.

4. *Daily Chronicle*, Jan. 19, 1883.

5. Ibid., Jan. 20, 1883.

"that salaried servants of the State waste their time in 'devilling' up the 'private' work and correcting the 'private' blunders of their Directors-General?" The writer further complained about the work of the Survey proceeding at a "snail's pace." Three days later, "Scholasticus" weighed in with a letter complaining that Geikie still had not repudiated the error of de Lapparent in giving him credit for Lapworth's accomplishments.[6] The correspondent referred to analogous charges of plagiarism against the director of the Botanical Gardens at Kew, and went on to make a general complaint that the claims for the moral force of science made by men such as Huxley were suspect indeed if this sort of thing were to be found in the "state-endowed priesthood of science." Indeed, the writer averred: "I doubt whether we are wise in making literary and moral culture give place to scientific instruction in our schools and colleges."

All this must have been extraordinarily embarrassing to Geikie, and he finally rose to his own defense with a short letter in *Nature*, published on January 26 (Geikie 1893). It was penned with Geikie's characteristic grace and skill, beginning: "In the kindly review of my work by Prof. de Lapparent, . . . there are one or two inaccuracies which I would have at once corrected had I not shrunk from drawing attention, even for purposes of rectification, to an article which I felt to be too eulogistic." But at long last Geikie was pushed into a position such that he had to give formal recognition to Lapworth.[7] In the long Northwest Highlands paper by Peach and Horne of 1888, it was true enough that a full statement of the work of relevant earlier workers in the northwest, including of course Lapworth, had been given, and Geikie had delivered this paper to the Geological Society, but he had never publicly stated in writing his own approbation of Lapworth's work. He now drew attention to the acknowledgments to Lapworth (and Nicol, etc.) in the Peach and Horne paper, and he concluded his letter by saying: "My friend Prof. Lapworth has no scientific comrade who had more frankly and practically acknowledged his great geological achievements than I have done." Geikie also acknowledged that he had had absolutely no part in bringing about the new view of the Northwest Highlands structure.

It is a nice question whether everyone should have been fully satisfied by this statement. For my part, I note the manner in which Geikie expressed his appreciation: there is no one as good as I am in giving praise to Lapworth, it seems to say. But on the whole, I suppose the statement, though extremely tardy, should have sufficed. The author of "A Geological Blunder" was, however, far from satisfied. In another column-long article published in the beginning of February, he returned vigorously to the attack.[8] He acknowledged that Geikie had supported Lapworth in his election to a fellowship of the Royal Society and the award of a gold medal for his work, but he insisted that

6. Ibid., Jan. 23, 1893.
7. It may be noted that Geikie failed to acknowledge Lapworth's contribution in the second edition of his *Text-book* (1885), which would have been an appropriate place for mention of Lapworth's work.
8. *Daily Chronicle*, Feb. 2, 1893.

Geikie had not given Lapworth credit at the right time—namely, back in 1884, when the surveyors' preliminary announcement of their results was given. And de Lapparent's essay actually suggested (in the eyes of the author of "A Geological Blunder") that Geikie was somehow personally responsible for the achievements of the surveyors in the northwest. (For my part, I cannot find that view expressed in de Lapparent's essay, but it is true that it put Geikie's contributions in the most favorable light.) The author further added that the Survey had only been stirred into action in the northwest as a result of concern expressed by overseas geologists.

The renewed attack upon the scientific establishment was reinforced by another letter from "A Field Naturalist," four days later.[9] He now widened the front by bringing up the question of the Southern Uplands debacle. Here in a way, the critics were on stronger ground, since the Survey had officially surveyed the Southern Uplands, had rejected the alternative views of Lapworth, and had eventually been forced, nonetheless, to redo the surveying work, whereupon they were obliged to acknowledge that Lapworth was offering a superior interpretation. Yet, said "Field Naturalist," as late as 1882, Geikie, in his *Text-book of Geology*, had condescendingly referred to Lapworth's alternative stratigraphy with an ironical exclamation mark (Geikie 1882a, 672).

This brought Lapworth into the discussion, though he did not take up the issue of the Southern Uplands work.[10] He merely stated that full acknowledgments of his Highlands contributions had been made, so far as he was concerned, in the paper of 1888, and that the Survey officers had been continuing their work in the northwest up to the present, producing results of the "very highest scientific value." He urged that the correspondents would bring more benefit to British science if they somehow aided publication of these results at "a price which would render them available for all British students of geology." Thus Lapworth was not to be drawn into a public slanging match, and his letter showed that he was in fact well disposed to the public support of science. This point of view was to emerge in his contribution to the debate about the role of the Survey within the British scientific community, which will be discussed later in this chapter.

All this may appear something of a storm in a teacup, and in a sense it was. But it represents the beginning of a swell of dissatisfaction with the affairs of the Survey that was eventually to lead to a public inquiry into its activities, and the termination of Geikie's period of office. We shall therefore examine the events leading up to Geikie's retirement, and endeavor to see what role was played in the discussions by the events of the Highlands controversy—which by the 1890s was "past history," so far as the "internal" history of science was concerned.[11]

9. Ibid., Feb. 6, 1893.
10. Ibid., Feb. 8, 1893.
11. I do not mean, of course, that all problems had been solved in the northwest by the 1890s: rather, that Nicol and Lapworth had been vindicated, the Murchison-Geikie theory had been repudiated, and the broad principles of the structure and the stratigraphical order in the northwest had been successfully established.

The Survey was undergoing all sorts of problems under Geikie's leadership in the 1890s. With the completion of the basic survey of the zone of complication from Eriboll to Skye, the surveyors began to try to extend their newfound knowledge into the main body of the Scottish Highlands: the Grampians. This, of course, was what Murchison and Geikie had sought to do back in 1860, with but little success. Indeed, Geikie realized at the time that the extrapolation from the Northwest to the Central Highlands was not working properly, though Murchison had been fully confident.

The Grampians form the large range of mountains and hills in the center of Scotland between the Great Glen Fault (running up Loch Ness from Fort William to Inverness) and the Highland Boundary Fault, which runs roughly parallel to this line and marks off the younger rocks of the Midland Valley of Scotland, where the chief industrial development is found. They are largely made of another series of metamorphic rocks: the "Dalradian Schists." It lies beyond the scope of the present study to attempt to give an account of the nineteenth-century work on the Dalradians. It is sufficient to say that the surveyors made a start on this work, but made disappointing headway with it, as was acknowledged by the Survey's own historian, Sir John Flett. Writing in 1937, he briefly described the efforts of the surveyors in the various parts of the Central Highlands:

> Much of the primary survey was of very unequal merit and probably the best was done by Clough and Barrow.[12] Geikie had no appreciation of the difficulties to be encountered in mapping Highland metamorphic rocks. There was no adequate inspection or correlation of the work. Howell, the Local Director, had no control over it. Geikie visited the geologists occasionally in the summer season but never really grasped the problems involved. There was no petrologist in charge. Teall was much too busy with the rocks of the North-West Highlands, though he made some interesting contributions at a later period. Moreover, the geologists were working at such distances apart that they had little opportunity of consulting one another. It is not to be wondered that this country remains one of the least understood parts of Great Britain and that the number of unsettled questions of its geology is still very great. (Flett 1937, 110)

The details of all this need to be established in some future study, but it is evident that the Survey was in difficulties and was coming into contact with a problem that was, if anything, more intractable than that of the Northwest Highlands. In the Grampians, just about everything was metamorphic. There were no nice clear marker beds like the fucoid beds; there was no help from

---

12. Charles Thomas Clough (1852–1916) was a major contributor to the mapping of the Northwest Highlands, as well as the Central Highlands. He was renowned within the Survey for the perfection of his map work and was awarded the Murchison Medal in 1906. On Barrow, see chapter 9.

fossils such as those to be found in the Durness Limestone; and so there was no established stratigraphical sequence. Indeed, one never knew which way up the rocks were to be regarded. Even today there are unsolved puzzles.

Geikie found himself unable to resolve these "intellectual" problems. His main research interests at that time were in truth elsewhere, namely, his ongoing studies of ancient volcanoes, which culminated in the publication of his magnum opus, *The Ancient Volcanoes of Great Britain* (Geikie 1897). The Survey was also in difficulties because of "manpower" problems. By the 1880s, much of the basic survey work in England, Wales, and Ireland had been completed, but there was still an enormous amount to do in Scotland. It was natural, therefore, that staff should be transferred to Scotland, where the greatest need was felt. But because of the perceived priority of the Northwest Highlands work, more and more resources were channeled into the study of that remote and economically insignificant part of the world; and Geikie's two best surveyors, Peach and Horne, were tied up there, or in the Southern Uplands (also a region of little economic importance), for most of the period of Geikie's directorate.

We find, then, that there was a growing public unease with the performance of the Survey during Geikie's period of office, as manifested in the exchange of correspondence in the *Daily Chronicle*. Dissatisfaction may also be seen in the correspondence files in the Survey archives for the 1890s. For example, we find reference to a question asked in parliament on June 8, 1891, about the state of the surveys of the coalfields in South Wales and Monmouthshire.[13] A letter dated September 3, 1891, from a Mr. C. B. Balfour, complained about the neglect of the survey of the Fife coalfield.[14] A letter dated May 30, 1893, from the North Staffordshire Naturalists Field Club and Archaeological Society, expressed concern about the lack of a six-inch map for the coalfields of their district.[15] On August 17, 1893, a Mr. Eustace Button, chairman of the Clevedon Water Works, wrote complaining of the inadequacy of the published one-inch map of his region for the company's needs; and he objected further that the locality where one of their wells was sunk in Magnesian Limestone (Permian) was mapped by the Survey as Mountain Limestone (Carboniferous).[16] A letter from a Whitehaven company, Kendall and Main, Mining and Civil Engineers, dated April 5, 1894, severely censured the local map of the Whitehaven area (on the Cumberland coast), claiming that a quarry of St. Bees Sandstone (Permian) had been colored as Coal Measures (Carboniferous); and at another spot Carboniferous Limestone had been mapped as slate.[17] On October 2, 1895, some of the inhabitants of Tregaron (in Central Wales) wrote to their M.P. requesting him to use his influence to obtain a mineral survey of their district by the Survey officers.[18]

13. British Geological Survey Archives, GSM 1/29, p. 61F.
14. Ibid., 25.
15. Ibid., 61. (The complaint was subsequently backed up by a letter from the local M.P., J. Heath, dated Sept. 1, 1893 [ibid., 69a].)
16. Ibid., 78.   17. Ibid., 91.   18. Ibid., 155a.

This sampling of the archives shows at once that there was substantial dissatisfaction with the performance of the Survey at a very practical level—quite apart from the high-minded objections that had been brought to bear against Geikie in the pages of the *Daily Chronicle* on the question of priorities in relation to the Highlands controversy. Moreover, the correspondence shows that Geikie was largely unable to accede to the requests made to his department, or to make immediate correction to the surveying errors to which his attention was drawn. He had insufficient staff to make revisions everywhere at once; and in any case a great many of his men were tied up in northwest or central Scotland, attempting to deal with questions that were of great interest and importance to pure science, but of negligible economic significance and but little concern to coal owners and the like. Moreover, it was a serious matter if misleading information had been published in official government maps for regions of economic importance. It would, for example, have been deeply embarrassing if some entrepreneur near Whitehaven had gone digging for coal in the St. Bees Sandstone on the strength of the information furnished in the official geological map of the district.

Geikie was pressed on yet another front. When he had taken over the Survey, it was with an explicit understanding that there should be an improvement on the rate of mapping that had been achieved under Ramsay's directorate. To this end, the surveyors had been instructed to record their results directly on one-inch maps, rather than six-inch maps. This may have led to some improvement in the rate of work in some areas, but it was obviously ludicrous to work in this way in the Northwest Highlands. The edict was highly unpopular with the surveyors and was soon abandoned, as a direct result of the events of 1883–84 in the northwest (Bailey 1952, 115). But more important, there was an expectation on the part of some government officials that the Survey would be terminated as soon as its basic job of mapping was accomplished. The Survey was not envisaged as a permanent institution—a permanent center for geological research in the country. On the contrary, the objective was to finish the survey work as quickly as possible, so that a drain on the public purse might be eliminated.

What Geikie's attitude to all this was is hard to tell. But we can learn something about his tactics and the accomplishments of the Survey's work by examining the manpower available to him, the manner in which he deployed it, and the results that the surveyors achieved, by examining the statistics furnished by the annual reports for the Survey during the period of Geikie's directorate. Some relevant information from these reports is abstracted in figures 10.1 and 10.2.[19]

Examining first the data on manpower, it should be emphasized that the

19. Data are available for the areas mapped, as well as the length of boundary lines. But the boundary lines give a better indication of the amount of work actually done. Obviously, in a complicated region, the ratio of boundary lines to area will be greater than in a simple area. However, the politicians and the public were very likely more interested in the areas surveyed and the number of maps published.

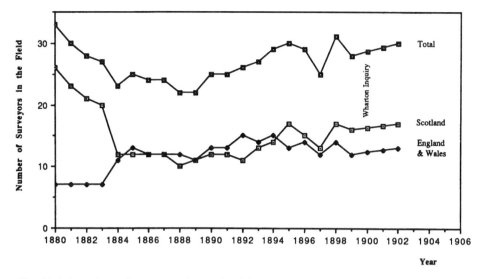

Fig. 10.1. Number of surveyors in the field for the period 1880–1902; (i) Total; (ii) England and Wales; (iii) Scotland.

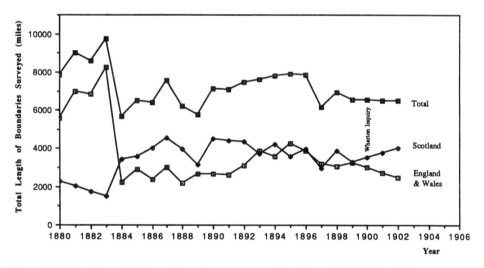

Fig. 10.2. Number of miles of boundary surveyed for the period 1880–1902: (i) Total; (ii) England and Wales; (iii) Scotland.

information refers to surveyors actually in the field, as reported by Geikie, rather than total Survey staff. Nevertheless, the graph seems to display some peculiarities, on first inspection. The first thing to notice is the significant decline in the number of men in the field in England and Wales at the beginning of Geikie's directorate. Was there then a mass of resignations because the staff found Geikie insufferable? This hypothesis is soon found wanting when we examine the information to be found in the Staff List appended to Flett's history of the Survey. The number of resignations in the early 1880s was not unusual. There were indeed a few retirements (one due to ill health), a transfer to the British Museum, a position taken up as a lecturer at the Royal College of Science, and James Geikie's appointment to the Murchison Chair at Edinburgh, recently vacated by his brother. But there was nothing out of the ordinary about such movements.

The problem, rather, was that Geikie did not secure sufficient replacements for the resignations or retirements. Thus the number of staff was allowed to run down. Whether this was Geikie's policy, or that of the department that administered the Survey (the Department of Science and Art) is a nice question. It may be that Geikie was, at the beginning, following government thinking. Or he may have felt that some retrenchment was not necessarily undesirable. However, it is the case that by the time that the government initiated an inquiry into the affairs of the Survey at the end of the century, Geikie found himself obliged to justify to the government the run-down in staff that had occurred.[20] At that time the total staff was counted as follows: 1881, fifty-two; 1890, forty-two; 1899, forty-one. Geikie explained how a number of the major resignations had come about, but it is clear from his document that he made little effort to ensure that replacements were made. Indeed, in one case, he actually urged that a vacancy arising from a resignation should not be filled. There was only one major senior staffing initiative, namely, the appointment of Teall as petrographer in 1888.[21] Thus one of the prime expectations of every government official—that the director of his or her section should work for the maintenance, or better the expansion, of the departmental workforce—was not fulfilled by Geikie. Murchison, by contrast, had been an outstanding director general so far as this side of his accomplishments was concerned.

Further problems arose in the matter of promotions and salaries, and again Geikie seems to have made little effort to protect the interests of his staff. As has been pointed out by H. E. Wilson (1985, 17), the senior staff were diminished almost to vanishing point during Geikie's reign, and a considerable number of men were kept waiting in junior or "temporary" positions, and were therefore without pension rights, even though they had given the best years of their lives to the Survey. It was really a deplorable state of affairs, and one may think, as Wilson obviously does, though without saying so explicitly, that promotions were blocked to senior positions because Geikie could brook no rivalry within his domains.

20. A. Geikie, "Statement in explanation of the reduction of the number of the STAFF of the Geological Survey," undated typescript, but carrying information up to 1899, British Geological Survey Archives, GSM 1/107. This document was prepared in response to a request for information from the Department of Science and Art. See page 322.
21. It is worth noting here that Geikie emphasized, as a special feature of Teall's case, that he had been appointed from outside the Survey. The usual expectation at that time was that the Survey would train its staff on the job, and perhaps in time provide personnel for other surveys or for university posts. The idea of bringing in experts from outside seemed to hold little attraction.

All this may not perhaps seem relevant to the Highlands Controversy, but these matters are indeed germane to the total situation that developed in the Survey in the 1890s, and on which the after-effects of the affairs in the North-West Highlands exerted a continuing powerful influence. Turning again to figure 10.1, it may be seen that Geikie shifted a considerable proportion of his "forces" into Scotland in the period 1883–1884. The surveyors John Dakyns, William Gunn, Charles Clough, and George Barrow were transferred from England and Wales to Scotland, and Hugh Miller, Jr., began work in Scotland at the end of 1884 (Geikie 1885a, 240). When Lionel Hinxman and Henry Cadell were appointed in 1883, they too were soon put to work in the Highlands, where they established sound reputations for themselves.[22] However, figure 10.1 also shows the effects of the complaints brought to Geikie's attention early in the 1890s from the various interest groups in England and Wales who were demanding revision of old maps and more information about regions of economic importance. I suggest that Geikie's response to these complaints can be discerned in the second half of the 1890s, as staff were increased in England and Wales once again, at Scotland's expense, even though Geikie well knew that the problems in the Central Highlands were far from resolution. It will be recalled that Flett stated that one of the problems in the surveying of the Grampians was that the surveyors were spread too thinly to be able to exchange ideas and information satisfactorily. Very likely, this followed from the decrease in the number of surveyors at work in Scotland resulting from the economic pressures from England and Wales.

Looking at figure 10.2, dealing with the surveyors' manpower and output, it will be seen that "producitvity" improved significantly during the early years of Geikie's directorate, which we might in fact have anticipated following the retirement of the aging and sick Ramsay. On the other hand, we see a slight drop in productivity in the closing years of Geikie's tenure, and it is not impossible—though hard to demonstrate—that this had to do with declining morale at that period. On the whole, however, there is a regular relationship between the number of men in the field and the amount of surveying work accomplished. (One must of course expect variations owing to seasonal weather conditions, when work in wild and remote country is being undertaken.)

The anomalous year in the statistics is 1884—the year of Geikie's "recantation" on the Highland question. A dramatic decline of work in England and Wales, only partly compensated by the additional performance in Scotland, is immediately apparent. It would appear that the transfer of a substantial number of surveyors to the north entailed a severe disruption of the program of survey, for the men would have had to come to grips with unfamiliar and difficult terrain, and some of the time of the established members of the Scottish staff would have been given over to assisting the newcomers. Nevertheless, the

22. Hinxman (1855–1936) was nevertheless one of those unfortunates who labored as a temporary geologist for a considerable number of years, only gaining a position as geologist in 1901, after the recommendations of the Wharton Inquiry were implemented.

surveyors' productivity in Scotland was actually somewhat higher than usual in 1884. This was, no doubt, in part due to the heroic efforts made that year in Sutherland by Peach, Horne, Clough, Hinxman, and Cadell, in response to the challenge set them by Lapworth. Also, it is known that the weather was exceptionally good for fieldwork in 1884, such that the surveyors in the northwest were able to cover 166 square miles of ground, mapping on a scale of six inches to the mile, in just two months (Geikie 1885a, 243). But the overall effect of the northwest concern on the Survey's work, particularly in England and Wales, was obviously severe, and the graphs suggest that it was never fully overcome in the period of Geikie's directorate. It should be stated, however, that the nature of the work carried out in England and Wales changed significantly at just that time, since the basic one-inch map was completed in 1883 (Geikie 1884a, 282). This meant that there was a lot of "indoor work" to be done in the completion of maps and memoirs; and thereafter attention in the field was given chiefly to the mapping of superficial "drift" deposits—important for agricultural purposes—rather than the basic solid geology (Geikie 1884a, 283; 1885a, 240).

Be this as it may, having shifted his resources into Scotland in 1883–84, Geikie then gradually had to shift them out again in order to meet the demands placed upon his organization by the economic and industrial interests in the parts of Britain to the south. Thus, I suggest, the effectiveness of the Survey as an agent for economic progress was significantly influenced by the events of the Highlands controversy. Indeed, this was one of the main factors contributing to the government's dissatisfaciton with the Survey, and leading to the establishment of the Wharton Inquiry, the findings of which precipitated Geikie's retirement from the scene as director general.

The public dissatisfaction with the affairs of the Survey and associated activities can be traced further in several petitions presented to the House of Commons in 1898 requesting that the Museum of Practical Geology not be moved away from its traditional home in Jermyn Street. This attracted the attention of the Department of Science and Art, whose permanent head, Sir John Donnelly, was an experienced and able civil servant.[23] From documents held in the

23. On Donnelly (1834–1902, see the anonymous memorial in *Nature* 65 (1901–2): 538–39, and Reeks (1920, 126–30). In the Reeks volume, Donnelly is described as "a first-rate man of business [who] would cut through red-tape entanglements and authoritatively make an end of official shilly-shallying over unimportant trifles (p. 127). He was also described in his obituary in *Nature* as having "an almost over-exacting sense of rectitude." Originally a military man who had served with distinction in the Crimean War, Donnelly became in time one of the most important government bureaucrats involved in the process of introducing scientific and technical training into the government-sponsored education system in Victorian Britain, and for the establishment of the complex of museums and teaching institutions at South Kensington. He was instrumental in bringing about the system of examinations in technical education organized by the Society of Arts, from which developed the important City and Guilds Institute for technical education and examination.

Public Records Office and in the Archives of the Geological Survey, we may see how events rapidly developed.

Donnelly asked one of his colleagues, Captain W. de W. Abney, to have the affairs of the Geological Survey looked into.[24] Abney requested another departmental officer, a Mr. P. G. Cooper, to examine the question in detail, so far as documents on the public record allowed, and to prepare a memorandum on the topic for the perusal of Donnelly. Abney also invited Judd to provide an informed outsider's view of the Survey's work. In a note attached to the document prepared by Cooper, Donnelly referred to the "recent applications for re-Survey of Coal-Fields" and "certain other circumstances that came to my knowledge" as his reasons for setting the government's wheels of inquiry in motion.[25] This indicates, I suggest, that Donnelly was responding to formal expressions of dissatisfaction with the Survey's work from the mining industry, and to things that he himself had heard or suspected.

The memorandum prepared by Cooper was submitted to Donnelly by Abney on July 5, 1898, and a copy was dispatched to Geikie for his comments on July 20. Cooper's document briefly surveyed the history of the Survey's work, giving particular attention to the inquiry that had been held in 1880 that led to Ramsay's retirement and the appointment of Geikie in his stead—at which time directions had been issued for the speedy completion of the survey work, with mapping to be done on the one-inch scale.[26] Both Ramsay and Geikie had said this was feasible. Cooper also quoted the estimates then made as to the time required to bring the survey work to a conclusion; and he compared the actual achievements since 1880 with the earlier sanguine estimates of Ramsay and Geikie. It emerged that at the current rate of progress it would take until 1932 to complete all the mapwork and issue "solid" and "drift" maps for Scotland—an estimate totally unacceptable to the government officials.[27] The report was furnished with statistics drawn from the Survey's own annual reports showing the work completed each year in terms of fieldwork and in the publication of maps, and it also gave figures for the staffing situation in the Survey, such as we have considered above.

In relation to the geological work in the Highlands, Cooper quoted Geikie as saying in his annual report that the establishment of a "new base-line" in

24. Abney (Sir William de Wiveleslie, 1843–1920) was another highly experienced bureaucrat in the Department of Science and Art and a scientist in his own right, being an authority on photography and the science of color vision. He was a Fellow of the Royal Society, and at different times president of the Royal Photographic Society, the Royal Astronomical Society, and the Physical Society. Over a hundred of his papers are listed in the Royal Society *Catalogue of Scientific Papers*, and he authored a number of books. His opinion would certainly have weighed with Donnelly.
25. Public Record Office, DSIR/9, file 13, 6/V4.
26. For Cooper's memorandum, see ibid. and BGS Archives, C & P/207.
27. A "drift" map shows "superficial" deposits such as river gravels and glacial detritus. Such information is usually of considerable importance for agricultural purposes, but it tells nothing about the basic geological structure of a region. Thus it is desirable to have two series of map. The term "drift" derives from the old notion that glacial deposits were emplaced by drifting and melting ice-bergs.

1884, and the reclassification of strata in 1891 as a result of the *Olenellus* discovery, constituted "an important scientific result" (see Geikie 1892b).[28] But, said Cooper: "It is . . . questionable whether the detailed examination which it involved was warranted by the principle laid down by My Lords namely 'general accuracy being secured without insisting too much on minute details.'" In a note attached to the copy of Cooper's report held in the Public Record Office, Donnelly stated that the investigation disclosed a "very serious state of things" and called for an inquiry. It can be seen, then, that the government, the Survey, and the wider geological community each had figured a significantly different order of priorities.

Geikie was slow in responding to Cooper's report, taking two months to prepare an eight-page reply, though, as he said, he had been in the field for most of that time.[29] His defense was self-assured and might have seemed entirely convincing to an outsider. He maintained that "the economic value of the work of the Survey ha[d] been constantly kept in view." But its work had to be carried out "on a scientific basis," which justified the emphasis given to surveying in the Highlands. In any case, there was hope that some metalliferous lodes might be discovered during the course of the Highland investigations. The petrographical work of Teall and Watts (which Donnelly had referred to in his letter of July 20 as "incidental employments") was properly defended as being essential to the success of the surveyors' work in the Highlands. Geikie denied that there had been any avoidable delay in the engraving of maps, and he pointed out that the resurvey of the South Wales coal field, which had begun in 1891, had already led to the issue of a number of new maps.

With respect to the Highlands work, Geikie emphasized the problems that attended work in this difficult country, with its complex geology. The natural difficulties of weather and terrain were compounded by the fact that the surveyors had to use hunting lodges for their accommodation: but these were unavailable during the shooting season. As a result, a surveyor had to work in two areas each year: one in high ground; the other in a lower region where he could work in the early summer and in the autumn. But perhaps more important was the significance of the Northwest Highlands to the understanding of the geology of the country as a whole:

> At first the mapping of the Highlands was begun in continuation of the survey of the Lowlands, but it was soon found impossible to make out the succession of the southern Highland rocks without recourse to the clearer evidence presented in the north-western counties. Accordingly some of the surveyors were detailed to the west of Sutherland and Ross, for the purpose of obtaining the key to the general geological structure of the Highlands. The whole tract of country where this

28. Geikie actually put it slightly differently: "The results obtained during the mapping of this part of the country [Dundonnell Forest] are the most important which the survey has achieved since the structure of the Durness-Eriboll area was worked out" (1892b, 384). He went on to explain how the new fossil find indicated that the Torridon Sandstone was pre-Cambrian in age.
29. Public Record Office, DSIR, 9/13, 6/V4, Geikie to Donnelly, Sept. 30, 1898.

structure is revealed has now been surveyed in detail, and the maps on which it is delineated are perhaps the most remarkable that the Survey has ever published. These maps have attracted close attention in foreign countries and have brought many distinguished men of science from all parts of Europe and America to the north-west of Scotland.

Now that the structure of this typical area has been made out, it is not necessary and it is not intended to continue the same style of detailed mapping over the rest of the Highlands. More rapid progress can now be made than for some years has been possible.[30]

Geikie also asserted that he planned to transfer some of the surveyors presently working in Scotland back to England. He concluded by pointing out that his staff was aging, which made it difficult to maintain the rate of progress of earlier years, though the experience of the older men was of course invaluable.

A series of comments on Geikie's statements in reply to Cooper's memorandum was made in an unsigned document attached to Geikie's letter, deposited in the Public Records Office.[31] The comments appear to have been made for the information of Donnelly, and the author may have been Abney, since it was he who sought further advice from Judd. The commentator doubted whether important mineral deposits would be discovered in the Highlands as a result of the survey work there. Geikie's assurances about the rate of progress of the map work were questioned. It was pointed out that forty-seven map areas were known to have been surveyed but the maps were awaiting publication. The commentator was skeptical as to how soon this work might see the light of day, given that it appeared that the current practice was for the field surveyors' work to pass through the hands of the petrographers Teall and Watts before the maps could be finalized for publication. He seemed to find this practice unacceptable, or at least he attributed much of the current delay in work to the intrusion of laboratory petrographical work into the mapping process: "Why a Surveyor of long experience, or a new Surveyor with high scientific attainments[,] is not considered equal to determining for the purposes of the map the rocks he meets with in the field seems inexplicable."

However, thinking no doubt that a more scientific opinion was required on the matter, Abney applied to Judd for his expert opinion on Cooper's memorandum and Geikie's reply. Judd's answer is highly pertinent to my present purpose, and some extended quotation is warranted. Judd pointed out that Cooper had correctly located the causes of the unfortunate delays in the completion of the preliminary survey of Britain as being in Scotland—as Geikie had tacitly acknowledged in his letter. So Judd launched into his own view of what had occurred in the mapping of Scotland:

Before the year 1881, the whole of the field-work in the Lowlands of Scotland . . . had been completed, the maps being either published or in the hands of the engravers.

30. Ibid.
31. "Comments on Sir A. Geikie's letter of 30th September 1898."

By the same year (1881) substantial progress had been made with the Survey of the Highlands, which was commenced in Perthshire about 1874. In 1881 it will be seen that much work had been done in the Highlands of Argyllshire, Perth and Forfar, the work being carried round through Aberdeenshire, Banff, Elgin, and Nairn as far as Inverness. It will be seen that the plan then adopted was to carry the lines surveyed in the lowlands directly into the Highlands, the work proceeding from South-East to North-West. By 1884, something like one third of the Highlands would appear, on the official report, to have been completed on this plan.

In the report for 1884 a complete change of plan was announced and the surveyors were transferred from the southern and eastern Highlands to the extreme North-West of Sutherland, and, in the fourteen years since that date, only a fringe of country in the Western Highlands has been completed and published.

It is suggested [by Geikie] that the delay in completing the Highland work is due to the fact that the Surveyors can only work there for a portion of the year. But this must have been clearly foreseen in 1881, when the estimate was made, for not only had large tracts of Highland been surveyed, but the present Director General had made a preliminary survey of the whole Western Highlands—one sufficient to enable him in 1861 and following years to publish (though not as official works) several maps in which the Highland geology is fully elucidated, and also a very elaborate memoir on the subject.

It is further stated that the large staff of Scottish geologists, when unable to work in the Highlands, were transferred to the Lowlands. But as we have seen, the whole of the lowland geology was announced as completed in 1881, and the maps were stated to be either actually published or in the hands of the engravers.

It is well known to all geologists, that the practical *re-survey* of the Highlands which was commenced in 1884 [*sic*] was shortly afterward followed by a *re-survey* of the whole of the Borderland [Southern Uplands]; maps which were announced as being ready for publication before 1881, being held back, and not published till twelve or fourteen years later. The classification and nomenclature of the strata adopted in the publications of the Scottish Geological Survey prior to 1885, have been entirely abandoned since that date, the necessary alterations being made in all the published maps.

Further than this, new surveys have been made in Fifeshire and other districts, so that the greater part of Scotland has been surveyed twice over.

Had the large staff of geologists employed on these new surveys been transferred to England, instead of the Lowlands during the periods of the year when work in the Highlands was impossible, there is no reason to doubt that the six-inch maps of the English Coalfields might have been completed, and the work of the Surveys brought to close within the estimated period.

There can be no doubt that, as suggested by Mr Cooper, a large part of the work undertaken by the officers of the Geological Survey during the last twenty years has not been of the kind contemplated when the Survey was first established—namely the promotion of our knowledge of the *economic* geology of the country. A reference to the general geological literature, during the period named, will show that much of the time and labour of the staff has been devoted to the solution of questions of a highly theoretical, often speculative, and sometimes polemical character—questions which, however interesting in themselves, ought to be left to private investigators.[32]

Judd's comments make interesting reading for the student of the Highlands controversy. Most of what he said was undoubtedly true; but his comments were partial. He might, for example, have drawn attention to the excellence and high scientific importance of the work of Peach and Horne in the northwest. He might have emphasized the essential nature of petrographical support for survey work in regions where metamorphic rocks were the chief objects of investigation. He might have referred to Geikie's major contributions with his studies of volcanoes. Instead, Judd deliberately drew the attention of the government officials to the debacle in the Southern Uplands and emphasized that the Survey had had to go over much of its Scottish ground twice. Yet in saying that the Highlands had been surveyed twice, he was being partly misleading, for no one could suppose that the rapid visit of Geikie and Murchison to the northwest in 1860 constituted a proper survey. There was, I suggest, something of a pattern in Judd's attitude. For example, in another publication, otherwise quite unconnected with the Highlands controversy— an introduction to Darwin's book on the geology of South America (1890)— Judd gratuitously drew attention to the controversy, and pointed to the repudiated paper of Murchison and Geikie of 1861 and the survey's abandonment of the old theory in 1884 (Judd 1890b, 276). Why he might have done this, we shall shortly see.

Following discussions in the Department of Science and Art, probably in conjunction with Geikie, a memorandum for the duke of Devonshire was prepared, recommending the transference of the Survey from the Department of Science and Art to the Ordnance Survey.[33] Letters from Geikie show that he made no strong objection to this proposal.[34] But in another memorandum in the file from Devonshire it was suggested that details of the transfer should be considered by a committee.[35] Thus events were moving toward the establish-

32. Memorandum of J. W. Judd, dated Nov. 18, 1898, DSIR, 9/13, 7/V4.
33. DSIR, 9/13, 7/V4, Jan. 11, 1899. (The document is initialled by Cooper.)
34. DSIR, 9/13, 7/V4, Geikie to Donnelly, March 6 and 13, 1899.
35. DSIR, 9/13, 7/V4, D[evonshire] to the Vice-President [of the Board of Education?], April 10, 1899.

ment of the Wharton Committee, which was set up to look into the affairs of the Survey, and which began its work the following year.

It is clear that this bureaucratic activity did not go unnoticed within the Survey, for we find that Henry Howell, the organization's second most senior officer, who had been nominally responsible for the Scottish work since 1884 and had several good reasons for being antipathetic toward Geikie, saw the moment as opportune to write, immediately prior to his retirement, respecting his role in the Highlands surveying and disclaiming responsibility for the slow results achieved in Scotland.[36]

The circumstances attending the dispatch of Howell's letter are of some interest. It appears that he wrote directly to the Department of Science and Art on May 22 and 23, but was advised by Donnelly that since the documents contained "certain grave reflections on your chief—the Director General" it was proper that they be submitted through Geikie.[37] As a result, Howell withdrew his original letters and prepared another document, which he submitted on May 30 to Geikie for transmission to the department. Geikie, however, failed to do as asked, only complying with Howell's request after some prodding and after having offered a feeble excuse for his inaction.[38]

These details are not perhaps so very important, but the contents of Howell's memorandum to the department are of surpassing interest, and are worth extended quotation, since they reveal so clearly the problems that had been encountered in the Scottish work as a result of all the things that had happened in the Southern Uplands and the Northwest Highlands. Howell wrote:

> When I took up my position as Director for Scotland at the Edinburgh Office in 1884, I found that the whole of the one-inch-maps of the Central [Midland] Valley and the Southern Uplands of Scotland were published, with the exception of three sheets viz. 8. 10. and 16. In the Index Map attached to the Annual Report for 1878 these three maps were entered as "Engraving,["] but this was probably an error, for when the inquiry was instituted by the Department 1880–81 into the progress of the Geological Survey at that date, and a Return was asked for in tabular form showing the state of the one-inch-maps, in the Return sent from Scotland the above three maps were entered as having been "sent to Southampton for engraving for the first time 9th October 1880." I have ascertained that this is the correct date at which they were sent to be engraved.
>
> In 1884 *Engraved* proofs of these maps had been returned from the Engravers, and the question was put before me as to their early publication.

36. On Howell, see the Geological Magazine's Eminent Living Geologists series (1899a); Smith Woodward (1916b); Greenly (1938, 2: 518–21). Howell started work with the Survey in 1850, at the early age of sixteen, having impressed Ramsay in the field.
37. DSIR, 9/13, 8/V4, Donnelly to Howell, May 26, 1899.
38. DSIR, 9/13, 8/V4, Howell to Donnelly, June 8, 1899.

Mr. B. N. Peach, District Surveyor, handed to me the following Memorandum:

Sheet 8.[39]—"Can't be published till Girvan ground is put right."
Sheet 10.[40]—"Line to trace between Upper and Lower Silurian in neighbourhood of Lockerbie. B.N.P. or Horne could trace this."
Sheet 16.—"Waiting till the Wrae Limestone and the black shale crops are mapped out in N.W. corner, Wrae Limestone is certainly interbedded among the Lowther Shales in this sheet. B.N.P. can trace these."

As regards sheet 10, the tracing of the line between the Upper and Lower Silurian was a comparatively small matter, and this map was published in 1885. But in the case of sheets 8 and 16, to put the Girvan ground right and to undertake a re-survey of these maps was a much graver matter, for it was seen that it would have a far reaching effect in the already published maps containing the Lower Silurian Rocks— covering an area not far short of 2000 Square Miles. So these two sheets were allowed to remain in abeyance for four years—the whole strength of the Staff then having been transferred to the Highlands.

In 1888 these two maps being still in abeyance Messrs. Peach and Horne were instructed to do the revision survey of the Lower Silurian Rocks in the Southern Uplands at such times of the year as field-work was impracticable in the mountainous districts of the Highlands. In this way, the revision survey extended over a period of ten years—from 1888 to 1897, and we were able to publish sheet 16 in 1889 and sheet 8 in 1893.

The revision survey of sheets 8 and 16 when published necessarily involved a revision survey of the adjoining already published maps to bring them into harmony with the new reading of the geological structure of the Silurian Rocks of the South of Scotland. As a matter of fact the published maps chiefly affected were 5. 7. 9. 15. and 24, and to a lesser extent 1. 2. 3. 4. A revised Edition of Sheet 7 has been published [in] 1895. . . .

The Highlands of Scotland.

In his reply to you dated 30th September 1898, the Director General reminds you that a Geological Survey to be of any practical value must be carried on upon a "scientific basis". The Department of Science and

---

39. This sheet covered the region inland from Girvan.
40. Sheets 10 and 16 covered important adjacent portions of the center of the Southern Uplands, Sheet 16 being the Moffat map, which included Dobb's Linn. Thus the long reach of Lapworth's work in the Southern Uplands can be discerned in this paragraph.

Art, I should think, would certainly expect that the Geological Survey of Scotland would be carried on upon a *true* scientific basis. But the disaster which overtook the Geological Survey of Southern Uplands, and the Highlands of Scotland was due to the fact that down to the year 1884 it was not carried on upon a true scientific basis.

When I came here in 1884, the basis upon which the Geological Survey of the Highlands was being carried on upon is set forth in the Annual Report of the Director General [Murchison] to the Department for the [year] 1860. . . .[41]

This basis which was said to have been established is further emphasised in the Explanation attached to the small Sketch-Map [of Murchison and Geikie] (published 1862 [*sic*]). . . .[42]

This . . . was the scientific basis established by Sir R. Murchison and the present Director General—the broad lines upon which the Geological Survey of the Highlands was to be carried on upon—the rank and file of the Survey to follow after and fill in the details, . . .

This basis held the field down to 1884—the year I came to Edinburgh as Director for Scotland . . .

From 1883 to 1885,[43] Messrs. Peach and Horne were carrying on the Survey work in Sutherlandshire. In July of the latter year, the District Surveyor—Mr. Peach—reported to me that the results of their field-work up to that date had proved that the basis said to have [been] established—the "ascending order of succession"—and upon which the detailed Survey of the Uplands had been commenced under the direction of the present Director General, was absolutely untenable and would have to be abandoned. This then largely accounts for the present backward state of the Geological Survey of the Highlands. It was found impossible to get the mapping done at the rate estimated in 1881.

There are now upwards of 8000 Square Miles of area remaining to be surveyed in Scotland—new ground absolutely untouched either for the fundamental Rocks or the Superficial (Drift) deposits.

In the interests of the Survey and the Surveyors a Departmental inquiry is most earnestly desired by the Staff. This found expression in a Memorial to the Director General and signed by the whole Staff of Surveyors in England and Scotland, and which I transmitted to him on the 4th March last [1899]. He acknowledged the receipt of it on the 6th March, and I was to assure the Staff that he would do his utmost to have the Survey and the members of the Staff placed on a proper footing.

41. There follows here a lengthy extract from Murchison's report of 1860.
42. There follows here a lengthy extract from the map's explanation. (It was published in 1861, not 1862 as stated by Howell.)
43. This should, I think, have read 1884.

Nothing has been heard of the Memorial since, and some members of the Staff—those most vitally interested in it—and their future prospects in the survey—officers who are now after from 15 to 20 years service still in the Temporary list of Assistants—are naturally anxious to know if it has been brought under the notice of the Department.

I have the honour to be,

Sir,

Your most obedient servant,

Henry H. Howell

Director.[44]

It is evident from Donnelly's annotation to this document that he had not seen the "Memorial" or petition to which Howell referred. We may conclude, then, that Geikie simply sat on it; and that Howell felt obliged to communicate directly with the department in order to make the surveyors' views known. I have been unable to locate the Howell memorial and it is possible that it is no longer in existence. However, it does not appear that Geikie attempted to stand in the way of a government inquiry.

Thus the government rapidly moved toward the establishment of a full-scale investigation of the Survey's activities, under the chairmanship of an M.P., John Wharton, rather than making a mere bureaucratic shuffle of the Survey from the Department of Science and Art into the Ordnance Survey. The members of the committee of inquiry were as follows (Wharton Committee 1900):

John L. Wharton, Chairman, M.P. for Ripon
Stephen Spring Rice, of the Treasury
T. H. Elliot, of the Board of Agriculture
General E. Robert Festing, of the Board of Education (Victoria and Albert Museum)
Dr. H. Franklin Parsons, Assistant Medical Officer of the Local Government Board
W. T. Blanford, formerly of the Geological Survey of India
Professor Charles Lapworth, of Mason University College, Birmingham
A. E. Cooper, of the Board of Education, South Kensington, Secretary.[45]

The Wharton Committee's terms of reference, laid down by the duke of Devonshire, were to inquire into the organization and staff of the Survey; to

---

44. DSIR, 9/13, 8/V4, Howell to Secretary of Department of Science and Art, May, 1899.
45. This was a printed document, but it bears no publisher's name or place of publication. A copy is held in the archives of the Geological Survey at GSM1/106. (The Wharton Committee's secretary, A. E. Cooper, was a geologist, and should not be confused with the Mr. P. G. Cooper in the Department of Science and Art, who had furnished information to Abney and Donnelly.)

report on its progress since 1881; to suggest arrangements for its improvement with a speedy and satisfactory termination of the work, having special regard to matters of economic importance; and to consider the desirability of transferring the Survey to another public department.

The committee, which met from May to July 1900, in Geikie's office in the Survey Museum in Jermyn Street (Geikie 1924, 319), took oral evidence from the director general and several members of his staff, and it received submissions from a number of scientific, educational, and economic interests, some in writing. Much of the committee's report consisted of statistics concerning the rates of progress of the Survey and the conditions of employment of the staff. Nine appendices were incorporated.[46]

Most of the statistical information need not concern us here, though it may be mentioned that the data assembled certainly displayed a sorry state among the surveying staff in regard to promotions, pay, and pensions, and that in these matters there had been a steady deterioration in conditions, especially for the more recently appointed members. Thus Aubrey Strahan calculated that the total value of an appointment in 1864 was £3,773 *more* than for one in 1875, for forty years service in each case.[47] And by the end of Geikie's reign, there were very few senior appointments remaining.[48]

So far as mapping was concerned, out of the 59,863 square miles in England and Wales, all had been surveyed and published in the "old series." All 32,177 square miles of Ireland had also been surveyed and published. But in preparation for the "new series" in England and Wales, 23,466 square miles were listed as having been surveyed on the six-inch scale; 16,280 were given as published; and 36,397 were unsurveyed. Of the 30,887 square miles in Scotland, 22,734 were surveyed; 15,828 were published; and 8,153 were not surveyed at all, either for an old series or a new. In Scotland, all of the Southern Uplands and the Midland Valley had been surveyed and published, and the

46. 1. A statement by H. B. Woodward on the staff, their dates of appointment and promotion (or their lack of promotion).
  2. A statement from Horne as to work remaining to be done in Scotland.
  3. A statement from Horne as to maps issued by the Scottish office up to the end of 1899.
  4. A statement from the staff member, Aubrey Strahan, as to the comparative value of appointments in 1875 and 1864.
  5. A statement from Geikie as to the work in hand on June 12, 1900.
  6. A statement from the president-elect of the Institute of Civil Engineers on the practical uses of geology.
  7. A statement showing the areas surveyed, published, engraved, and unsurveyed, as of July, 1900.
  8. A memorandum from committee member Dr. Parsons on the role of the Geological Survey in relation to public health.
  9. A paper from Lapworth on the uses of geology and the Geological Survey.
47. Strahan (1852–1928) himself joined the service as a temporary assistant geologist in 1875, in which insecure position he remained until 1896. Thereafter, he rose quite quickly, eventually becoming director of the Survey in 1914.
48. In 1899, following Howell's retirement, there was only the director general, two district surveyors (Woodward and Peach), and one senior geologist (in Ireland).

eastern border or margin all the way up to Wick and Thurso in the far north; also the area of the zone of complication running from Durness-Eriboll down to (but not including) Skye. But the Western Isles, Orkney and Shetlands, and all the mountainous regions of the center of Scotland awaited survey and publication. Thus the government had some reason to be displeased, considering that the Survey had been established as long ago as 1835. Geikie estimated that it would take another fifteen years to complete the survey work in Scotland, but he no longer suggested that the work could be done properly on one-inch maps. That point had finally been settled as a result of the Highlands controversy.

One major question that had to be determined by the inquiry was whether the Survey should be wound down or wound up, so to speak. In this matter, the opinion of Lapworth doubtless weighed heavily. In fact, Lapworth spoke strongly in favor of the continuation of the Survey, and he argued for its economic, scientific, and educational importance.[49] Likewise, there were representations from industry and from the medical profession in favor of the continuation of the Survey's activities. The primary question was thus readily settled in the Survey's favor. Moreover, the committee was readily persuaded of the need for a significant improvement in the pay and conditions of the Survey staff, and recommendations were made to that effect. It was also recommended that the Survey be maintained under the aegis of the Board of Education.

Perhaps the major suggestion for the improved working of the Survey was that a definite "programme of work be framed and adhered to" (Wharton Committee 1900, x). Doubtless this recommendation was made with the events of 1883–84 in the Northwest Highlands in mind, when Geikie suddenly changed the whole program of survey work in Scotland and was thereafter forever moving staff about like pawns on a chessboard. However, there was no overt criticism of Geikie in the report, and so far as the surveyors were concerned it was stated that the committee was much impressed with the staff's work and enthusiasm (ibid., v).

Thus the Wharton Committee did not really make any major criticisms of the Survey's work. Rather, it made positive suggestions for improved organization and for a plan of work that would best meet the economic needs of the community. There was absolutely no mention of the Highlands controversy, though as we have seen this was a major factor leading to the establishment of the Wharton Committee, through the memoranda of Judd and Howell. In fact, the only point where geological controversy was mentioned in the committee's report was in relation to outside consulting work that the Survey staff might engage in from time to time: "We do not consider . . . it to be consistent with the purposes or dignity of the Survey that its officers should give expert

---

49. His views were furnished in an appendix to the Wharton Committee's report, and this essay was also published as a separate pamphlet: "On some of the uses of geology and the Geological survey—Communicated by request to the Geological Survey Committee," 1900. Lapworth delivered another paper on the same theme on Dec. 13, 1901 (Lapworth 1902).

evidence in contested cases as scientific witnesses in geology" (ibid., vii). It seems to me possible that Lapworth had a hand in the wording of that part of the committee's recommendations.

Although Wharton Inquiry made no adverse remarks about the director-ship of Geikie, it seems likely that he was led to understand that it was high time he retired (though there is no documentary evidence, so far as I am aware, that he was forced to retire). Geikie was sixty-five on December 28, 1900, so there was no difficulty about this, sixty-five being then the normal age for retirement in the civil service. Thus he was able to retire without loss of face as a result of the inquiry, and in fact he actually hung on until March 1901, finishing his annual report for 1900. I may say, however, that there had been no talk of Geikie's retirement prior to the Wharton Committee's work, and I suspect that had there been no inquiry he might have sought to con-tinue into old age, as Murchison had done.

In fact, at sixty-five Geikie was still extraordinarily vigorous (full of intel-lectual vigor). He did not sever his work with the Survey entirely, since he published a major official memoir on the geology of Fife (Geikie 1902); he cooperated with Gunn, Peach, and Harker in a memoir on Arran, Bute, et cetera (Geikie et al. 1903); and there was of course his contribution to the edit-ing of the Northwest Highlands memoir of 1907, discussed in chapter 9. He was also very active in his capacity as secretary of the Royal Society from 1903 to 1908 and as president from 1908 to 1913. In his old age he did much valu-able writing in the history of geology, and it may well be that this work was, in the last analysis, his most enduring contribution to scholarship, for he was ex-ceptionally well suited to historical work, with his lucid style of writing and his intimate knowledge of the history of geology. Thus the continued smooth up-ward ascent of Geikie's career was barely interrupted by the events of the Wharton Inquiry.

As to the evidence that was gathered during the inquiry, we have rather little information other than the statistical material furnished in the report it-self. The public record concerning the activities of the Wharton Committee is unfortunately incomplete—that is, most of the transcripts of the "interroga-tions of witnesses" appear not to have been preserved. The only one that I have been able to locate is that of Horne, who by 1900 was acting director of the Scottish branch of the Survey. (He was appointed director the following year.) The transcript of Horne's evidence is held in the archives of the Scottish branch of the Survey, rather than in the Public Record Office; so, for what-ever reason, all the information collected from witnesses was not preserved in the government archives.[50]

My impression from reading this transcript is that the questions were not at all "aggressive." They seem to have been framed simply to elicit information

50. BGS, Edinburgh Archives, SOA 1/331. (This file also contains papers that Horne prepared in advance of his questioning by the committee, with various points that he wished to make. He seems to have been successful in getting most of these across. The papers include a letter from Macconochie, who rightly supposed that this was a possible opportunity to get his case heard.)

about the general workings of the Survey—as to how manpower was deployed, how decisions were made on mapping "tactics," the conditions of work at the Edinburgh office, the rate of mapwork and whether delays in this might be due to the Survey or the engravers of the Ordnance Survey, the arrangements employed for the display of specimens in the museum, the consulting work done by staff, the sale of maps and memoirs, the relationship between the work involved in mapping and the work involved in the preparation of descriptive memoirs, the economic significance of the Survey's work, and so on. At no time did a questioner show any disagreement or dissatisfaction with Horne's answers.

Horne himself had obviously come well prepared to make known some of the specific grievances held by the staff. For example, he pressed the case of the fossil collector, Arthur Macchonochie, who had been stuck on a salary of £182 p.a. for twelve years, even though he had performed wide-ranging duties and had made some important discoveries, as for example the *Olenellus* remains in the Northwest Highlands. It seems that representations had been made to Geikie to do something to assist Macconochie's case, but apparently it had not been taken up with the department. A very unsatisfactory situation was revealed in relation to traveling allowances. The government regulations allowed for a certain sum to be expended per day for accommodation, but this had to fit into a fixed annual budget (£900) for the whole Survey. The size of this budget, not the amount of money actually expended, determined the amount of reimbursement a surveyor received. So the surveyors were sometimes out of pocket, in contravention of the government regulations. Moreover, since it appeared the travel vote was to be reduced to £750 in 1900, it was clear that further problems were foreseen by the staff. Again, we find that Geikie had not been protecting the interests of his subordinates.

The question of the work in the Northwest Highlands came up in Horne's "interrogation," but no information was elicited that the reader will not already know. Horne did not seek to attach any blame to Geikie for all that had occurred in the northwest, or in the Southern Uplands. He did, however, say that if he had his way the surveyors in the Highlands would work closer together, so that consultation between them might be more effective. Lapworth asked very few questions. He showed interest in the amount of detail in the surveyor's work in the Northwest Highlands and the Southern Uplands, and wondered whether such detailed work was to be required in the mapping needed for the completion of the survey of Scotland. Horne thought not. It is not clear to me what Lapworth's motives were in asking these questions. He may simply have been wishing to provide Horne with an opportunity to justify all the effort in these two areas in the eyes of the committee members. At one point Horne mentioned that the rate of survey work was now slower than had been the case twenty years before, since "zonal" techniques of collecting were now being employed. (No doubt Lapworth smiled inwardly at that point.) The committee asked Horne no technical questions about geological matters.

It would be a delight to the student of the Highlands controversy to have available a transcript of the questions posed to the director general, and of his replies, but unfortunately a copy of that interesting document does not appear to have been preserved. We turn, then, to the steps taken after the completion of the committee's work to implement its recommendations.[51] With Geikie's (1924, 321) support, the directorship went to Teall. Horne and Woodward were named as deputies, with responsibility for Scotland and England and Wales respectively. Seven district geologists were named and there were to be thirty geologists, which required the appointment of five additional men. All professional staff obtained rises in salary, according to their years of service and perceived abilities. Besides Geikie, three men retired because of their age, one because of ill-health, and one was compulsorily retired because of inadequate performance.

Also, following the reorganization of the Survey, a committee was established by the Board of Education to oversee the work of the Survey on a regular basis—but only in an advisory capacity. Four of the members of the Wharton Committee (Wharton, Blanford, Parsons, and Lapworth) formed the nucleus of this new committee, and they were joined by two additional geologists (Judd, and the Oxford professor William Sollas) and three non-scientists. Teams of from three to five surveyors were to be established, with responsibility for particular areas where work was required to be done.[52] It can be seen from the proposed disposition of forces that attention was to be given particularly to areas of economic importance, especially mining regions. But in Scotland work was to be continued with the primary surveying of the Highlands and the Hebridean islands. This plan seemed to work satisfactorily, so far as one can tell from the three books on the history of the Survey (of Flett, Bailey, and Wilson). Never again did the Survey become traumatized, as occurred under the directorship of Geikie. But discussion of these later developments takes us beyond the scope of the present study.

It is a nice question as to the nature of the personal relationships that existed between Geikie and his subordinates. He has been accused of shameless plagiarism of the ideas of his staff. Thus Wilson has written that Geikie "apparently regarded the right to publish the findings of his staff as a kind of droit de seigneur," and that there were "few tears when he finally retired" (1985, 17, 18). However, it is not possible at this distance in time to establish exactly what the specific acts of plagiarism were, and Wilson sites no concrete cases. It remains, then, a matter of tradition within the Survey, though doubtless with some truth in it.

We know that Lapworth had a poor opinon of Geikie on a personal level. So too did Teall to some degree, but he accepted a job under Geikie and was later recommended by him as the one to fill his place, over the heads of Wood-

51. Public Record Office, DSIR, 9/78, 4/5.
52. DSIR, 9/15.

ward and Horne. As will be discussed below, Howell and Judd, who both spoke against Geikie's conduct of the Survey prior to the establishment of the Wharton Committee, each had their personal reasons for acting as they did. Bonney was not overly impressed by Geikie.[53] But Geikie's retirement was marked by a highly successful complimentary dinner, and he was presented with a "beautifully illuminated address bearing the signatures of the Staff of the Survey in the three kingdoms" (Geikie 1924, 323).[54] In his address in reply, Geikie spoke of the esprit de corps of the Survey and his honor and pleasure at being placed at its head. However, when he went on to claim that he had striven to the utmost in his power to ensure the welfare of his staff, this must have struck his subordinates as a piece of humbug, when we consider how readily the government acquiesced to improvements in the pay and conditions of the surveyors once the Wharton Committee looked into the matter properly.

I have no evidence of any tension between Geikie and Peach and Horne. Indeed, there is evidence to the contrary. As we have seen, Horne emphasized to Lapworth that Geikie changed his views over the Highlands without quarreling with his subordinates. And we know that Horne made no attempt to use the opportunity of the Wharton Inquiry to denigrate Geikie. The day before Geikie retired, Peach made a point of writing him an extraordinarily generous letter, which Geikie was proud to publish in his autobiography. It is worth quoting a few sentences here (Geikie 1924, 321–22).

Edinburgh, 28th February, 1901.

My dear Sir Archibald,

    I cannot allow the day to pass without writing to express my regret that our official connection ceases tomorrow. I feel how much I am indebted to you, not only for all I know of geology, but for the constant support I have felt in having you as a friend. I trust that this latter sense will long remain with me, and that as long as you are alive I shall

---

53. On one occasion, Geikie wrote to Bonney: "You were perfectly entitled to criticise my paper as severely as it seemed to you to deserve: but a general charge of inexperience, unsupported by proofs, is not, in my opinion at least, fair criticism" (Geological Society Archives, E32, Geikie to Bonney, April 18, 1893). There is no published work of Geikie from early 1893 to which this letter might be "fitted." We may assume, therefore, that the paper received such a hostile reception that Geikie did not choose to publish it.

54. The chief guest, Lord Avebury, proposed the toast. Among the numerous distinguished guests were most members of the Wharton Committee including Lapworth, and Watts and Bonney. But Judd, Peach, and Horne were not present. Presumably Peach and Horne were away in the field somewhere.

The address bore the following words: "We desire, upon the close of tenure of your office, to express our sense of the high value of the services which you have rendered to these Institutions; we proudly recognize the high position attained by you in the scientific world and gratefully acknowledge the beneficial influence of your example. That you may live, after more than forty-five years in the public service, to enjoy your freedom from official cares and to enrich geological literature with your luminous writings in our warmest desire" ("Dinner to Sir Archibald Geikie," *Nature* 64 [1901]: 34–36).

still look up to you, and if I survive you, I shall still have the sense of
my obligation to you.

   . . . Long may you live to enjoy your well-earned leisure, which I
know you will employ for the instruction and delight of others.

I am, Yours sincerely,

B. N. Peach

Peach could hardly have written more warmly, and this surely demonstrates
the excellent personal relationship that he had with Geikie. Also, the obituary
of Geikie published by Peach and Horne (1925) was entirely without animus,
though the two old geologists did mention some of the mistakes that Geikie
had made during the course of his career.[55]

I should like to conclude this chapter by attempting to gauge the motives of
Howell and Judd, whose actions appear to have precipitated the establishment of the Wharton Committee, and hence the retirement of Geikie and the
reorganization of the Survey. The case of Howell is relatively straightforward.
Howell and Geikie were acquainted for a very long time. Indeed, as we have
seen, Geikie started his career in the Survey as an assistant to Howell back in
1855, and they cooperated with one another in the publication of a memoir
on the geology of the Edinburgh district (Howell and Geikie 1861). Howell
had a slow but steady career in the Survey. His main work had to do with the
mapping of coalfields in the English Midlands, the north of England, and
Scotland, but he also did work on the survey of Mesozoic rocks, particularly
Jurassic. His chief claim to fame was his recognition of Scottish equivalents of
the three major divisions of the Carboniferous, previously established in
England.

While Howell was an expert on the economically important coal-bearing
rocks, he had no special expertise in metamorphics, and it is not obvious that
he would have been an outstanding success in investigations in the complex
geology of northern Scotland. Nevertheless, by steady application, and with
the help of his seniority, Howell was appointed to the directorship of the Scottish branch of the Survey in 1882, soon after Geikie became director general.
Howell was at that time still busy with surveying the coalfields of Northum-
berland and Durham and was resident in Newcastle, so he did not move to
Edinburgh to take up his Scottish post until 1884. Meanwhile, Geikie continued to exercise control in Scotland, as he had done for many years. But he
now operated from London, rather than Edinburgh.

There was, no doubt, good reason for Geikie to leave Howell in the north
of England to finish his work there. But to others in the Survey, including
Howell himself, it must have seemed as if Geikie had been reluctant to relinquish control in Scotland; and indeed this was partly the case. Thus, even

---

55. By contrast, the obituary written by Sir Aubrey Strahan (1925, lx) made some significant
negative comments about Geikie's personality. It is doubtless relevant that Strahan was one of the
several surveyors whose careers were impeded during the latter part of Geikie's directorate.

when Howell did eventually take up his position in 1884, he never had anything more than nominal responsibility for the work in the Highlands. This sensitive work was maintained under the control of Geikie, who insisted on doing all the field inspection of the surveyors' work (Greenly 1938, 2: 519)—but he was too remote and too busy to do justice to the tremendous complexities of that large region. (It is, of course, questionable whether Geikie had the intellectual capacity to deal with the problems of the Scottish Highlands. It was, after all, Lapworth, Peach, and Horne who sorted out the stratigraphical relations of the Northwest Highlands; and the Grampians, et cetera, did not really get sorted out during the nineteenth century.)

In 1888, Howell became director for Great Britain. That is, he became officially the second-in-command in the Survey. But this made little difference to his actual responsibilities. He remained in Edinburgh—("immured in the Sheriff's Court," as Wilson (1985, 16–17) has put it—and only had effective command over the southern portions of Scotland and the north of England. His promotion had as much to do with Geikie's gradual winding-down of the number of senior positions in the Survey as to his appreciation of Howell's administrative and intellectual capacities.

It can be seen, then, that Howell had no special reason to be grateful to Geikie. He was also perfectly justified in claiming that he had no responsibility for the debacles in the Highlands and the Southern Uplands, even though he was nominally responsible for the conduct of the Scottish branch of the Survey. Moreover, it appears to me that his analysis of the effects of the Northwest Highlands and the Southern Uplands work on the progress of the Survey during Geikie's tenure as director general was entirely correct. I do not know whether Howell's bitterness toward Geikie over the way in which he was permitted (or not permitted) to conduct the affairs of the Scottish Survey prompted him to write to Donnelly in the terms that he did. It may well have been so. But it is unquestionably the case that Howell was justified in disclaiming responsibility for the Highlands "fiasco" and all that that entailed.[56] Thus Howell's motives in writing to Donnelly as he did are not difficult to comprehend. Moreover, he had every right to feel aggrieved if, as appears to have been the case, the staff's "Memorial" was blocked by Geikie.

The case of Judd is more complex, rather less certain, but more interesting. As we have seen, Judd was a member of the Survey staff in his earlier days (1867–71), but to my knowledge nothing occurred then that would have led him to cross swords with Geikie. However, as we have seen, Judd was an important member of the "amateur" party that gathered round Lapworth, and he was very keen to vindicate Nicol when the second meeting of the British Association at Aberdeen occurred in 1885. Why, apart from the perceived jus-

56. This "fiasco" was, of course, a triumph of a kind. The 1888 paper of Peach and Horne was one of the most important contributions to geology from Britain in the nineteenth century. And the two great memoirs of 1899 and 1907 were perhaps the major achievements of their kind in Britain from that period.

tices of the causes of Lapworth and Nicol, might Judd have taken the position that he did, and why, when his opinion was solicited by Abney, might Judd have acted as he did, helping to precipitate the Wharton Inquiry?

It is, of course, always difficult for the historian to comprehend motives and intentions accurately, but in the case of Judd I think we may see how it came about that he took such a negative attitude toward the Survey and the Murchison-Geikie theory of the Highlands. The reason, I suggest, is that Geikie and Judd had become involved in a furious disagreement over the interpretation of some very important igneous rocks in Skye. It is worth our while to give some account of this controversy, since it throws further light on the general relationship between the members of the geological community in Britain in the nineteenth century, and it provides a kind of subplot to the story of the Highlands controversy, even though it had to do with a set of rocks that were distinct from those that form the zone of complication from Durness-Eriboll down to Skye.

As a young man, surveying Mesozoic rocks in the English Midlands, Judd had already found some errors in work previously performed by officers of the Survey, and this had doubtless led him to realize that its work was not always beyond criticism.[57] On resigning from the Survey, Judd subsequently examined the patch of Mesozoic rocks near Brora in Sutherland on the northeast coast of Scotland (in which Murchison and Sedgwick had taken an interest early on), and he then moved over to the western side of Scotland to look at the rather similar deposits on Skye and the adjacent islands. These rocks are associated with large masses of igneous rocks, including the Cuillins (chiefly gabbro) and the Red Hills (chiefly granite); and so Judd (1874) was led to examine, and publish on, these igneous rocks as well. But in so doing, he was beginning to intrude into territory that Geikie might have felt was specially his own, for Geikie had done some of his very first mapping in the Mesozoic rocks of Skye; and one of the most important pieces of his work was the examination of the British volcanic rocks, which culminated in his great treatise on the topic (Geikie 1897).

In fact, Judd's work paralleled that of Geikie in another way, since in 1875 Judd was invited by the aged geologist George Poulett Scrope to travel to Europe at Scrope's expense to make investigations of a number of European volcanoes, and to report back the results. It will be recalled that Geikie performed a similar task for Scrope in 1870, but the work was cut short by Geikie's severe illness. Judd, however, carried through his work successfully, and it resulted in the publication of a considerable number of papers on the volcanoes of Italy and the volcanic rocks of Hungary and the Alpine regions. Thus Judd would, with reason, have begun to think of himself as one of the leading experts in Britain on volcanoes both ancient and modern.

In 1871 Geikie had read a paper that included material on the igneous rocks of Skye before the Geological Society of London (Geikie 1871a), and

57. See the Geological Magazine's Eminent Living Geologists series (1950, 387–88).

according to an account that he gave subsequently (Geikie 1894, 213) he withdrew from publishing further papers on the topic since he felt that his further efforts were rendered superfluous by Judd's paper of 1874 on essentially the same topic. Even so, according to Geikie's reconstruction of events, he was not satisfied with the interpretations of certain phenomena proposed by Judd; and on reconsidering the matter in the light of observations that he made during the course of his journey to the United States in 1879, Geikie returned to the topic from time to time and presented a paper on his new views to the Royal Society of Edinburgh in May 1888 (Geikie 1890). The following year, Judd (1889) presented a paper to the Geological Society opposing Geikie's interpretations, without having gone back to Skye to reexamine the controverted ground. In the next few years, Judd (1890a, 1893) returned to the attack, until Geikie was eventually goaded into reply with a paper delivered on February 21, 1894, that was much more polemical than he was accustomed to present in public (Geikie 1894). He followed this up later that year (on June 6) with another paper (written in conjunction with Teall) developing his ideas (Geikie and Teall 1894).

This exchange generated a good deal of heat. In his paper of 1874, Judd put forward the idea that there were five great ancient volcanoes in the Inner Hebrides (of Mull, Ardnamurchan, Rum, Skye, and St. Kilda), from which had been formed the large igneous mountains and the extensive sheets of plateau basalts such as are found in the northern part of Skye. He also stated his belief that there were three main periods of igneous activity: the first producing chiefly acidic (quartz-rich) rocks such as granites and felsites; the second chiefly basic (quartz-deficient) rocks such as gabbro and basalt; and the third chiefly minor eruptions of materials of variable composition. The idea that Geikie picked up in America was that the eruptions could occur along linear fissures, as much as from separate central vents, and he sought to apply this hypothesis to the volcanic deposits of the Western Isles.[58]

We do not, perhaps, need to go into all the details of the controversy that developed between Geikie and Judd. But the issues settled down to the question of which was older: the Cuillins or the Red Hills of Skye. As visitors to Skye will know, the black, coarse-grained gabbro of the Cuillins seems very slightly weathered, and the mountains have a remarkable jagged appearance. By contrast, the adjacent, lighter-colored granitic Red Hills have a smoother and more weathered appearance, and on first examination might be thought to be the older structures. The question as to which was the younger and which the older mountains might seem to be a relatively simple one, to be determined by absolutely straightforward geological principles. That is, all that had to be done was to show which set of rocks intruded into which. The ones

---

58. His account of his conversion to this view, as a result of his examination of the layered volcanic deposits dissected by the Snake River in Idaho, is given in two works (Geikie 1880a; 1882b, 239–49). The idea of fissure eruptions was originally due to the German geologist Baron Ferdinand von Richthofen.

forcing their way into the others should be younger, just as any common dyke cutting across a series of sedimentary rocks would be known to be the younger.

This was one of the first principles of Huttonian geology, on which Geikie had been born and bred as a young geologist in Edinburgh—principles that he had publicly expounded on numerous occasions, and had used by way of illustration as to how a keen-eyed observer might read the geological history of an area from examination of its rocks. It was, one might think, an issue on which there need be no controversy. But it was important to get it right. Needless to say, if Geikie could not read a suite of volcanic rocks properly—a topic on which he was supposed to be the leading authority in the country—he would really be beginning to lose all credibility as a geologist, particularly after the disaster of the Northwest Highlands.

But matters were, needless to say, by no means simple and straightforward in the volcanoes of Skye. It is now believed that a whole intermingling and hybridization of the acidic and basic rock materials occurred at the junctions of the Cuillins and the Red Hills, particularly at a mountain called Marsco— which has in fact given its name to a special rock type, marscoite, produced by such an interaction of acidic and basic materials (Harker and Clough 1904, 175–87). And this interaction of acidic and basic material made the whole problem much more complex than a simple application of Huttonian principles would be able to handle.

Judd did not hesitate to use highly polemical language in his several addresses to the Geological Society, and it is clear—even from the relatively sanitized versions of the debate that were published in the society's *Journal*—that tempers were running high on the issue. Indeed, one has the impression that the whole question as to who was to count as the source of geological authority in the matter of the igenous rocks of Britain was at stake. Judd, from a fairly secure power base at the Royal College of Science, was challenging Geikie as the chief spokesman and source of authority on the geology of igneous rocks in Britain.

By 1893, the debate had settled down to a kind of "crucial experiment." The chosen locality was a ridge on a hillside, called Druim an Eidhne ("ridge of ivy"?), a little to the west of a track that runs up Glen Sligachan between Sligachan and Loch Coruisk in the mountains of Skye. At this point the boundary between the acidic granite (or granophyre)[59] and the basic gabbro may readily be seen on the relevant geological map (Institute of Geological Sciences 1952). Judd described the ridge as "a great intrusive sheet of gabbro"; and *within* the basic rock he reckoned he could see a "number of patches of pale granitic rock" (Judd 1893, 182). However, the patches of pale rock were more like a rhyolite than a granite,[60] and they were sometimes almost

59. A fine-grained porphyritic granite with a characteristic texture of the material of the groundmass, having inclusions of quartz in a feldspar matrix, sometimes spherulitic.
60. A rhyolite is a fine-grained, acidic igneous rock, formed by fairly rapid cooling. Spherulites are small, aggregations of radiating, needle-shaped crystals.

glassy and also had a spherulitic texture in places. In thin section, Judd found crystals of quartz and feldspar in a more or less devitrified glassy base. This, he thought, provided conclusive evidence as to the order of emplacement.

In commenting on the paper, however, Geikie referred to the existence of granitic veins that might be seen "proceeding from the mass of granite and traversing the gabbro." Why, he asked, had Judd not mentioned these? As for the supposed "inclusions" of pale-colored granitic matter, they might be ascribed to an earlier series of acidic rocks, which otherwise had only been found as ejected fragments. They need not be assumed to belong to the main mass of the granite of the Red Hills.

Judd tried to defend himself with a note added as a postscript to his paper. He said that he had not mentioned the claimed veins proceeding from the granite into the gabbro because of a "desire not to complicate the very definite issue raised in the title of the paper" (ibid, 194). Certainly he had seen some white veins traversing fragments of basic rock, but these he regarded as "segregation veins," rather than "apophyses of the granite." He denied that he could find any such veins as were referred to by Geikie.

This was Geikie's great opportunity. He took up the challenge with vigor (Geikie 1894) and demonstrated that such veins *could* indeed be found, and he figured them in a highly convincing manner (ibid., 222). Indeed, he went so far as to photograph them, display the photographs (ibid., plate 14), and describe the exact locations of the veins. They were, he said, to be found occupying three gullies in which rivulets ran off Druim an Eidhne—which in itself suggested that the granitic material was softer and weathered more easily than the gabbro, even though it was younger. The veins could be seen to cut the gabbro's banding, and there was no doubt that they were younger than the gabbro. Since they connected up directly with the main mass of granite, this furnished proof that the rounded granitic Red Hills were in fact more recent than the beautifully crystalline, rugged Cuillins, even though the latter had the general appearance of being much fresher and younger rock. Incidentally, this was certainly *not* a case where Geikie helped himself to the observations of his subordinates without acknowledgment, for the original sketch for the three granitic veins cutting the gabbro may be found in his field notebook for 1893–94, held in the archives of the University of Edinburgh.[61]

Judd tried to defend himself in the discussion following the presentation of the paper, saying that there was evidence that the veins referred to by Geikie also cut the granite, and thus would be younger than both the granite and the gabbro. He also continued to insist that there were inclusions of granitic material in the gabbro. In reply, Geikie claimed that the special spherulitic texture observed in Judd's "inclusions" could also be seen in Geikie's "veins" as well as in the main body of the granitic material.

The issue was not formally decided there and then, at the meeting of the

61. University of Edinburgh Archives, Gen. 521/1, A. Geikie, Field Notebook QQ. 1893–94, p. 2.

Geological Society. But in fact Geikie won this battle. Indeed, it was perhaps
the greatest intellectual "victory" of his career (though whether it was achieved
single-handed I have no means of knowing). Judd retired from the field, severely wounded, and published nothing further on the topic. By contrast,
Geikie was able to detail the arguments at length in his authoritative *Ancient
Volcanoes of Great Britain,* where he reproduced the figure from his 1894 paper
showing the granophyre veins piercing the mass of gabbro (Geikie 1897, 2:
439). In a footnote, Geikie referred to his paper, and the earlier one of Judd,
and indicated that he had had the last word on the subject. It was not, of
course, the last word that geologists were to have on the igneous rocks of Skye.
But when the Survey eventually published its fine memoir on the topic in
1904, by the master petrologist Alfred Harker (Harker and Clough 1904), the
authority of this work was placed fully behind Geikie, both in the matter of
the idea of linear fissures for the outpouring of the basalt sheets of the Hebrides and in the question of the relative ages of the Cuillins and the Red Hills
(ibid., 12, 286).[62] In a helpful and clear section (ibid., 93) the authors showed
the structure of the cliffs of Druim an Eidhne, with the gabbro *overlying* the
granite like a lava flow, and hence having the apearance of being younger, but
there were veins of granite piercing the gabbro. The surveyors had nothing to
say against Judd's petrological work; but it was clear, nonetheless, that he had
mistaken the field evidence.

This point was rammed home by Geikie in his *Ancient Volcanoes,* where he
stated that the general appearance of the gabbro (dark, almost black) and the
granite (pinkish) was such that the geologist could easily trace the contacts
over the countryside by eye. The implication, therefore, was that he might
easily be seduced into making hasty judgments. The easy inference for any
geologist was that this was what had happened to Judd. In fact, I am inclined
to think that Judd *had* failed to find Geikie's dykelike veins of granophyre in
the gabbro. He had not searched hard enough; he had not used his legs sufficiently to examine every nook and cranny of the gabbro-granite junction before he pronounced on it. Thus we may understand the significance of the
following passage in Geikie's *Ancient Volcanoes:*

> No geological boundary is more easily traced than that between the
> pale reddish granophyre and the dark gabbro. It can be followed with
> the eye up a whole mountain side, and can be examined so closely that
> again and again the observer can walk or climb for some distance with
> one foot on each rock. That there should ever have been any doubt
> about the relations of the two eruptive masses is possibly explicable by
> the very facility with which their junction may be observed. Their con-
> trasts of form and colour make their boundary over crag and ridge so
> clear that geologists do not seem to have taken the trouble to follow it
> out in detail. (Geikie 1897, 2: 391)

62. Harker wrote major treatises on igneous and metamorphic rocks (Harker 1909, 1932). He
also wrote what was the standard textbook on petrology in English for many years (Harker 1895).

In the eyes of the Geological Society members, this would have been something difficult to overlook or forgive.[63]

From this we can move to an examination of Judd's motivations in offering such negative advice to Abney and Donnelly when his opinion was sought as to the workings of the Survey. It does not mean that Judd's advice was necessarily invalid. But we can at least understand why it may have been that Judd wished to paint the Survey and its director general in an unfavorable light. There can be no doubt that Judd lost face severely in the geological community as a result of the dispute with Geikie about the igneous rocks of Skye.[64] By contrast, Geikie had for once entered into a fierce controversy and proved himself the winner.

It should be noted that the heated dispute between Geikie and Judd concerning the rocks of Skye was in full swing at the time that the exchange of correspondence concerning the Geological Survey took place in the pages of the *Daily Chronicle,* as described at the beginning of this chapter. It is a nice question whether Judd was one of the anonymous participants in this debate. He certainly had the motivation to denigrate the Survey's work. But he would not, I think, have made some of the mistakes that appeared in the original critical note of de Lapparent's essay. Judd could possibly have been "Scholasticus," but such a speculation cannot be substantiated.[65] In any event, it may not have been a coincidence that the attack on Geikie and the Survey occurred at about the same time that the Skye debate was in full swing. Also, we may note that the beginnings of the Skye debate between Geikie and Judd went back to the 1870s, well before Judd's address to the British Association meeting at Aberdeen in 1885, when Nicol's work was posthumously vindicated. It is very likely, then, that the long-standing differences of opinion between Geikie and Judd helped to shape the presentation of Judd's historical account of the Highlands controversy on that occasion.

There is one other speculation that I wish to pursue in conclusion of this chapter. One may wonder whether Geikie ever got to know that Judd was one of those whose influence led to the establishment of the Wharton Committee. Although the commission did not formally find against Geikie's administration, and it does not seem to have had any lasting ill-effect on his career, it must nevertheless have been of considerable embarrassment to Geikie. I have

63. Of course, we do not know that Geikie himself actually discovered the granophyre veins or dykes at Druim an Eidhne. This could well have been a case where he relied on a discovery of one of his subordinates.

64. It must be said that the problems of the igneous geology of Skye were, of course, far from over as a result of Geikie's work. For example, I have referred to the hybridization of magmas such as one finds at Marsco. This difficult problem had to be elucidated. There are also "multiple dykes" (with successive intrusions of magmas of different chemical composition), which required careful examination and have been the subject of a good deal of subsequent research. But on the essential question of the relative ages of the gabbros and the granites, Geikie has been vindicated by later work.

65. In fact, my suspicion is that this correspondent was a classicist, judging by his chosen pseudonym and his apparent antipathy toward certain aspects of science.

no documentary evidence to offer under this head. It may be noticed, however, that when it came to Judd's retirement in 1905, he did not receive the
pension to which he would have been entitled. According to his obituarist,
Professor Bonney (1916, 192), Judd joined the Royal School of Mines as a
professor on a full-time basis in 1876, but full-time attendance to the pro-
fessorial duties only became obligatory in 1881 upon the creation of the Royal
College of Science; and the pension was calculated from this latter date. One
wonders whether the injustice done to Judd, on the basis of a kind of bureaucratic quibble, was in any way due to Geikie. I have no evidence to be able to
confirm or deny this speculation. Certain it is, however, that Judd's health began to decline seriously after his retirement. He suffered from deafness and
vertigo, and although he did some valuable writing after his retirement,
chiefly historical work on Darwin and Darwinism, he had no major acheive-
ments in the latter part of his career. By contrast, Geikie continued to carry all
before him, doing extensive lecturing and writing, returning to his old classical interests with great success (he became president of the Classical Association in 1910), editing the great Northwest Highlands memoir, rising to the
dizzy height of the presidency of the Royal Society and gaining the prestigious
Order of Merit, having a contented retirement in Surrey (where he gave a
strong helping hand to the Haslemere Educational Museum), and tidily completing his autobiography in the last year of his life while still in full possession
of his faculties.[66] Geikie was always the winner.

66. Geikie's house of retirement still stands on a wooded hill at the southern edge of Haslemere.
It is a large house that would have been ultra-modern at the time that he had it built for himself
(1913). At present, it is divided into two homes and has probably lost some of its former architec-
tural attractiveness.

11

Issues: Methodological, Epistemological, and Social

To conclude this study, I should like to examine some of the broader issues raised by the historical work in which I have been engaged, considering particularly certain important problems concerning the nature of scientific knowledge, in relation to the social system of science within which such knowledge is generated. The discussion will be undertaken in the light of contemporary writings on these matters, but I should like to open up the issues by considering what nineteenth-century geologists had to say on methodological and epistemological matters. It will be convenient to focus attention on Geikie once again, partly because he wrote more widely on general matters than any of the other geologists that we have examined, and partly because his views may be taken as representative of what the British geological community thought about such matters in the second half of the nineteenth century. I shall also make some remarks about the general character of the work in which the Survey officers were engaged.

An entree to the problems that I shall address may be found in Geikie's more popular writings and his textbooks. The latter were *Physical Geography* (1873), *Geology* (1874), *Outlines of Field-geology* (1876), *Elementary Lessons in Physical Geography* (1877), *Text-book of Geology* (1882), *Class-book of Geology* (1886), *The Teaching of Geography* (1887), and *An Elementary Geography of the British Isles* (1888). They were among the most successful science textbooks published in Britain during the nineteenth century, and although Geikie candidly admitted in private that he wrote them chiefly to supplement his income, this motive did not detract from the merits of the works or prevent them from shaping the view of the geological enterprise for students in schools and universities in Britain, and for numerous adult amateur geologists. Thus the view of geology Geikie offered in these books had considerable social significance. Geikie, of course, influenced professionals as well as amateurs, for the views expounded in his more popular writings were partly based on the ethos of the Survey to which he devoted his professional life; and serving in senior positions in the Survey for many years, his views inevitably shaped this ethos too.

We know from Geikie's autobiography that it was chiefly Hugh Miller who inspired Geikie's love of geological science and determined him to make it his calling. We know too that Geikie was a devoted follower of the ideas and methods of his Edinburgh forebear, James Hutton. The third chief inspiration was of course Sir Roderick Murchison. From Miller, Geikie imbibed his enduring love of nature, captured in his fine prose and in charming watercolors. Miller also taught Geikie something of the romance of geology, and of the relationship of geography and geology to human history. From Hutton, Geikie learned the basic theory of geological science and uniformitarian methodology. From Murchison, I suggest, Geikie learned the finer points of how to succeed in the scientific community, though he seemed to need rather little instruction in this, for he had had early successes, even as a teenager in Edinburgh, before he met Murchison. Murchison also showed Geikie how to think and act quickly, to accomplish a lot in a limited time, so that as much might be packed into life as possible. In addition, Geikie took over from his mentor the idea of geological research as a kind of military campaign, with territory to be subdued, opponents overcome, and with secrets to be wrested from Nature. Nonetheless, geology was also regarded as a morally uplifting enterprise, and healthful besides. It could be warmly commended to the large readership of Geikie's more popular writings, and this elevation of the subject in public esteem at the semipopular level was surely one of Geikie's outstanding contributions to his science. (Naturally, however, ways to climb the geological social ladder were not mentioned in his textbooks!)

Geikie's philosophy of science was not particularly sophisticated, even by the standards of his day. He was a thoroughgoing empiricist and believed that with careful observation, application of the maxim that "the present is the key to the past," and restraint placed on unwarranted speculation, it was possible for the geologist to reveal the structure of the surface of the globe and its geological history, these being the chief goals of geological science. This view of the geological enterprise was formed early in Geikie's career. Thus, even in his very early efforts at geological writing, such as his description of fossils from the Burdiehouse Quarry near Edinburgh (see chap. 5), one can see these elements as constitutive of his geological method. It was all well set forth in his first book, *The Story of a Boulder* (Geikie 1858).

In this interesting volume, Geikie explained that geologists were, so to speak, "interpreters of hieroglyphics, and historians of long-perished dynasties" (ibid., xi–xii). The larger boulder, which he invited his readers to contemplate with his assistance, might be regarded as "a curious volume, regularly paged, with a few extracts from older works" (ibid., 4).[1] It was, "as it were, a quaint, old, black-letter volume of the Middle Ages, giving an account of events that were taking place at the time it was written, and containing on its earlier pages numerous quotations from authors of antiquity. The scratched surface, to complete the simile [might] be compared to this old work done up in a modern binding" (ibid., 5). Thus, knowing how to read the "text" of the boulder with the help of modern analogies, the geologist would soon recog-

---

1. The "few extracts from older works" were, I take it, xenoliths—that is, inclusions of some "foreign" rock within an igneous rock body.

nize it as a large lump of rock, worn by ice action, and deposited by a floating iceberg. All one had to do was "proceed cautiously, reasoning from positive facts, and striving as far as possible to exhaust what Bacon call[ed] the 'negative instances.'" Working in this way, "our deductions [should] possess all the certainty of truth" (ibid., 181).

This example was, however, an unfortunate one for young Geikie (he was only twenty-three in 1858), for five years later we find him following geological opinion and publicly repudiating the "ice-raft" theory to account for glacial erratics and adopting the land-ice theory of Louis Agassiz (Geikie 1863b, 74). Yet nowhere in Geikie's writings is there, to my knowledge, any statement of the fact that observations are customarily made in the light of some theory or other, that they may be influenced by the theoretical views that are held, so that reading the "book of Nature" is very much a theoretical matter and hence problematical. If he had been called upon to account for his change of theory on glaciation, Geikie would no doubt have acknowledged that it was due to changes in the views of the more senior members of the geological community; but he would have held that *they* changed their minds as a result of more careful and extensive observations. A twentieth-century philosopher of science may find Geikie's empiricism naive. But on the whole, his method of using the present as the key to the past served him reasonably well, and he took care to warn his readers that the particular geological epoch in which we live may not be characteristic of the whole of geological history. Thus the maxim should be just a "working hypothesis" (Geikie 1882a, 3). In illustration of his method, Geikie (1880c) made an interesting attempt to gauge the rate of rock weathering for different types of rocks by examining the rate of decay of date-bearing tombstones. His argument here showed in a practical way how one might seek to use the present as a key to unlock the past.

The emphasis Geikie placed on observation as the way to success in geological research appeared in many of his writings. Indeed, accurate observation was held to mark the road toward moral virtue. For example, in his inaugural lecture at the University of Edinburgh in 1871 he concluded with the words:

> Let us turn from the lessons of the lecture-room to the lessons of the crags and ravines, appealing constantly to nature for the explanation and verification of what is taught. And thus, whatsoever may be your career in the future, you will in the meantime cultivate habits of observation and communion with the free fresh world around you—habits which will give a zest to every journey, which will enable you to add to the sum of human knowledge, and which will assuredly make you wiser and better men. (Geikie 1892c, 271)

In his historical writings, Geikie conspicuously singled out for approbation geologists who were thought to be particularly adept at observation. William Smith was described as "eagerly scrutinizing every field, ridge, and hill" (Geikie [1905] 1962, 385). Sedgwick had "eagle eyes [that] seemed as if they

must instantly pierce into the very heart of the stiffest geological problem"
(ibid., 433). Murchison, though "not a profound thinker" in Geikie's estimation, had a "patient and sagacious power of gathering and marshalling facts,"
such that one "could hardly find a more keen-eyed and careful observer"
(Geikie 1875, 2: 346).

How did the always observant geologist of Geikie's day go about his work?
How were facts collected and synthesized? To put it another way, what did the
fieldworker of the Survey actually *do?* Geikie's *Outlines of Field-geology* provides
some useful information on this head. The fundamental role of the Survey
was, of course, to furnish geological maps of the country, and to provide information concerning the geological structure and history of Britain in accompanying descriptive memoirs, such that the country could gain economic
benefit from this knowledge. We learn from *Field-geology* that a surveyor
would average ten to fifteen miles of walking a day and would spend about
two hundred days in the field each year. A square mile of mapping might require twenty to twenty-five miles of walking, though considerably more in the
Highlands (Geikie 1900, 47–48). All relevant information—rock types, fossils, dips and strikes, faults, stratigraphical boundaries, et cetera—had to be
collected and recorded in notebooks and on field maps. The preliminary information, collected and entered on a topographical map, might, then, look
something like what is shown in figure 11.1.

The information then needed to be synthesized to yield a complete map
and theoretical section, as is shown in figure 11.2.

The drawing of sections would proceed as shown in figure 11.3, starting
from a known set of elevations above sea level along a line of traverse, then
marking in different rock units on the upper surface of the section at the correct angles of dip, and then completing the section so as to show the supposed
structure below the ground.

Needless to say, such procedures involved a considerable amount of theory
and extrapolation of the evidence, over and above keen-eyed and sagacious
observation. The establishment of the main classificatory units of the stratigraphical column was a matter of difficulty and controversy, as has been
shown by the studies of Rudwick and Secord. At the more local level, the surveyor would constantly have to make minor decisions as to the boundaries
that he would adopt in his mapping procedure and which line on a rock surface corresponded to a particular (theory-based) boundary. Sometimes the
boundaries would have to be made on lithological criteria; but by the second
half of the nineteenth century, it was generally agreed that lithology would
have to defer to the evidence of fossils, wherever it was available.

As modern students of geology well know, by reason of the "mapping exercises" that they are required to undertake during their training, there are established techniques for determining geological structures underground, on
the basis of surface outcrops. Thus, for example, if a plane bed outcrops at
certain elevations known by the contour lines of the topographical map that

Fig. 11.1. Data entered on a field map, before completing boundaries (Geikie 1900, 110).

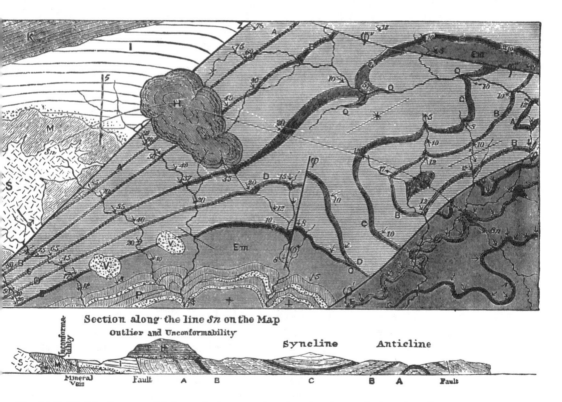

Fig. 11.2. Finished map, with boundaries completed and a theoretical section drawn (Geikie 1900, 116).

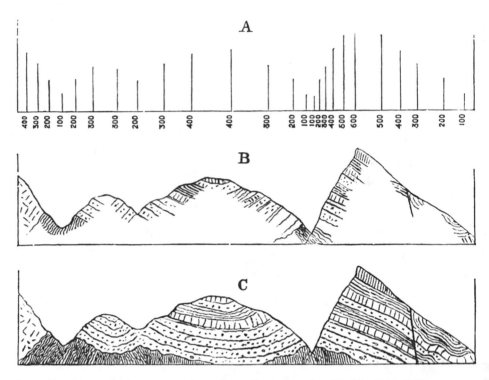

Fig. 11.3. Stages in the construction of a geological section (Geikie 1900, 216).

the geologist is using, then the lie of the bed through the three-dimensional structure of the ground can be estimated. The outcrop of a fault plane can be plotted similarly, and the underground extension of the fault can be gauged; likewise the plane of an unconformity. With more difficulty, the structures of folded rocks can also be worked out geometrically on the basis of the surface exposures. Thus the structure of very complicated areas can be inferred with reasonable success from surface indications. A combination of deductive and inductive reasoning is required; and although the structures that may be obtained, illustrated in suitable sections, are never strictly deducible from surface observations, by the plotting of extrapolated strike lines for beds under the surface of the ground one may work out in a semi-routine way what the structures are likely to be. Once the structure is known, the sequence of events leading to that structure may be hypothesized.

However, for much of the period that we have been discussing, the geologists in the Highlands were only carrying out reconnaissance surveys, not even with the rigor indicated in figures 11.1 to 11.3. Sometimes the technique was simply to walk down a valley, simple observation of the exposures on the valley walls serving as the main evidence for interpreting the subsurface structure of the area. Or one might sail along a coastline. We may notice the binoculars hanging round Murchison's neck in the cartoon of him by Prosper Mérimée (plate 4). They were, I suggest, a vital piece of equipment for Murchison, who usually maintained a hectic pace during the course of his field-work.[2] Also, in his *Field-geology* Geikie pointed out how the geologist could, with practice, "identify rocks and formations even from a distance by their outlines, colour, character of vegetation, or other distinctive trait" (1900, 44). This, of course, is perfectly true. But these methods should be borne in mind when we consider the quality of the evidence adduced during the early stages of the Highlands controversy.

As mentioned, however, another technique was to observe rocks along a line of traverse in order to prepare a section (idealized in figure 11.3). This technique was used by most of the participants in the Highlands controversy. But there was always some doubt, for the readers of early geological papers, exactly how much of the published line of section had been walked over, and there was often uncertainty as to the precise positioning of the section. Only with Bonney do we find the first publication of a geological map of one of the important sites for the controversy (fig. 7.3), and that was a very primitive effort. As we see from figures 11.1 and 11.2, the ideal procedure was to map first and then draw a section. But the usual method was somewhat as follows: reading of previous literature on the area; observation; theoretical interpretation of the structure; section; presentation of description of area and section at meeting of Geological Society or some other venue; defense of interpretation in subsequent debate at meeting; publication or withdrawal; acceptance, rejection, or modification in subsequent work.

It scarcely needs emphasis, but it should be mentioned that the procedures involved in the making of geological maps, as well as sections, are highly theoretical, and that maps of the same area made according to different theoretical presuppositions may look quite different (Harrison 1963). It is well known that scientific observations are "theory-laden" (Hanson 1961, chap. 1), and this applies to observations made in the course of geological fieldwork as much as any other area of scientific research. Thus one does not simply *see* things in the field; one *sees them as* something or other. And one's system of classification of rocks and strata is always a theoretical matter. There is no universal "natural" system of geological classification, either for stratigraphical boundaries or for rock or mineral types, and even fossils can be problematical. There is always room for disagreement among geologists as to the "best," or most natural, or most convenient, classificatory system. Classifications are always revisable and open to negotiation, or may be the source of controversy and debate, as has been amply demonstrated in the present study. There being no manifestly perfect or natural system, the actual observations are always open to a number of different competing interpretations. The stratigraphical column can always be divided up in different ways. One geologist's rock is igneous; for another the same rock is metamorphic; and for another

2. This was particularly so for Murchison's long journeys in Russia. We learn, for example, that "mounted on a light calèche, sometimes with five or six horses harnessed to it, he rushed through the country, over sand, boulders, and bogs, at the rate of often as much as ten or twelve miles an hour (Geikie 1875, 1:296).

the issue is indeterminable on the evidence available. One geologist's line of thrust is another's bedding plane. Some may see a regularly ascending stratigraphical sequence while others may suppose they have evidence for large-scale inversions and faultings. One geologist may think a gabbroic mountain range has been intruded by granite, but another may suppose that the granite is traversed by gabbro. An interpretation of a region that has been agreed upon for many years may be revised when new evidence comes to light, when a new prevailing theory is established, or when the balance of power within the geological community shifts. Thus, as is commonly said, theories are "underdetermined" by the evidence. No empirical evidence can be so solid that it is impossible to revise, for the theory is not strictly deducible from the evidence.

All this is commonplace; but the problem of underdetermination is, I suggest, particularly well demonstrated in historical geology, which rarely has the capacity to test its claims by direct experiment.[3] Yet there is a further point that should be emphasized, as has been done by Rudwick in his study of the great Devonian controversy (Rudwick 1985, 431). Not only may we think of observations as "theory-laden," we should also think of them as *controversy-laden.*" This term may need a little elucidation, and the introduction of further vocabulary.

We have seen how the activity of the geological community is rather like a field of contest. Indeed, many of the nineteenth-century geologists we have been considering regarded their subject as a kind of battlefield, at two levels. At one level, the geologist was in a kind of battle with Nature: ground had to be traversed and "conquered" or "subdued." But, as the geologists well realized, they were also in competition with one another. Each of them was trying to make *his* view prevail, perhaps by gathering supporters so as to advance and enhance a common set of interests. Thus was important, since if one's views prevailed, they would bring in their train professional esteem, public recognition, honors, money, power, and influence. Thus knowledge and power become interdependent. Thinking along these lines, one may liken the scientific community to an "agonistic field"—a field of contest—as has been done by Bruno Latour and Steven Woolgar (1979), developing ideas previously suggested by Pierre Bourdieu (1971, 1975).

It should be remarked that the field of science, as envisaged by Bourdieu and his followers, is *not* just a field of contest, like an athletics field, where there are clear-cut winners and losers. A better analogy is that of an electromagnetic field, in that activity anywhere in the field subtly influences every other region of the field. As we have seen, scientists seek to make their views of reality prevail, and to accomplish this they need to establish appropriate positions of power for themselves within their community. Then it is more

---

3. This is not to say that laboratory-based experiments are necessarily decisive in situations of scientific controversy.

likely that the world will be seen according to their preferred interpretations. But the case is intriguing, for so far as the interpretation of the geological history of the earth (for example) is concerned, there is no source of authority external to the community of scientists endeavoring to write that history. There is, as Bourdieu puts it, "no judge who is not also party to the dispute" (1975, 25). The data collected by the observer do not, therefore, in themselves, unequivocally determine the outcome of a scientific controversy. The scientist is forever involved in a series of feed-back loops (either positive or negative) within the agonistic field of the scientific community: nothing succeeds like success; and nothing fails like failure. Thus it is the scientific community that is the arbiter of what *counts* as truth at any given point in the history of science. It is for this reason that scientists (or for that matter, historians of science) seek to become *authorities,* so that their views will prevail within their agonistic field. For one's views to prevail, one has to acquire an appropriate field-strength. This study has sought to illustrate in a concrete way how this might be accomplished—in the cases of Murchison and Geikie, for example. The consequences of an insufficient field-strength to support one's theoretical views are illustrated by the case of Nicol. Needless to say, many factors are usually at work in the allocation of a particular scientist's field-strength: recognized skills in observation, imaginative grasp, and reasoning; skill as a writer and speaker; draftsmanship; and so on. Sometimes it may be a case of Machiavellianism at work. But this is not necessarily or typically so.

In considering these matters, I should also like to draw on the work of the sociologist of science, H. M. Collins (1981), who has introduced the useful concept of the "core set." According to this notion, there can be, for any scientific specialty, a group of authorities whose opinions count, in determining the outcome of a scientific controversy, particularly in the matter of the empirical techniques that are favored—or, in the case that we have been considering, the field observations and techniques that are to be preferred. Thus there can be a group of key players in the field, whose collective judgment determines the outcome of a controversy, and who will make a judgment as to whose techniques, results, or observations are to be relied on. However, in Collins's view, "functioning" core sets are not so very common in science. They only emerge when some significant controversy develops—some "hot spot" as Collins calls it. And then, when the controversy is settled, they may die away again, to emerge once more with a somewhat altered membership when a new controversy breaks out within the specialism. For much of the time papers are written, refereed, and published, and are utilized by other scientists, without serious controversy erupting. This is the characteristic "normal science" of Thomas Kuhn (1962), when research proceeds smoothly and without challenge. But when a hot spot emerges, the behavior pattern changes significantly. Heated discussion occurs; correspondence flows; there is an increased intensity of experimentation, field observation, and publication; and there may be some behind-the-scenes skullduggery or acrimonious public debates. Then when the core set has made up its collective mind on the issue, the debate dies down and the research returns to a condition of "normalcy." This, I suggest, was what happened in the case of the Highlands controversy. So although I have used for convenience the Kuhnian term "paradigm" in chapters 7 and 8, it should be made clear that the Highlands controversy was more of a hot spot than a scientific revolution in Kuhn's sense. Nevertheless, the controversy was of the greatest importance to the history of British geology: it had a profound effect on the Survey, and it provides us with an excellent example for studying issues in the philosophy and sociology of science.

In order to track the emergence of the "hot spot" of the Highlands controversy, I offer in figure 11.4 a graph of the number of publications pertaining to the geology of the Northwest Highlands, plotted over time, from the first publication of Sedgwick and Murchison in 1827 to 1902, the last year for which publications are listed in the Survey memoir of 1907. The data are taken from Horne's bibliography published in that memoir, with a few additions derived from my own researches. Published maps are included in the data.[4]

From this graph, one may see in outline the major "topographical" features of the Highlands controversy, and it will be more clearly apparent that there were two distinct phases to the debate—or two hot spots. Serious work began in the northwest following Charles Peach's discoveries of fossils at Durness in 1854, and there was active research and debate for about ten years thereafter, with the views of Murchison and Geikie prevailing and with Nicol effectively retiring from the field. There followed a period of consensus in the 1870s, with the Murchisonian theory being generally accepted and received into the textbooks. Then following Hicks's reopening of the controversy in 1878, there was even more active debate culminating in Geikie's abandonment of the old view in 1884. Thereafter, it was a question of the Survey consolidating the new interpretation. With the publication of most of the relevant maps in 1892, the interest in the field as an active area of research died down, though there were increasing numbers of geological visitors to the Northwest Highlands toward the end of the century, eager to see for themselves the marvels of what Callaway, Lapworth, and the Survey officers had accomplished. The manner in which publication declined after the meetings of the British Association at Aberdeen in 1859 and 1885 is worthy of note.

4. The data refer to all publications from 1827 to 1902 that had to do with the geology of the Northwest Highlands, except for those pertaining to the so-called *Eozoon* controversy. About 1871, following claims originally made for the Laurentian rocks of Canada by the Canadian Geological Survey, there was interest in what appeared to be possible fossil remains in the corresponding gneisses of northwest Scotland. These seeming organic remains, which were named *Eozoon Canadense*, gave rise to a number of papers that were included in Horne's bibliography. The *Eozoon* controversy was an interesting episode in the geology of that period, and it acted as a kind of "background noise" to the Highlands controversy. However, I have not thought it sufficiently important for the purposes of the present study to include a discussion of the matter in this book. If the papers on the *Eozoon* controversy are deleted from the data, we have a more accurate "profile" of the Highlands controversy, as it has been discussed in the present study. For details of the *Eozoon* controversy, see King and Rowney (1881) and O'Brien (1970). (The first announcement of organic remains in the Laurentian was made by William Logan, director of the Canadian Geological Survey, in 1858.)

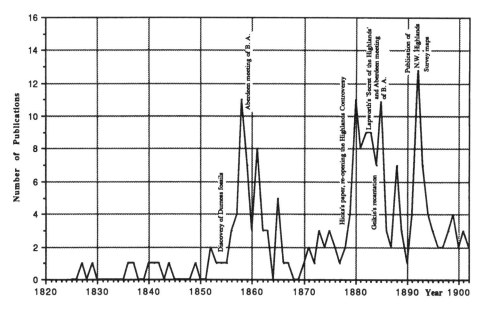

Fig. 11.4. Annual number of papers pertaining to the geology of the Northwest Highlands for the period 1827–1902.

If we would ask who were the members of the "core set" for the two phases of the controversy, I would suggest the following lists. The geologists are ranked with asterisks, according to my estimate of their relative standings in the agonistic field of the geological community with respect to the debate concerning the Northwest Highlands, not with respect to their overall status as geologists.

Phase 1	Phase 2
Murchison*****	Archibald Geikie*****
Ramsay*****	Lapworth*****
Nicol****	Benjamin Peach*****
Archibald Geikie****	Horne*****
Sedgwick****	Bonney*****
Lyell***	Judd*****
Harkness***	Callaway****
Scrope**	Hicks****
Salter**	Teall***
Jones*	Hudleston***
Nicholson*	Heddle***
Phillips*	Hull**
Darwin*	Nicholson**
Smyth*	James Geikie**
Horner*	Blake**

Charles Peach*   Howell**
Duke of Argyll*   Clough**
John Miller   Cadell**
                                  Woodward**
                                  Sorby**
                                  Bristow**
                                  Etheridge**
                                  Prestwich*
                                  Marr*
                                  Hughes*
                                  Rudler
                                  Topley
                                  Jolly

These lists are, of course, drawn up on the basis of my qualitative judgment of the relative "field strengths" in the British geological community at the time of, and in relation to, the Highlands controversy. Another analyst would probably have two somewhat different lists, but I would hope that they would agree in broad outline. Thus we may see how a geologist at the time of the Highlands controversy might have assessed the weight of opinion of those whose views needed to be taken into consideration, if he was seeking to gauge the merits of the opposing cases. The opinion of any given geologist would "count" according to his general standing and influence within the community and according to his known familiarity with the particular issues under debate. A distinguished geologist such as Lyell might be very influential, even though he took a negligible part in the controversy and never visited the Northwest Highlands. On the other hand, a geologist such as Nicol, who might seem a relatively minor figure in nineteenth-century geology by historians of science, and whose reputation suffered severely in his own lifetime as a result of his "defeat" by Murchison and his allies, must nevertheless stand high in our list, for he would have been recognized as having done much fieldwork in the north; he also had a university chair, was a former friend and colleague of Murchison, had previously been assistant secretary and librarian to the Geological Society, and had managed to develop an ingenious theory to account for the structure of the Highlands.

It will be seen from these lists that there was a considerable changeover in the personnel between the two major episodes of the Highlands controversy. I would suggest that this was one of the major reasons why the outcomes were so different at the two meetings of the British Association at Aberdeen. Whereas Murchison wielded all the power in the 1850s and 1860s, by the time the debate was reopened by Hicks, Murchison was dead and it was left to Geikie to carry the flag for the Murchisonian theory almost alone, though the Survey officers had begun their fieldwork in the northwest with that theory in mind.

It is, I hope, sufficiently clear from our study of the Highlands controversy that judgments about the points at issue were not always made on the basis of empirical evidence alone, or simply upon the cogency of the arguments adduced, the prevailing philosophy of science notwithstanding. Each geologist had his own specific social and academic interests to pursue, which existed side by side—indeed were part of—the efforts to achieve a true understanding of the geological history of the Northwest Highlands. Murchison was interested in gaining more territory for his Silurian kingdom; Geikie wanted a chair at Edinburgh; Lapworth wanted to demonstrate that "amateur" geologists' views should be taken into account, as well as those of the "professionals" in the Survey; and so on. Thus one should think of such social concerns as entering into the very fabric of the debate, constantly modulating the work that was done—even though all concerned genuinely wanted to know the history of the rocks of northern Scotland.

As a geologist engaged in controversy looks at an outcrop, he will not only be seeing the rocks through the spectacles of his theory. He will also be thinking about how the observations he is making may help or hinder the controversy (or theoretical debate) in which he is engaged. He will be thinking about how the data collected will be received by the other members of the field; and he will specifically look for evidence that will assist him in his controversy and tend to diminish the views of his opponents in the eyes of the core set. Occasionally, one can find direct traces of this thinking in the historical record, as, for example, the derogatory remarks about Nicol that may be found in Murchison's field notebooks. (But Geikie's notes included no such remarks.) In this way, a social dimension enters into the research right at the rock face, as it were, even if there is no other geologist present to form a miniscule, transient geological "community."

The same consideration applies to the historian of science, of course. While collecting material for this book, I was all the time thinking how the information discovered would or would not fit into the argument that I was gradually developing. Having regard to the views of the colleagues who might act as referees for the press, I was constantly thinking of the impression—favorable or otherwise—that something I might write would produce in their minds. Then looking ahead, I had to try to envisage the things that might be said by critics in the history of science journals. Nor was I unmindful of the positive or negative effects that the book might have on my reputation as a historian of science! Thus it is indeed the case that there is a social dimension to the knowledge claims that I have been endeavoring to advance. This does not, of course, mean that there is no empirical component to the text. But it does mean that I have failed to write a history "exactly how it was." The history has been constructed within the community of the historians of science and bears (I hope) something of the mark of the concerns of contemporary historians of science. Likewise, the history of the Northwest Highlands constructed by Lapworth, Peach and Horne, and others bore the mark of its social origins.

The Survey had by this time a well-established procedure for the construction of that knowledge, and I should like to examine this at some length. Geikie himself described this procedure in his annual report for 1897:

> The collectors are placed under the supervision of the field-officers. The assistant-geologists are promoted, as vacancies occur, to the ranks of geologists, who supervise the work of a number of geologists or assistant-geologists in a wide district. The district-surveyors report to their director, who takes general charge of the work in his own kingdom. The Director-General is the head of the whole organisation and is responsible for its conduct. He personally visits the officers in the field in each of the three countries, and is thus enabled to see that the work is being everywhere conducted on the same lines, and that the results obtained harmonise. It is his duty to bring the experience gained in one kingdom to the elucidation of difficulties met with in another, and to decide from time to time when the surveyors of one branch may usefully be sent to see the work in progress by another branch. It will be understood that to these duties in the field are added the general correspondence and administration of the whole service, and editorial labour connected with the issue of various publications. (Geikie 1898, 9)[5]

The whole social structure of the Survey was like that of a military unit, in keeping, perhaps, with the view of the geological enterprise as one with some resemblance to a military campaign or the actions of a colonial power.[6] Information was brought in from the field, analyzed and synthesized, and gradually digested, passing through a number of hands until it eventually emerged in the form of official maps and memoirs. In this way, a great mass of empirical information—the result of many weary hours of effort in wind and rain—would be boiled down to a single map, a few sections, and a small pamphlet describing the geology of the area of the map. The task of the director general was immense. He had to assimilate the information collected each year and see that it issued forth satisfactorily in digested form in the Survey's publications. He had to organize the day-to-day running of the organization, oversee its expenditure, administer the Survey's museum, and ensure that everyone was kept up to the mark. He had to determine the deployment of his "forces" and—using Geikie's word—he had to "harmonize" the results obtained. In addition, he had to keep abreast of the developments in geology worldwide and ensure that there was "harmonization" with the work of surveys in other countries. And further, to maintain some credibility as a geolo-

---

5. It is interesting that Geikie brought out this argument as and when he did. The Survey was, at that stage, in something of a state of siege, and it is not unlikely that the document was considered by Geikie to be a weapon for the defense of his position as director general during this siege.
6. It may be mentioned that the Survey in fact had its origins in the British military establishment, and in the early days the surveyors were required to wear quasi-military uniforms—with buttons bearing crossed hammers.

gist, it was desirable that he should keep up some research of his own. Geikie did all this to a remarkable extent, particulary when we recall the scope of his other activities. One can hardly wonder that he occasionally made mistakes.

I mention these matters for two reasons. First, they have a bearing on the manner in which Geikie comported himself as director general. He had to rely, to some degree, on the work of his subordinates. But he also had to back them, and the Survey, once an "official" view had been determined on any specific quesiton. If all had gone well, the work would have emerged regularly, pumped out of the system, so to speak, in the manner indicated by Geikie in his annual report above. Everyone would have been happy. The government would have felt that it was getting its money's worth. So too would the landowners, mining companies, water companies, etc. The role of the amateurs would have been either to amuse themselves in a healthful and morally uplifting activity, adding bits and pieces of information that were too insignificant for the surveyors to worry about, and seeking help and advice from the professionals as required. Or they would have done specialized research in the universities, and would perhaps develop some high-level geological theory that could be utilized by the Survey when required. Also they would train the third and fourth generations of surveyors. (The surveyors of De la Beche's generation were often largely self-taught or were trained by the director himself. Those such as Peach and Geikie gained most of their knowledge by a kind of apprenticeship within the Survey, assisting an experienced surveyor before going on to do independent work. By the end of the century, there were appointments such as Teall, who had been formally trained in the universities.)

It may be understood, then, why Geikie might be sensitive to any threat to his authority and that of the survey. If the Survey's work were questioned by the public at large, it would bode ill for the organization, for British governments in the nineteenth century were always anxious to curb expenditure and would be quick to sense any public dissatisfaction with the Survey's performance, even if the politicans and bureaucrats did not know the first thing about geology themselves. Of great importance, of course, was the coloring of the maps, which depended upon the choices made for the subdivisions of the stratigraphical column. Thus there arose the great battle between Murchison and the Survey over the Devonian system, and the battle between Lapworth and the Survey over the admission of the Ordovician system into the stratigraphical column. It was imperative for the Survey to be sovereign in that aspect, if no other; or so it seemed to Geikie.

So although one may feel that Geikie was insufficiently generous in his response to Lapworth's views, one can well understand why he was so unsympathetic to suggestions for the revision of the geology of the Southern Uplands, the Ordovician system, and the reinterpretation of the structure of the Northwest Highlands. Here was a local schoolmaster, an "amateur," asking the geological community to reject the ideas of the official Survey, to redraw most of the maps that had been prepared in the south of Scotland, and to recolor much of Wales. The outside world was confused as to what was going on. Thus we find the distinguished American geologist J. D. Dana writing to Geikie in 1892 to inquire whether he did or did not intend to adopt Lapworth's Ordovician system.[7] Dana indicated that he was willing to accept Geikie's judgment, but he informed him that many American geologists were beginning to go over to Lapworth's classification. Earlier, we find Barrande writing to Geikie congratulating him on his attempt to reestablish the authority of the Survey in the matter of the interpretation of the rocks at St. David's (see chap. 8). Barrande said that he had always considered the Survey to be the "sovereign authority" in all matters pertaining to British stratigraphy and that he was glad to see this authority being exercised.[8]

The difficulty, of course, was that the Survey did not have a monopoly of talent, and amateurs could sometimes examine in detail an area the Survey officers had only looked at in their usual manner; and then the amateurs might well find fault with the official interpretations. In the nineteenth century, amateur geologists could understand the issues involved and make substantial contributions to the progress of the science in a manner that is hardly possible today. Indeed, it was part of the rhetoric of books such as Geikie's *Field-geology* that the amateur *could* make useful contributions to geological science. It was thus almost inevitable that tensions should develop somewhere along the line between the amateurs and the professionals, in a way that was much less likely to occur in the physical sciences, which were already beginning to utilize specialist laboratories. Obviously, some amateur geologists had little of significance to say to the professionals. It was, however, foolish in my judgment for Geikie to tend to lump all amateurs together and fail to recognize that men of the caliber of Callaway and Lapworth might produce results as good as or better than those of the Survey; and that university geologists such as Bonney and Judd might have opinions that were every bit as well informed as those of Geikie himself.

An interesting feature of the Highlands controversy is worthy of note here. In the case of the Southern Uplands, Geikie had been performing his regular role (as director of the Scottish branch of the Survey). One of his staff, Logan Jack, had made a mistake in a general area that Geikie knew well, and where he had himself done some of the surveying. But Geikie thought that Jack's ideas were tenable, and so he backed him, putting the weight of the Survey's authority behind the proposed interpretation, indeed (according to Lapworth) building the Survey's theory on the basis of Jack's data and ideas. In the case of the Northwest Highlands, things stood rather differently. Geikie did not know the area particularly well, having examined it in a great rush in his tour with Murchison in 1860, and apparently he did not visit at least one of the key areas until 1884. More important, perhaps, the Survey did not pretend to have surveyed the ground properly at the time that Hicks, Callaway, and Lapworth went into the field in the northwest.

7. Edinburgh University Archives, Gen. 1425/57, Dana to Geikie, April 28, 1892.
8. Ibid., Gen. 1425/19, Barrande to Geikie, June 29, 1883.

So while Geikie stood by the Murchisonian theory of the Northwest Highlands until the last possible moment, it was not in fact a doctrine formulated as an outcome of the Survey's official mapping program. Geikie certainly backed Murchison's ideas; and he had good reason to do so for the advancement of his career, both in the matter of the directorship of the Scottish branch of the Survey and the Edinburgh chair. But there is little doubt that Geikie was uneasy about the extrapolation of the doctrines of the Northwest Highlands to the region of the Central Highlands; and when it came to the push, he was able to change his views, without too much loss of face, with talk about sending out the surveyors with the "special injunction . . . to divest themselves of any prepossessions in favour of published views and to map the actual facts in entire disregard of theory" (Geikie 1884b, 29). As we have seen, given this injunction, the surveyors made full use of their freedom and soon only Geikie was holding out. He reluctantly changed his mind in the face of the evidence the surveyors showed him at Loch Eriboll—and with the knowledge that no other geologist whose opinion really counted, and who was familiar with the rocks of the northwest, was supporting him any more. Geikie was in the awkward and, in truth, unusual situation of having to defend an interpretation that was not one the Survey had "formally" adopted as a result of its map work. Nevertheless, a line of three successive directors general had adhered to the notion of a regular ascending sequence of rocks in the Northwest Highlands.

I would refer again to the general manner of working of the Survey: the gradual analysis and synthesis of empirical information so as to yield a structure and geological history of the United Kingdom. In an interesting recent volume, Bruno Latour has referred to the data collected in the old voyages of discovery and then carried back to what he calls the scientific "centers of calculation" in Europe (Latour 1987, chap. 6). Specifically, he refers to the case of the French explorer La Pérouse and his determination of whether Sakhalin, to the north of Japan, was or was not an island. When La Pérouse first visited the area, the locals knew the answer to this question but he did not. However, the information he gathered was transmitted back to Europe and entered on European maps of East Asia. In consequence, the knowledge relationship between the Europeans and the East Asians was significantly altered by this event. For La Pérouse's voyage, the locals knew more about the topography than the visitors. But for the second European visit, the voyagers were already armed with the information previously gathered by La Pérouse; and as visit succeeded visit, the balance of knowledge and power between the Europeans and the Asian fishermen gradually altered. For not only did the Europeans have superior technology, they also had a superior system for garnering, abstracting, and systematizing knowledge, and this enhanced their power vis-à-vis the local inhabitants. (Thus knowledge and power were interrelated.)

This is a simple yet profound point. Latour goes on to develop a theory of the way in which fundamental ideas about space and time emerge from the abstraction of all the written knowledge carried back to the centers of power. I have serious doubts about his arguments on this head (Oldroyd 1987b), but

I think he is unquestionably right about the manner in which power relations may be altered by the successive accumulation of written items of information and its abstraction to ever higher levels in the "centers of calculation." Moreover, we can see all this operating on a small scale, even *within* the United Kingdom. The Geological Society, with its magnificent buildings and library in Burlington House, Piccadilly, was a leading "center of calculation." So too was the Survey in Jermyn Street, and in a more systematic fashion than the Geological Society.

Thus we can see that when the first surveyors, such as Macculloch, visited the Northwest Highlands, in some respects they would have known less about the rocks than the local crofters—for example, as to where limestone might be found. But this situation soon shifted, so that each successive generation of visitors could come armed with ever better maps, books, and prepared information. The knowledge relationship rapidly altered, until today the average Highlander's local knowledge of rocks is virtually irrelevant to the work of the geologist. I would like to suggest that the fate of Nicol illustrates this principle in some measure. When he and Murchison first did their work together, their power relationship was not so very different, as far as their views on the geology of the Northwest Highlands was concerned. But once they had parted company in their views, the balance of power rapidly shifted. Though Nicol ended up knowing a great deal more about the rocks of the Northwest Highlands than Murchison ever knew—and though his views were at least as sound as those of Murchison, if not better—Murchison's views were the ones that prevailed, for Murchison was at a "center of calculation." Indeed, as director general of the Survey he was in effect a center of calculation on his own account.[9]

Thus Murchison sought to be a world repository of information about, and a governing authority on, Lower Palaeozoic rocks. Through the issue of the successive editions of *Siluria*, he sought to ensure that his views were accepted. He had every advantage. He had money and influence, and patronage to bestow; and a succession of geologists such as Harkness, Ramsay, and Geikie found it in their interests to adopt the Murchisonian theory of the Northwest Highlands. He produced mammoth books expounding his views, and as we have seen, he took good care to ensure that they were favorably reviewed in influential places. Also, he could take pleasure in the fact that the government-sponsored Survey used *his* system as the basis of its map publications, with a far-flung Silurian kingdom. All this was possible, in part, because the information that was gathered was systematically transferred to Edinburgh and London.

9. Here one may make a point that Latour does not stress. There was a *two-way* flow of information between periphery and center. For example, Geikie, in Jermyn Street, was directing and coordinating work in the field—at the periphery. The information was gathered at the "edges" according to the instructions of the spider at the center of the web, so to speak, rather than in an uncoordinated fashion. That is, the information was gathered according to the theories that were being developed at the center, as well as according to ideas in the minds of the actual observers—the peripheral fieldworkers.

The case was quite otherwise with Nicol. He had little patronage to bestow. He carefully gathered up information, year by year; but all he could do with it was take it back to Aberdeen and abstract and synthesize it privately. Certainly, his ideas found some favor in Aberdeen. But this city was not a real center of calculation. It was a British analogue of Sakhalin. Even at the meeting of the British Association in Aberdeen in 1859, when the scientific world came to see what was going on in that remote corner of Britain, Nicol could not get his views across, and he was easily swamped by the Murchisonian onslaught against him, with its strong public relations campaign (evidenced by the attention given to Murchison's views by the newspapers). Later, Nicol found his efforts to rebut Murchison's arguments in the pages of the *Quarterly Journal of the Geological Society* blocked by the society's council. Edinburgh (both in the Survey and at the University) was in the hands of the enemy. The best that Nicol could do was to present his views in the form of lectures, and publish them privately as a slim volume (Nicol 1866). There would be no market for a lavish publication putting forward the theories of a discredited geologist. Even for these lectures, Geikie was instructed by Murchison to "hammer" poor Nicol. If ever there was a case of the "Matthew effect" (Merton 1968) in operation in science, this was it![10] By contrast, the interests of Geikie and Murchison were conveniently linked, to the mutual advantage of both warriors. For his part, Geikie deliberately nailed his colors to Murchison's mast, with prospects of preferment in view. On the other hand, Geikie was useful to Murchison as the standard bearer for the Murchisonian doctrine into the next generation of geologists, and with his dextrous pen the ambitious young geologist wrote in praise of *Siluria*. The two of them thus in combination formed a formidable force, whereas Nicol acted against his own best interests in falling out with Murchison.

There have been some interesting discussions in the recent literature in philosophy of science as to what happens, and how, when a controversy is terminated or brought to a close. In particular, one may ask what are the relative parts played by epistemic factors (whether they are empirical, methodological, or theoretical) and social factors. One may also wonder whether such a distinction can be made in a convincing manner. This literature is important, both for the historian and for the epistemologist, since it is when a controversy is terminated—when a scientific community has made up its mind on an issue—that we may say that knowledge has been constructed. While the controversy is still being pursued, there is uncertainty in the minds of all concerned and the knowledge cannot become bedded down in textbooks or university lectures. How, then, are controversies concluded, and what factors operate in effecting a conclusion?

In the growing literature on this topic, I have found the suggestions of

10. Cf. The Gospel according to St. Matthew, chapter 13, verse 12: "For whosoever hath, to him shall be given, . . . ; but whosoever hath not, from him shall be taken away even that he hath."

Ernan McMullin (1984, 1987) particularly interesting, and it may be worth-while to see how far they may be utilized in the present inquiry, without neces-sarily endorsing them fully. McMullin (1987) distinguishes between *epistemic* and *nonepistemic* factors, influencing the course of a scientific controversy; and between factors that the analyst may regard as *standard* and *nonstandard*. Fac-tors such as observation reports, hypotheses, reasoned arguments, appeals to logical principles, or (what are deemed to be) principles of method may be called epistemic. Also there may be personality traits (e.g., laziness or ambi-tion), institutional pressures, political pressures, nationalistic feelings, chance events, et cetera, which may be regarded as nonepistemic, but which undoubt-edly influence the course of scientific controversies. This distinction has the appearance of being moderately clear-cut, and it approximates the older and better-known distinction in the historiography of science between internal and external factors; but we shall want to know whether or not the distinction is valid or useful.

McMullin does not suppose that the epistemic-nonepistemic distinction is sufficient to do justice to all the complex issues involved when scientific con-troversies are terminated. He also introduces his second dichotomy: between standard and nonstandard factors. The distinction arises according to the views of the analyst (or historian) as to what does or does not constitute scien-tific thinking. So the boundary between standard and nonstandard can shift according to changes in opinion as to what constitutes scientific rationality. It is an evaluative distinction. For example, in the seventeenth century it was "standard" to attempt to weave together geology and theology, giving the "physico-theological" systems of authors such as Burnet (1684) and Wood-ward (1695). Today such efforts would not (perhaps) count as scientific; but that is our judgment, not that of Woodward or Burnet. Physico-theology was "standard" in the seventeenth century, whereas it is "nonstandard" today. Working with McMullin's two dichotomies, then, we have four possibilities:

> Standard epistemic
> Nonstandard epistemic
> Standard nonepistemic
> Nonstandard nonepistemic

But a nonepistemic factor would necessarily make it nonstandard also, for al-most any analyst of science at any period. So, says McMullin (1987, 62), one need only consider the following possibilities:

> Epistemic
>   a. Standard
>   b. Nonstandard
> Nonepistemic

He further distinguishes between controversies of *fact*, *theory*, and *principle*, the last of these having to do with questions of method and of fundamental ontology. There can also be *mixed* controversies, which involve more than one of these types of controversy.

Armed with these distinctions, then, McMullin discusses three ways in which scientific controversies are typically brought to an end. A controversy may, he suggests, be *resolved*, when the participants in the debate agree among themselves that the issues have been decided satisfactorily. The participants will agree that either one view or the other is correct, or they find a mutually acceptable middle ground. When a controversy is settled in this way, it will (according to McMullin) be accomplished by means of epistemic factors; and these factors will appear to be standard to the participants, though they may appear nonstandard to some later commentators.

A controversy may also be brought to an end by *closure*, with the application of nonepistemic factors such as political pressure, declining interest due to illness or old age, loss of research funds, biased reviewing, and so on. In such cases the disagreement may persist, but for some reason one side chooses to (or is forced to) give up.

The third possibility canvased by McMullin is *abandonment*. In such cases the scientific community as a whole loses interest. The controversy is neither resolved nor closed, but the participants can find no way of settling the matter and choose to get on with something else.

It should be noted that McMullin's mode of analysis allows a strong role for rational critical discussion in the settlement of scientific controversies, though his system acknowledges that there may be more to a scientific controversy than the use of observation and "sweet reason," and that nonepistemic factors may play an important part, which the historian must seek to identify and describe if he or she would give a satisfactory account of the course of scientific debate and progress. McMullin would transcend the old "internal/external" dichotomy in the historiography of science, and would acknowledge the role of the nonepistemic and what appears to us to be nonstandard as being potentially of essential importance in the history of science. But he maintains nonetheless the preeminent role of the empirical and the rational in scientific debate, and in effecting the resolution of controversies.

Let us see, then, how we fare if we deploy McMullin's categories in the case of the Highlands controversy. There is, I suggest, no need to invoke the distinction between standard and nonstandard factors. So far as I can see, the issues were standard, both by the criteria of nineteenth-century philosophy of science and those of the present day. Nowhere do we find appeals to God, the Bible, transcendental intuitions, or anything of that kind, which might lead one to regard the debate as in any way nonstandard. We may therefore leave that aspect of McMullin's analytical apparatus aside. (It may be objected that this is a cavalier attitude on my part. But as I understand the matter, it is the decision of the analyst/historian that is to count, for McMullin, in the assessment of what is and what is not "standard."[11])

11. It might, however, be suggested that the social factors involved in the debate that we have discussed in the present study would appear standard to a twentieth-century analyst, whereas

It is immediately clear that the two main phases of the Highlands controversy were significantly different: seemingly the first was a case of *closure,* the second a case of *resolution.* (In neither case was the debate simply abandoned.) In both cases, the issues were ones of fact and theory, rather than principle. Murchison and Geikie clearly "won" the first phase of the debate, and Nicol effectively retreated from the battle, maintaining his views in private, with a small body of support in Aberdeen. He certainly did not acknowledge that the Murchison-Survey opinion was correct. So far as Nicol was concerned, both geological theory and observational evidence supported his view of the structure and history of the Northwest Highlands, as expounded in his late publications. Murchison "won," not because of the superiority of his observations or his arguments, but chiefly because he argued from a position of greater strength in the agonistic field of the geological community. As the writer in the *Daily Chronicle* wrote in 1893, Murchison successfully "bounced" everyone into accepting his theory!

It does not follow, however, that the debate was won by Murchison *solely* because he had a position of greater power within the geological community. Some of his arguments against Nicol were cogent. He could point to the difference in strike between the "western" and "eastern" metamorphic rocks of the Northwest Highlands. He could refer to sections at places like Knockan Cliff and argue that they manifested a regular ascending sequence. This being so, it appears that we cannot say with certainty that the events following the first meeting of the British Association in Aberdeen constituted a situation of either pure closure or pure resolution. We can say, perhaps, that with all the "hammering" that went on, it was much more a case of closure than resolution, in McMullin's terms. But here a somewhat subjective decision on the part of the historian/analyst has been made.

In any event, it is clear that not everyone was satisfied with the outcome of the 1859 British Association meeting—even though the "authorities" had apparently settled the matter, with Murchison's "bouncing" having been successful. In his old age Nicol was able to persuade Judd of the merits of his case, when the two were together in the field. Hicks's interest was also aroused, and he began to look into the question. Thereby the interest of the geological community as a whole was rekindled. Heddle, Hudleston, Bonney, Callaway, Lapworth, and others began to do fieldwork in the northwest, and soon the controversy was raging again. The Survey fought a rearguard action, as for example, with the publication of Hull (1882–84), but despite all its "field-strength" the organization failed to prevail on the second occasion. Of the earlier generation of geologists who had espoused the Murchisonian theory, only Geikie was left to carry on the contest, Ramsay being in declining years by the time and Harkness having died in 1878. Peach certainly had long-standing obligations to Murchison, and his father had been an important figure in the establishment of the Murchisonian paradigm. But it was not the Peaches' map

---

they would have been deemed wholly alien to scientific reasoning in the nineteenth century, and should have been excluded as far as possible by the historian.

work or their theory that was in question; and Peach and Horne had been specifically enjoined by Geikie *not* to feel under any special obligation to uphold the Murchisonian doctrine. Moreover, they were well aware of Lapworth's abilities as a field geologist, from their knowledge of his work in the Southern Uplands. And further, as they worked in the Durness-Eriboll region in 1883–84, they would have known that Lapworth and Callaway had been there before them and gone over the ground with meticulous care (whereas I believe Geikie had never even been to Eriboll before 1884).

It is a great misfortune to the student of the Highlands controversy that we do not know exactly what happened when Geikie met Peach and Horne on the north coast of Scotland in October 1884. We know that the weather was bad, that Geikie showed extreme reluctance to concede, and that his face was a pitiable sight when eventually he did so. We also know that even then he dragged the surveyors down to Ross-shire, to show them the field evidence further south, which Geikie, walking in Murchison's footsteps, had found sufficiently convincing on previous occasions. It was only when exposures of Torridon Sandstone were found *east* of the presumed thrust plane near Loch Coulin that Geikie fully and finally yielded.

Given our lack of knowledge of what happened at Eriboll in October 1884, I shall beg an "indulgence of the understanding" on the part of readers and ask them to contemplate a historical thought-experiment. Let us suppose that Peach and Horne showed Geikie how to walk along the outcrop of the "lower" quartz rock until he found himself on what Murchison had called the "upper" quartz rock, without ever getting off the easily identifiable, brilliant white, hard quartzite.[12] It will be recalled that it was by this procedure that Lapworth succeeded in persuading Teall and Blake that the claimed distinction between upper and lower quartz rock was spurious and that they were one and the same unit; and so it was that Blake and Teall became enlisted on Lapworth's side of the debate. It might be thought that one could hardly have evidence that is more "empirical."

Was it the case, then, that Geikie changed his mind simply by virtue of this piece of empirical evidence, or something like it? If so, it would be an important argument against the proponents of the sociology of knowledge, in its application to the study of the history of science. It might seem that a scientist can be swayed, or have his or her views formed, primarily or solely on the basis of concrete empirical evidence—there before one's very eyes and not requiring the use of some esoteric (and hence problematical) scientific instrument in order to be apprehended.

The example is a simple but important one. For one thing, study of the quartz rock had for some time before 1884 come to be accepted as a Baconian *experimentum crucis.* If there were two quartz rocks, then one might plausibly envisage a regular ascending sequence as commended by Murchison. But if it could be demonstrated conclusively that there was only one quartz rock, this

---

12. This requires some stretch of the imagination. A good deal of the ground is, of course, covered by vegetation.

would mean that the appearance of there being two such rocks must arise from folding or faulting. In which case, the notion of a regular ascending sequence of the strata was inadmissible.[13]

In fact, as has frequently been emphasized elsewhere, it is rare for the participants in a scientific controversy to reach a position from which they can agree that the matter can be settled neatly by an *experimentum crucis*. By the time this happens, the debate is already close to resolution, so that the crucial experiment is symbolic rather than determinative.[14] This seems to have been so in the case that we have been considering hypothetically. That is to say, resolution of the controversy was imminent in October 1884. Was Geikie's change of opinion based *solely* on the empirical evidence that he saw before him at Eriboll? Certainly, the evidence that he found or was shown there was at least in part empirical. But he would have been well aware that the social tide was running against him; he was by then familiar with some of the recent Continental work, such as that of Brøgger and Heim, which Lapworth had utilized; and he would have read the work of Callaway and Lapworth and was perhaps already partly convinced by it.[15] The change of heart cannot, I think, be attributed wholly to the empirical evidence that was now brought to Geikie's attention. However, in presenting the revised view to the public in *Nature* in November 1884, he represented matters as if it were simply "the evidence" that made him change his mind. Thus he could maintain his empiricism—indeed, he might gain some measure of credit in the public eye by acquiescing to the dictates of empirical information.

Could that have been the whole story? Surely not. For we may carry our thought-experiment a little further and imagine Peach and Horne watching Geikie carefully to see his reactions as he walked along the outcrop of the quartz rock, placing him in a very firm social context. And we may be confident that Geikie's thoughts were not directed exclusively to the quartz rock on which he was walking. He must surely have been thinking of the consequences for himself (his standing in his new position as director general) and for the Survey as a whole (its position vis-à-vis the general public, the amateur geologists, and the government paymasters). If he changed his mind, he would have to bear the indignity of having been "proved wrong" in a matter that had been under discussion for thirty years. But if he didn't change his mind, he would perhaps look even more foolish, insisting that the ideas of his deceased chief (to whom everyone knew he was deeply beholden) were correct, when

13. If one wished to be pedantic, one might, I suppose, say as an ad hoc hypothesis that there *could* be a metamorphic rock lying on top of a sedimentary sequence that was somehow folded back on itself in an S-shape. Thus in a sense the existence of one or two quartz rocks need not have settled the larger debate unambiguously. But it was agreed to be a crucial point by the disputants.

14. In any case, the resolution of a controversy rarely occurs like that, outside sanitized versions of history, or accounts specially devised for public consumption, such as TV expositions.

15. Geikie also knew Lehmann's work, which was similar to views developed by Lapworth, though it will be recalled that Lapworth maintained that he worked out his ideas concerning metamorphism independently of Lehmann.

even his own staff had now turned to the opposition. He would not wish to be seen, so to speak, as a shag on a rock. So there was, I suggest, necessarily a strong social dimension to his thoughts, as well as empirical and theoretical considerations.

If we follow McMullin's arguments, the epistemic and the nonepistemic factors operating in Geikie's mind would presumably be distinct and in principle separable—certainly by Geikie himself, and perhaps by the historian if blessed with enough documentary information. It seems to me, however, that the epistemic and nonepistemic factors cannot be fully disengaged from one another. Certainly, Geikie would have looked at the quartz rock and would have seen something very like what anyone else might see, the theory-ladenness of observations notwithstanding. (You can hit the white rock with a hammer, and it has a characteristic bounce and ring about it.) But its minute subdivisions, such as were instituted by Peach and Horne for mapping purposes, do not manifest themselves to the cursory observer, or even to the eye of the experienced geologists. (The Continental visitors to the Northwest Highlands who looked round the exposures under the guidance of Peach and Horne recorded their wonder at how the Scottish geologists could perceive and rely upon such minute differences for the different subdivisions of the quartz rock.) Just recognizing quartz rock as quartz rock came no way near to solving the riddle of the Highlands rocks. After all, Nicol and Murchison, both highly experienced geological observers, could not agree on the evidence of the "trumpet pipes" as to whether the rocks were or were not inverted. Thus, as we have seen more than sufficiently, the whole issue was very complex, and by the time the issues had been boiled down to whether there was one or two quartz rocks the controversy was close to settlement. The "walking experiment" merely served to clinch or symbolize a decision that was close to resolution. On the other hand, the exposure of quartz rock was certainly not irrelevant. If Geikie had *not* been able to maintain a continuous walk from the old "lower" quartz rock to the "upper," perhaps he would not have yielded. But either way, was he thinking *only* about quartz rock as his decision to yield was made? No; this is not plausible.

So, strictly speaking, should we say that the controversy concerning the geology of the Northwest Highlands was "closed" or "resolved" after 1884? After Geikie's recantation, it was "all over bar the shouting." After 1884, there were no geologists harboring a secret belief that there was a regular ascending sequence in the Northwest Highlands and carrying on their research on that assumption. Unlike the case after the first Aberdeen meeting of the British Association, there was (so far as I am aware) no arm-twisting, no "arrangements" of anonymous reviews, no personal "hammering." On the contrary, the geological community went in for self-congratulation, with talk about letting the evidence speak for itself; and that the community regarded the matter as resolved can scarcely be doubted.[16] In particular, Lapworth was more

16. Lapworth evidently recognized this when he entitled his 1885 paper "The Close of the Highland Controversy." His term "close" meant what McMullin and others would call "resolution."

than satisfied that the contributions of the "amateur" party were now recognized. The controversy did not simmer on. All parties could take the view that the dispute was settled the way scientific disputes ought to be settled—by one side climbing down in the face of empirical evidence.

Thus if we are ever to find a case of a "resolution" of a controversy, the termination of the Highlands controversy after 1884 should serve very well, and in fact I am happy to regard the term as appropriate in this case. Nevertheless, I would not wish to say, as would McMullin (1987, 77), that it was a case of resolution rather than closure, simply because there were no significant nonepistemic factors at work. As suggested above, we can envisage some significant social considerations, having to do with power relations in the geological community, entering into Geikie's thinking as he paced along the quartz rock. And in any case, the emergence of a consensus was something that occurred within the agonistic field of the geological community. Even after Geikie's recantation in the field in 1884, the issue still had to be reported in the pages of *Nature* and argued out at the meetings of the Geologists' Association, the Geological Society, and the British Association, and in the published papers; and the geological community had to *decide* whether or not a satisfactory resolution of the controversy had been achieved. In this, the very fact of *Geikie's* recantation would have been as persuasive for many geologists as what might be seen in the field. Indeed, most geologists did not get to the Highlands; so their "witnessing," such as it was, was entirely vicarious.

But this acknowledgment of the continuing role of the geological community does not mean that there were no epistemic factors involved (no observations, no theorizing), or that a significant advance in understanding the geology of Scotland was not made when the Murchisonian theory was finally abandoned. Quite the contrary. New empirical information was gathered, and there was development or application of new theory and new methods of work, in the period between Hicks's investigations and the second Aberdeen meeting of the British Association. As I say, I am happy to call the second collapse of the Northwest Highlands "hot-spot" a case of "resolution" rather than "closure." It would in fact appear that the turn of events in 1884–85 was dependent chiefly on epistemic factors. But I do not believe that it rested on them alone.

In any case, the resolution of the controversy did not mean that all problems on the geology of the Northwest Highlands had been settled for all time. (Indeed, almost immediately after the Survey "officially" accepted Lapworth's doctrine, Barrow expressed doubts about the theory of the origin of the Moine Schists, according to that doctrine.) But it did mean that the acrimonious and controversial elements of the research receded, at least for the time being.[17] The work carried out by the surveyors in the latter part of the 1880s, for example, was what might reasonably be called "normal science." Everyone (except perhaps Barrow) was happy with what was being done, and with the

17. In the twentieth century, there have been bitter debates concerning the interpretation of the Lewisian Gneiss, and in particular the so-called Scourie Dykes.

results being obtained, though it was obvious that there were still questions to be answered. As we have seen, disagreement developed between Peach and Horne in the latter part of their careers concerning the origin of the Moines, but this did not degenerate into controversy; it was, rather, a strong difference of opinion that could be discussed between them amicably—though it was a stimulus to further research. This suggests that controversy is as much a social phenomenon as an intellectual matter. The disagreement between Peach and Horne was quite different in quality from that which broke out between Murchison and Nicol. Peach and Horne in no way threatened each other's interests, and they were great friends.

It should be emphasized that in the Highlands controversy, the resolution of 1884–85 involved the decision that neither Murchison nor Nicol was found to be wholly in the right. The theory developed by Callaway, Lapworth, Peach and Horne, and their co-workers was not that of Nicol, even though he was posthumously awarded the victor's crown by Judd in 1885. The later theory held that the eastern metamorphic rocks were indeed emplaced by faulting, and that the western and eastern gneisses were essentially one and the same. But the faulting was not that conceived by Nicol, and he was (we believe) incorrect in regarding Logan Rock as igneous. There was no upwelling of diorite or syenite along the supposed line of fault in the manner that Nicol envisaged, and the Aberdonian had no sense of the way in which the Moines might have been *formed* by the processes of earth movement.

On the other hand, Murchison was right in saying that the macroappearance of the western and eastern metamorphics was quite different, and that they had different strikes. He was also quite right to say that at a place like Knockan Cliff there was every indication of a regular ascending sequence. It appears, then, as it customarily the case in scientific controversy, that the final outcome—when a theory was eventually constructed to everyone's satisfaction—was a kind of amalgam of the two previously contending theories or a *via media*. It took elements from, and accepted certain observational evidence from, both doctrines; but it also added something of its own. Obviously, in the case that we have been considering, Lapworth adduced ideas about faulting and folding from authors such as Heim and Brøgger, and he developed his own ideas about metamorphism, perhaps with some help from Lehmann's work. Without the Continental ideas, Callaway was only able to construct a partially adequate theory—a kind of Tychonic system, heralding Lapworth's Keplerian system. (Of course, Lapworth did not base his argument on Callaway's data.)

So when Judd (1886) more or less suggested that Nicol, long before, had given a satisfactory account of the structure and history of the Northwest Highlands, this was not an accurate interpretation,—though I'm sure it gave old Mrs. Nicol a lot of pleasure, and for that reason alone I'm not at all sorry that Judd did what he did! But why did he do it?

The reason for his posthumous eulogy of Nicol is not, I think, hard to discover. As we have seen, Judd and Geikie were engaged in their own private battle concerning the volcanoes of Skye. It would therefore be advantageous to Judd's position, and prejudicial to that of Geikie, to show that Geikie was associated with a discredited theory, and had, perhaps, been doing some "hammering" of the deceased geologist who had been on the right track all along. This perspective allows us to make sense of Judd's action. We must remember, however, that Nicol succeeded in enlisting Judd to his cause not long before he died, as he showed him round the outcrops at Loch Maree. It was not the case, therefore, that Judd spoke well of Nicol simply because of his antipathy toward doctrines with which Geikie was associated. He had empirical considerations in view.

Again we may discern epistemic and nonepistemic factors in harness. They may be distinguished by the historian's analysis, but one must always expect to find them acting in concert. The problem is to know which element should be assigned precedence in any given case. Needless to say, I do not think it possible to assign "percentage points," as it were, for the two components of a person's thinking. The case is somewhat like that of the IQ controversy, though even less likely to be resolved satisfactorily, for the situation will alter within a given scientist's career from one case to the next, or even from moment to moment, and there seems to be nothing like twin studies that might (in principle) help to solve empirically the problem of the epistemic-nonepistemic relationship. The best that we may hope for is that the historian/analyst will be sensitive to the possible role of empirical, theoretical, methodological, and social factors and will incorporate all these in the historical account, according to the evidence available.

Can there ever be *pure* epistemic or *pure* nonepistemic factors at work? Can we, for example, suggest a factor for the Highlands controversy that might be said to be wholly nonepistemic? There was political pressure on Ramsay and Geikie to speed up the surveying, and hence they thought it appropriate to direct that map work be done on the one-inch rather than the six-inch scale. This must have been a great inconvenience and doubtless hampered the surveyors' efforts significantly. However, we know that they resisted by buying their own six-inch topographical maps, and then handing in the work after reducing it later to the one-inch scale (Bailey 1952, 84). Thus the epistemic effect of the government's edict was negligible, though the progress of work was doubtless slowed. Insofar as the edict was relevant, it would, I suppose, have to count as an essentially "pure" nonepistemic factor. However, I have thought it to be a relatively insignificant issue, and hence I did not introduce it into my account in chapter 8 or chapter 9.[18]

As another putative example, one might suggest that the movement of surveyors out of the Northwest Highlands in the 1890s, as a result of demands for resurvey of coalfields, and so on, in the south, had an effect on the work in

18. Just conceivably, that was a mistaken judgment on my part. It appears that the geological community tended to down-grade Callaway's contributions because he worked with one-inch rather than six-inch maps at Eriboll. See Bailey (1952, 114). (Bailey's opinion was probably based on a remark by John Hulke, the president of the Geological Society of London for 1883–84, in his presidential address for that year.)

the Highlands. This is no doubt true, but it should be remembered that the controversy had been resolved by that date. We cannot, therefore, offer it as an example of a political nonepistemic factor influencing the events leading to the resolution of the controversy. In fact I am unable to think of a purely nonepistemic political factor that may have played a manifestly significant role in the debate.

We are not in much better case when we look for other kinds of "pure" nonepistemic factors. For the sake of discussion, I offer the suggestion that Nicol happened to live in Aberdeen and hence had ample opportunity to work in the Northwest Highlands. Geikie, by contrast, was living in London after 1882, and was unable to visit Durness and Eriboll; and so he maintained his old stance, even though it was perhaps foolish for him to do so. Or again, it might be a piece of luck that Lapworth landed a job in Galashiels. Thereby he came to play an important role in the Southern Uplands debate, and hence his position vis-à-vis Geikie became settled. In this light we do have nonepistemic factors playing a part in the course of the Highlands controversy. The chance meeting of Lapworth, Teall, and Blake in the Northwest Highlands in 1883 might serve as another example. But such factors are marginal in the overall story, though essential for its understanding. They would not have been operative in people's minds when thinking over the merits of the opposing arguments in the controversy. I am thus hard put to find something that was a significant pure nonepistemic factor in the final resolution of the controversy in 1884–85. This is not to say that pure nonepistemic factors were wholly absent, or that they could not have been at work. It is simply to say that in the particular case before us, they do not seem to have been important. And I would agree with McMullin that if it were the case that a controversy came to an end simply as a result of the action of nonepistemic factors, it would have to be counted an example of closure (or perhaps abandonment) rather than resolution.

What, then, about the possibility of there being some purely epistemic factors at work? This is a trickier question. I have said earlier that we may expect to find an interweaving of social considerations in the very heart of the debate and controversy, for the observations, their interpretations, and the reasonings about them are made within the inevitable context of the controversy. But were there any "pure" observations or arguments? I do not think it would have been possible for any geologist to have done work in the Northwest Highlands in the 1880s in disregard of the controversy. At the very least, any geologist at work in the northwest would have to have read the previous literature (or perhaps, like Peach, would have been told about it), in order to be able to make a contribution to the subject under debate. So even if one were not already a participant, one would inevitably be caught up in the controversy to some degree and be aware of its social ramifications in some measure.

Thus we may still ask whether the observations and arguments could have been made in such a way as to transcend or override the social dimensions of the debate. It seems clear, in fact, that this *was* possible, for it happened in the case of Geikie. That is to say, the sum of all the information at his disposal by October 1884 was sufficient to make him realize that the Murchisonian doctrine could no longer be sustained. Even so, I don't think it was possible for Geikie to free his mind from the social considerations. He could not withdraw from the controversy without resigning from the director generalship. He could not be the only geologist in Britain left adhering to the Murchisonian doctrine; yet he could find no empirical or theoretical arguments that would sway the forces now ranged against him, in the form of Lapworth and of Geikie's own two best surveyors, who (we may suppose) were there with him in 1884 urging him to recant and directly drawing his attention to field evidence in favor of Lapworth's theory. As for Peach and Horne, they entered the ground at Eriboll knowing that a brilliant geologist, Lapworth, had gone before them very recently and had repudiated the Murchisonian theory. Lapworth himself went north with a strong presumption that the theory was wrong and a determination to show that the Survey had blundered, as it had in the Southern Uplands.

I suggest, then, that there would not have been, and could not have been, any thinking pertaining *solely* to epistemic factors, as characterized by McMullin, during the final stages of the resolution of the controversy. This does not mean that there were no unbiased observations, no careful reading of clinometers and compasses, no genuine entering of boundaries of rock units on field maps to the best of the geologists' abilities, or that there was no careful and unprejudiced reading of the literature, or attempts to apply theoretical ideas therein to the situation in the Northwest Highlands. All this, I suggest, is not precluded by the presence of the social factors at work in the geologists' thinking. So even if we offer the observational evidence for there being a single quartz rock as a candidate for a pure epistemic factor, we must admit that this factor would not have been operative alone and free from social considerations in Geikie's mind. This would have been so even at the point when he finally changed his mind.[19] But whether it was the view of the quartz rock, the urging and prompting of Peach and Horne, or the thought of the government paymasters that finally tipped the balance it is impossible now to say. However, it is the historian's task to do what he can to display the construction of the arguments and their public formulation. This I have attempted to do for the Highlands controversy, showing the manner in which the empirical and theoretical were interwoven with the social. For we can at least be confident that the empirical and theoretical considerations did not act in isolation from the social.

But as I say, one cannot be sanguine as to achieving some exact allocation of the relative strengths of each kind of factor operating in a person's mind, and some factors can hardly be represented as being definitely epistemic or nonepistemic. So a total "Collingwoodian" reenactment is impossible, no matter

---

19. I take it that McMullin's position is not that a factor become epistemic merely by virtue of being a consideration in a scientist's mind.

how closely one follows in the footsteps of the early geologists. One cannot enter *fully* into the mind of another person, now long deceased, and tease apart all the factors at work therein. And this shows the near impossibility of settling what might at first seem a relatively straightforward question: did Geikie recant for epistemic or nonepistemic reasons? I have gone as far as I might reasonably go with a Collingwoodian reenactment, having to resort to a thought experiment in the end, rather than some obliging document. At the last moment I am baffled and find myself resorting to guesswork. Sadly, it isn't much use saying what *I* would have done in Geikie's boots!

But perhaps we are becoming just too ambitious. So let us step back from this argument in order to make a further clarification. In my inquiry, and in the foregoing discussion of the roles of epistemic and nonepistemic factors, I have implied that scientists' interests can and do play an important part in the process and progress of science. They may, for example, from time to time exert influence or have a determinative role when a scientist is making a choice between two or more contending hypotheses or theories. Also, I have shown more than sufficiently how scientific debate involves many considerations over and beyond the straightforwardly empirical and theoretical. But this does not mean that scientific knowledge is, as it were, a puppet on the strings of ideological forces. Under normal circumstances (there are notorious exceptions such as the Lysenko affair), scientific knowledge about the world *does* grow over time and become more "verisimilitudinous." So in the particular case that we have been examining, we might expect that the account of the Northwest Highlands given by Lapworth, Peach, and Horne was a significant improvement on the views of either Murchison or Nicol. But how can we feel confident about this if, as Bourdieu (rightly) suggests, scientists are both judges of and participants in their own disputes?

Furthermore, according to David Hull's (1988) analysis of the community of zoological taxonomists, bias, personal commitment, and self-interest are the norm in science. Scientists are often "hammerers," and some are devious or Machiavellian. Why should truth emerge from all this? The question is pertinent, for it seems to me that to a striking degree the Highlands controversy was not so very different socially from the controversies that Hull has detailed. Actually, Hull goes somewhat further than I think is required and seems to hold that controversy is *necessary* for the successful prosecution of science. I do not think that it is essential, but it is commonplace. So why does it not obstruct rather than facilitate the progress of science? Why may we suppose that the theory of Lapworth "limned" reality better than did the theories of Murchison or Nicol?

In a sense, we cannot be certain about this at all. It may be that future science will decide to take up and develop further the theories of Murchison or Nicol. But this is extremely unlikely, and one is entitled to ask why this is so. In discussing the Devonian controversy, Rudwick offers the analogy of individual empirical observations—a marine fossil here, a terrestrial plant impression there—being like the individual black dots of a newsprint picture. As more and more of the dots are emplaced, the picture emerges and can be construed successfully by the geologist, who is, in Rudwick's view, "constrained" by the empirical information, but not determined by it. So long as there are only a few bits of empirical information available, a considerable number of alternative "pictures" may be legitimately suggested; but as more empirical information comes to hand, more and more of the possible pictures have to be eliminated, and eventually perhaps only one may seem "consensually persuasive" (Rudwick 1985, 455).

This "newsprint model" offers a useful analogy, even though it does not allow for the fact that the location of the "dots" in the picture may not be "fixed," because of the theory-ladenness or controversy-ladenness of the observations. Even so, I think that it is true to say that the addition of new items of empirical information can, and often does, serve to eliminate alternative interpretations. I would add, further, that if the observations are made with greater accuracy and precision, then one may proceed more effectively and certainly toward a consensus as to how the data should be construed, for improvements in accuracy and precision lead to the "ruling-out" of certain hypotheses or generalizations. Thinking in this way, we can, I believe, see how progress toward more *truthful* theories is achieved.

"Truth," however, is a word about which philosophers are very wary. There are two main contenders as to what is meant by truth: the correspondence theory and the coherence theory. Most people believe there is a real world "out there" (ontological realism); that we gain access to it with the help of our sense organs, instruments, and experimental investigations; that there is *correspondence* between our scientific theories and the way things exist and occur in the world; and that our "theoretical entities" (atoms, thrust planes, mylonites, or whatever) *refer* to things that actually exist (epistemological realism). The commonsense view is that our theoretical pictures "limn" reality—with greater or lesser success. When they do so, we have truth: correspondence of ideas and the way things are. Or, as Aristotle put it in his characteristically pedantic (and not particularly illuminating) way: "To say that 'what is' is not, or that 'what is not' is, is false; but to say that 'what is' is, and 'what is not' is not, is true" (1961, 201).

Truth can also be characterized by the so-called coherence theory. When the parts of a discourse are congruent and consistent, or hang together, we say that they are coherent with one another. This is characteristically so for a branch of mathematics. One might say that the meaning and truth of all propositions in arithmetic are bound up with one another: they all cohere: $2 + 2 = 4$; $2 = (8)^{1/3}$; $2 = 4 - 2$; and so on. One cannot and does not have the meaning of "2" varying from one arithmetical proposition to the next. All the propositions of arithmetic hang together or cohere, so that our knowledge of arithmetical truth is wholistic, for the truth of one arithmetical proposition emerges from its congruence with all other possible arithmetical propositions.

Thinking in terms of this arithmetical model, Leibniz imagined the world to be one in which all its component parts—a system of monads—were logically related to one another in a coherent system, analogous to the numbers of arithmetic.[20] F. H. Bradley imagined the true propositions about the world all formed one grand cohering system of knowledge.[21] But we can hardly expect to achieve some godlike understanding of all knowledge, and it seems implausible to suppose that truth *is* coherence. Rather, one may say that knowledge is true when we achieve correspondence between our descriptions of the world and the way the world is; and that we believe we have this correspondence when the information we have coheres with knowledge from other parts of science. However, this does not provide a route for the complete eradication of nonepistemic factors and a wholly rational science: what *counts* as coherence is, like everything else, something that has to be determined within the agonistic field of the scientific community.

The difficulty, of course, is that so far as science is concerned, what we know of the way the world *is* is discovered by means of our scientific researches. We have no independent way of getting to know the world other than through science. So if we are looking for correspondence between our scientific (theoretical) descriptions of the world and the way the world is, we cannot look at our description of the world and the world itself and see how they match up. One might suppose, therefore, that a correspondence theory of truth is useless. Indeed it is, if we want to use correspondence as a criterion of truth. On the other hand, it is perfectly serviceable as a statement of what we *mean* by truth: a statement is true when what it says about the world corresponds with the way the world is in fact. As to how we know that we have achieved correspondence (and hence truth), I suggest that we may deploy the notion of coherence. We use it as an *arbiter* of truth.

For example, we may find that the measurement of some physical quantity (the age of a rock, say) by a number of logically independent methods gives the same result in each case, within the limits of experimental error.[22] If this is so, then we may be confident that the value we give for the age is indeed true, within those experimental limits. The results are consistent, the knowledge coheres, and hence the information can be regarded as true. However, coherence does not give a one-hundred-percent guarantee of truth. For example, the Ptolemaic astronomy cohered internally and with the wider system of Aristotelian philosophy, but it was, we believe, false. The same might be said of

20. The best analogy, perhaps, is to think of the system of Leibnizian monads as a set of solutions to some grand "cosmic equation."
21. Bradley (1914, 233) wrote: "Truth is an ideal expression of the Universe; at once coherent and comprehensive. It must not conflict with itself, and there must be no suggestion which fails to fall inside it. Perfect truth in short must realize the idea of a systematic whole. And a whole . . . possesse[s] essentially the two characters of coherence and comprehensiveness."
22. For example, the potassium/argon method, the fission-track method, the palaeomagnetic method, and the evidence from palaeontology. For my account of these methods, and an elaboration of some of the arguments canvassed here, see Oldroyd (1987a).

the system of the world and the theology of St. Thomas Aquinas. At a less exalted level, two different stories, both apparently cohering, may be told at a trial, and in consequence the jurors may have the greatest difficulty in deciding which account to prefer. In science, there will certainly be room for social negotiations in deciding what degree of precision in physical determinations is to be sought for or is acceptable. Nevertheless, the use of logically independent empirical methods for determining physical quantities does offer a most valuable test for the verisimilitude of scientific statements. The "mesh" of knowledge in physics, chemistry, biology, and so on, is indicative of the "truthlikeness" of all these sciences, which is further substantiated by their practical efficacy and predictive capacities.[23]

I should like to suggest that this consideration of coherence is very much a matter of concern to geologists, just as it is to other physical scientists. It was certainly recognized as such by Geikie, as well as by the other participants in the Highlands controversy. We notice, for example, his mention of the task of the director general to harmonize the results of the surveyors in their fieldwork in different areas (Geikie 1898). A rudimentary example of this would be the requirement that the boundary lines drawn on adjacent maps linked up when one reached the common edges of the maps. Likewise, it was needful that the stratigraphical divisions laid down should be of general application, and not just restricted to some local region. Thus it was that Murchison desired his Silurian system to have worldwide application and not be restricted to use in the type area in Wales. Murchison had a perfectly "honorable" reason to wish to see his kingdom expand, over and above his manifest "territorial" ambitions that I have emphasized in this study.

Theories, likewise, should be mutually compatible with one another and cohere. The theories of palaeontology, sedimentary petrology, tectonics, geomorphology, etc., all have to lock into one another—and ideally with any grander metaphysical beliefs that one might have. In the case that we have been considering, however, Murchison's theory of a regular ascending sequence in the Highlands was not compatible with the then current theories of metamorphism. Nicol could rightly say that the idea of schist lying on top of unmetamorphosed sediments such as the fucoid beds made no sense. Murchison might say that the materials out of which the schists had been made were somehow susceptible to metamorphism, whereas the fucoid beds were not. But this was manifestly implausible. There was no independent reason to believe that the fucoid beds should, for some reason, be resistant to metamorphism. Thus Murchison's view on this did not cohere well with other geological ideas.

---

23. The important caveat should be entered here that what *counts* as practical success is itself subject to social negotiation. In science, pragmatic success, as well as coherence, may be used as a criterion of truth. During the course of the history of science, it may sometimes be pragmatic success and sometimes coherence that is the preferred criterion of truth; and whether it is pragmatism or coherence that is to rule is again something that will have to be fought out within the agonistic field of the scientific community. In the Highlands controversy, however, pragmatic considerations seem to have been subordinate to consistency and coherence.

By contrast, Cadell was able to make Lapworth's idea cohere with the results of practical experimental models for mountain building.

I suggest, then, that the account of the history of the Northwest Highlands offered by Lapworth cohered much better with other geological knowledge than did the story offered by either Nicol or Murchison. Lapworth's theory meshed with the work being done on mountain building in Europe by geologists such as Heim and Brøgger. It gave a description of events that worked at both Assynt and on the far north coast of Scotland (and that later could be extended satisfactorily all along the zone of complication as far as Skye). It cohered with the ideas about metamorphism being propounded by Lehmann. And when the Assynt Series was suggested to be Cambrian (by reason of the *Olenellus* fauna), this allowed a correlation between the Torridonian in Scotland and the pre-Cambrian rocks of Shropshire. Altogether, Lapworth offered a story that meshed with other geological knowledge; and thus it was acceptable to the director general, even though Geikie had all sorts of social reasons for being opposed to the theory. Needless to say, Lapworth did not claim to have had the last word on the geology of the Northwest Highlands. Indeed, he was well aware that the theory of the formation of schists and gneisses was in a very provisional state. But although there was evidently much to do, there was good reason to regard the theory as an acceptable basis for further research.

I would therefore suggest that although the geologists after 1885 had no reason to believe that the story they were telling about the Northwest Highlands was true in every particular, they did have reason to think that it was truer than the old Murchisonian theory. Thus, while it is correct to say that the Lapworth/Peach/Horne theory was socially constructed within the agonistic field of the geological community, it is also correct to say that it was more "verisimilitudinous" than the Murchisonian theory: there was greater correspondence, albeit imperfect, between what was said in the (official) theoretical geological history of 1885 and the actual history of the portion of the globe now known as northwest Scotland than was the case for the (official) theoretical history of 1882. This correspondence could be recognized by the coherence between the elements of theory and observation mentioned in the preceding paragraph. This was perhaps the major epistemic factor leading to the *resolution* of the Highlands controversy.

I hold that scientific theories about the world and its geological history are socially constructed, but that the world and its history are not. Also, we know more about the world and its history as the years pass: there is increasing coherence of scientific theories, and increasing correspondence between theory and "the world." These statements may seem banal, and they are, of course, the generally received view of science. But I believe they need to be said, for some recent commentators within the school of "social constructivists," such as Latour and Woolgar, seemingly would have it otherwise. (And it is possible that some of my earlier remarks may lead readers to suppose that I am a camp follower of their work.)

In a recent volume of Woolgar's, for example, it is maintained that "objects are constituted in virtue of [their] representation[s]," and that "the representation gives rise to the object" (1988, 56, 65). It is true that thinking about science in this manner may, in some measure, give rise to a host of interesting questions and insights. But it does not follow from *that*, that the world is somehow constituted (in an ontological sense) by its scientific representations. To suggest this is a misuse of language and a source of confusion. It is the scientific representations that are socially constructed or constituted within the agonistic field of the scientific community. The representations can be known to be more truthful as they increase in coherence and as they facilitate pragmatic success or control. Within the scientific community, there is a host of actors at work, each seeking to get his or her view of reality accepted by the other members of the field. But there are rational arguments being deployed, involving *inter alia* reference to observations (albeit ones that are theory- and controversy-laden). People do have rational reasons for changing their views, which interweave with their social concerns and interests in a complex manner. The waters into which the scientist peers (and the historian also) are often deep and muddy, and truth may not shine forth immediately. I should like to think, however, that this study has revealed at least something of what was at the bottom of the pond in the matter of the Northwest Highlands, as well as something of the mud that obscured the vision of reality. Curiously, the stirring of waters during controversy seems to be one of the processes by which we come to form a clearer picture of reality. At the very least, it induces people to look more attentively, and when the mud has settled a bit they may see things more precisely and more truthfully than before.

Glossary of Geological Terms

Acidic/Basic: Igneous rocks may be classified according to their content of silica (silicon dioxide). Those composed of more than two-thirds silica are said to be "acidic." Those low in quartz (less than half) are said to be "basic" or "ultrabasic." The others may be classified as "intermediate."

Actinolite: A green metamorphic mineral of the amphibole family—a calcium/magnesium/iron silicate.

Almandine: A variety of garnet—an iron/aluminium silicate.

Amphibole: An important family of dark-colored ferromagnesian minerals, of variable composition, the most common being hornblende. The amphiboles characteristically have two cleavage systems, approximately at 120° to one another, a fact useful in the diagnosis of hand specimens. (Cf. pyroxenes.)

Analyzer: The polarizing device (formerly a Nicol prism, but nowadays a Polaroid filter) in a petrographic microscope that resolves the polarized light after it has passed through a thin section of the rock or mineral under examination. (Cf. polarizer, polarized light, polarizing microscope, and Nicol prism.)

Anticline: A fold in strata such that the convex side is uppermost.

Archaean: A term meaning "ancient," generally applied to early pre-Cambrian rocks; pertaining to the earliest geological time.

Arenaceous: Sandy; rocks containing or derived from sand.

Augen: German for "eyes." Applied to eye-shaped lenticles of minerals, or aggregates of minerals, found in metamorphic rocks such as gneisses.

Augite: A common green, brown, or black member of the pyroxene family of ferromagnesian minerals.

In preparing this glossary, I have made considerable use of Howell (1957).

373

Basalt: A dark, fine-grained igneous rock commonly forming dykes or lava flows. Basalts contain calcic plagioclase (a type of feldspar) and pyroxene (commonly augite); olivine is also present in many basalts.

Bed: The smallest depositional division of a series of strata, differentiated from its neighbors above and below by more or less well-defined division planes.

Bedding: Depositional layering in sedimentary rocks.

Bedding plane: The division planes that separate the individual layers or beds of sedimentary rocks.

Borolanite: An igneous rock, formed at depth, being a particular type of syenite. (First recognized at Loch Borolan in the Northwest Highlands.)

Breccia: A rock made up of angular fragments cemented together.

Bronzite: A member of the pyroxene family of minerals—a magnesium/iron silicate.

Calcareous: Containing calcium, usually as calcium carbonate.

Caledonian Orogeny: A large-scale series of earth movements that occurred during Lower Palaeozoic times and led to the formation of a mountain system extending from Ireland through Scotland to Scandinavia, as well as mountains along the Atlantic coast of North America.

Cambrian: The oldest of the Palaeozoic systems (beginning about 590 million years ago), containing the first surviving fossils of organisms with hard parts.

Carboniferous: One of the Palaeozoic systems, between the Devonian and Permian and beginning about 360 million years ago. Most British coals are of Carboniferous age.

Carboniferous Limestone: The lowest lithological division of the Carboniferous system in Britain, and an important and easily recognizable stratigraphical unit. Formerly called the Mountain Limestone.

Catastrophism: The doctrine that some past geological changes have occurred as a result of sudden catastrophes.

Chlorite: A common group of greenish flaky minerals, being hydrous silicates of aluminium, iron, and magnesium. Common in metamorphic rocks (e.g., "green schists") and in igneous rocks as a result of alteration of ferromagnesian minerals.

Clastic: Consisting of broken fragments of older rocks.

Clay slate: A slate derived from a shale.

Cleavage: A tendency to split along definite, parallel, closely spaced planes. Found in both minerals (such as micas) and rocks (such as slates).

Colonies: In this book used to refer to the theory of Barrande that precursory forms of fossils could be found in a stratigraphical sequence in advance of the large-scale appearance of the type.

Conformable: Beds or strata lying one upon another in unbroken parallel sequence, with no evidence of disturbance or erosion having occurred during the process of deposition of the strata, are said to be conformable to each other.

Conglomerate: A sedimentary rock consisting of rounded water-worn pebbles cemented together.

Coprolite: Petrified excrement.

Cross-bedding: An arrangement of the laminations in a stratum such that there are subsidiary laminations oblique to the main bedding planes. Such arrangements can arise when sediments are deposited in moving water, but also in wind-blown deposits such as desert sand dunes. Also known as current-bedding or false-bedding.

Denudation: The removal of strata by erosion due to water, ice, etc., to expose an underlying set of rocks. In the nineteenth century, it was customary to speak of "a denudation" as some specific occurrence.

Devonian: The Palaeozoic system lying between the Silurian and the Carboniferous and beginning about 408 million years ago.

Diorite: A coarse-grained igneous rock, formed at depth, and consisting typically of intermediate plagioclase (feldspar) and hornblende, with quartz, biotite (brown mica), or a pyroxene in small amounts.

Dip: The maximum inclination of a stratum or other planar feature to the horizontal.

Dolerite: A family of dark, medium-grained, intrusive, igneous rocks containing minerals such as calcic plagioclase, augite, and olivine, and in many cases with ophitic texture. Dolerite is equivalent to basalt in composition.

Dolomite: Calcium magnesium carbonate (as mineral), or a magnesium-bearing limestone (as a rock).

Drift map: A geological map, showing superficial deposits such as river gravels, glacial debris, etc., which obscure the underlying solid formations. (So called because of the early theory that materials such as "boulder clay"—today thought to have been emplaced by the action of glaciers or ice sheets—were deposited by floating icebergs.)

Dyke (or dike): A mass of igneous rock filling a fissure in strata and cutting across them. So called because it may appear as a "wall" of rock after the surrounding strata have been stripped away by erosion. If the material of the dyke is weathered more easily than the surrounding rock, a trench may be formed.

Eozoon: A name given to what for a time in the nineteenth century was believed to be the remains of life in Archaean rocks, such as the Lewisian Gneiss.

Epidosite: A rock formed chiefly of epidote with some quartz.

Epidote: A greenish or yellow-green mineral commonly found in metamorphic rocks, and in igneous rocks as a result of alteration of calcic plagioclase—a hydrous silicate of calcium, aluminium, and iron.

Extinction angle: The angle through which a crystal in thin section must be rotated on a petrographic microscope, from a known crystallographic plane or direction, to the position of maximum darkness (extinction), when viewed between "crossed" polarizer and analyzer. An important quantity for mineral identification.

Facies: The general appearance of a sedimentary rock body, with respect to composition, type of bedding, fossil content, etc., indicative of the conditions of its formation. Also used in relation to igneous rocks where some part of the rock mass differs from the rest. For example, a granite might be said to show a porphyritic facies as its margins. (The term "metamorphic facies," denoting particular temperature-pressure regimes in metamorphism, was not current in the nineteenth century.)

Fault: A fracture along which there has been displacement of the two sides relative to one another and parallel to the fracture.

Feldspar (formerly often written *felspar*): An abundant group of rock-forming minerals, being alumino-silicates of calcium, sodium, and potassium. The main types are orthoclase (potassium aluminium silicate) and plagioclase (a series from calcium aluminium silicate to sodium aluminium silicate, divided, arbitrarily, into the minerals anorthite, bytownite, labradorite, andesine, oligoclase, and albite). (More rarely there are forms with barium instead of calcium.)

Felsite: A loose field-term used to refer to a fine-grained light-colored igneous rock.

Flaggy: A term to describe strata with beds between about 10 and 100 millimeters in thickness. (Such rocks are often used for paving-stones.)

Floetz series: The name given by Werner (see chap. 2) to parallel-layered, approximately horizontal, rocks. Roughly equivalent to the sedimentary strata, from the Old Red Sandstone upward.

Foliation: The laminated or banded structure found in metamorphic rocks arising from the segregation of minerals into layers of different composition. The term is also used more loosely to refer to the parallel fabrics (schistosity) of metamorphic rocks.

Fossil: Originally anything dug from the ground; but in the modern meaning, the term refers to the remains of animals or plants preserved in the earth's crust by natural causes.

Fucoid: The term formerly referred to what were believed to be the fossilized remains of seaweeds; but it came to be used loosely for various indefinite markings on the surface of a sediment, thought to be caused by living organisms but that could not be referred to any recognized type.

Fundamental Gneiss: A term suggested by Murchison for the gneiss rocks at the bottom of the stratigraphical column in Britain, exposed in the northwest of Scotland. Correlated with the "Laurentian" rocks of the Canadian shield. Also called "Lewisian Gneiss" or "Hebridean Gneiss," Lewis being the major island of the Outer Hebrides.

Gabbro: A dark, coarse-grained, basic igneous rock, equivalent in composition to the fine-grained basalt. Contains calcic plagioclase and minerals such as augite and olivine.

Garnet: A family of cubic minerals found mostly in metamorphic rocks; some garnets are used as abrasives and some as semiprecious gems.

Geognosy: The term favored by Werner (see chap. 2) for the scientific study of the earth. It was intended to be used in contradistinction to the term geology, which was considered more speculative. In reference to the work of Werner's school, it indicates an emphasis on mineral and rock studies and attempts to develop stratigraphy on lithological rather than palaeontological principles.

Gneiss: A coarse-grained metamorphic rock, roughly granitic in composition, with layers or lenticles of granular minerals alternating with layers of lenticles of platy materials. Some gneisses are marked by alternating light- and dark-colored layers, respectively rich in quartz and feldspar and in dark minerals such as biotite (mica) or horneblende (amphibole).

Granite: A family of generally light-colored, coarse-grained, igneous rocks, formed at depth and rich in quartz and potassic/sodic feldspar.

Granulite: In modern usage, the term refers to foliated rocks of granular texture that have undergone high-grade (granulite facies) metamorphism. As used by Scottish geologists in the nineteenth century, the term referred vaguely to a rock made of quartz and feldspar and some accessory minerals, with some foliation. Initially thought by Nicol to be igneous, such rocks were later regarded as metamorphic, and both schists and gneisses were sometimes called granulites.

Graptolites: A group of extinct invertebrates found as fossils in Palaeozoic rocks. The organisms were colonial animals, with individual polyps living in small cups (thecae) arranged along a common stem. In appearance, the fossils may look like part of a fretsaw blade. They were originally so called because they sometimes resembled writing on a slate.

Greenstone: An old field-term for a dark, fine-grained, basic igneous rock affected by metamorphism.

Greywacke: A loose term derived from the German word *Grauwacke,* referring to a dark sandstone or grit with angular fragments of quartz and feldspar in a more fine-grained "clay" matrix.

Grit: A coarse sandstone.

Groundmass: The fine-grained crystals between the larger crystals in a porphyritic igneous rock.

Hebridean Gneiss: See "Fundamental Gneiss."

High-grade (or rank) metamorphism: Metamorphism produced by high temperatures and/or pressures.

Hornblende: An important member of the amphibole family of minerals, commonly found in igneous and metamorphic rocks, and occurring as dark brown, black, or greenish-black monoclinic crystals or grains. It is a complex alumino-silicate of calcium, magnesium, iron, and aluminum, with variable composition.

Hypersthene: A brown to black ferromagnesian mineral of the pyroxene family; similar to bronzite.

Hypersthenite: A dark, coarse-grained igneous rock consisting wholly or almost wholly of hypersthene. Minor amounts of other pyroxenes, olivine, or plagioclase may be present.

Igneous rock: A rock "formed by fire"—i.e., by solidification from a molten or partially molten state.

Imbricate structure: Refers to tabular masses that overlap each other like the tiles on a roof. In the context of the present book, a series of thrust sheets dipping in the same direction. Also called *Schuppen* structure.

Intrusion: A body of igneous rock that invades older rock; or the process of forming an intrusion.

Jasper: Red, brown, or green, slightly translucent, poorly crystalline, impure quartz.

Joint: A fracture or parting, usually vertical, that interrupts the continuity of a rock mass and along the plane of which little or no relative movement has occurred.

Jurassic: The highly fossiliferous Mesozoic geological system between the Triassic and the Cretaceous, beginning about 213 million years ago.

Klippe: An isolated block of rocks separated from the underlying rocks by a fault.

Kyanite: A triclinic aluminium silicate, usually blue, found in aluminous metamorphic rocks subjected to high-pressure regimes.

Lamprophyre: A porphyritic igneous rock typically found in dykes. Its dark minerals are found as phenocrysts and in the groundmass, while the lighter constituents are found only in the groundmass.

Laurentian: See "Fundamental Gneiss."

Lewisian: See "Fundamental Gneiss."

Lias: The lower division of the Jurassic system.

Limestone: A sedimentary rock, consisting chiefly of calcium carbonate.

Lithology: The science of the mineral character of rocks (not their fossil contents); or the study of rocks according to their appearance in hand-specimens, not thin sections.

Logan Rock: A provisional, nontheoretical name suggested by Heddle for certain exposures of Hebridean Gneiss, not then recognized to be such and previously described in several different and incompatible ways by different observers.

Magma: Molten or mobile material, generated within the earth, from which igneous rocks are thought to have been formed by cooling and solidification.

Marble: A metamorphic rock produced by the action of heat and pressure on limestone or dolomite.

Melanite: A black, titanium-bearing variety of the garnet andradite—a calcium/iron silicate.

Mesozoic: An era of geological time, made up of three systems: Triassic, Jurassic, and Cretaceous.

Metamorphic rock: A rock formed by the agencies of heat and pressure or the chemical environment, producing alterations including the formation of new minerals, without the occurrence of melting. Schists and gneisses are important examples of products of metamorphic action on a regional scale.

Mica: A group of minerals with sheetlike structures arising from perfect cleavage, particularly biotite ("brown mica") and muscovite ("white mica").

Mylonite: A fine-grained laminated rock produced by the "milling" of rocks on a fault surface, with little or no growth of new crystals, though the fragments are welded together by frictional heat.

Nappe: A large body of rock, moved forward from its original position by overthrusting.

Nepheline: A sodium aluminium silicate of the feldspathoid family, commonly found in some syenites and related rocks.

Neptunism: The doctrine associated with the theories of Werner (see chap. 2) that all the rocks of the earth's crust were originally deposited from water, either by chemical precipitation, or by erosion and deposition of sediment derived from the supposedly precipitated rocks.

New Red Sandstone: A series of red sandstones formed under desert conditions, constituting much of the Permian and Triassic systems in Britain.

Nicol prism: A device invented by William Nicol to produce plane-polarized light. A crystal of calcite is cut and then cemented together again, in such a way that only the "extraordinary ray" arising from the double refraction of the calcite is transmitted; and this ray is plane polarized. The nineteenth-century petrographic microscope used two Nicol prisms, one as polarizer and the other as analyzer. (In modern instruments, the Nicol prisms are replaced by Polaroid filters.)

Normal fault: A fault in which the displacement of strata is downward on the fault plane. This arises when the overall maximum stress is vertical rather than horizontal. (Cf. reverse fault.)

Old Red Sandstone: A succession of conglomerates, red shales, and sandstones forming the continental facies of the Devonian system in Britain. Well known for its fossil fish remains.

Olivine: A group of green, brown-green, or yellow-green orthorhombic minerals (iron/magnesium silicates), with glassy fracture, commonly found in basic igneous rocks.

Oolite: A limestone made up of small cemented spherical granules of calcium carbonate, somewhat resembling fish roe in appearance.

Ophitic texture: A term applied to the texture of dolerites in which feldspar (plagioclase) crystals are enclosed within grains of pyroxene, chiefly augite.

Ordovician: The Palaeozoic system between the Cambrian and the Silurian that began about 505 million years ago.

Orogeny: A period or process of mountain-building.

Orthoclase: Potassic feldspar; potassium aluminium silicate.

Palaeo-environment: An ancient environment; the environmental conditions at some period in the geological history of a part of the earth's crust.

Palaeontology: The scientific study of fossils.

Palaeozoic: The period of geological time following the pre-Cambrian, encompassing (in Britain) the time spans of the Cambrian, Ordovician, Silurian, Devonian, and Permian systems, and beginning about 590 millions years ago.

Pegmatite: Coarse-grained igneous rocks, usually occurring as dykes, found associated with large masses of igneous rock of somewhat finer grain. Pegmatites may contain large, well-formed crystals, sometimes of rare minerals.

Permian: The geological system following the Carboniferous, beginning about 286 million years ago.

Petrography: The branch of science dealing with the systematic description and classification of rocks.

Phenocrysts: The larger, generally well-formed, cyrstals, set in a fine-grained matrix or groundmass, of porphyritic igneous rocks. The phenocrysts are the first crystals to form as the melt cools during the formation of the rock.

Photomicrograph: An enlarged photograph of a microscopic object such as a rock in thin section, taken by attaching a camera to a microscope.

Pipe rock: The name given to a type of quartzite found in the Northwest Highlands of Scotland, containing vertical cylindrical bodies believed to result from the in-filling of worm tubes.

Plagioclase: The series of sodium/calcium aluminium silicates forming an important group of feldspars: albite, oligoclase, andesine, labradorite, bytownite, anorthite.

Plutonic: Relating to the earth's depths, Plutonic rocks, such as granite, are formed by slow cooling and crystalization at depth, with the consequent formation of coarse-grained rock.

Plutonism: A term given to the theories of Hutton (see chap. 2), which held that rocks such as granites were formed by the cooling and consolidation of magma deep under the surface of the earth; also that basalt is igneous in origin. (Cf. Neptunism.)

Polarized light: Light in which the electromagnetic vibrations are constrained. Thus, in plane-polarized light, the vibrations are confined to a single plane.

Polarizer: The Nicol prism (or Polaroid filter) of a petrographic microscope through which light passes before reaching the thin section being examined. It is that part of the microscope which produces the polarized light required for the examination of minerals in thin section.

Polarizing microscope: A microscope used in petrography, provided with polarizer and analyzer, and with a rotating stage on which thin sections can be mounted for examination. Minerals capable of doubly refracting light display characteristic colors and appearances when viewed in this way and these features can be diagnostic.

Porphyry: A medium- or fine-grained igneous rock consisting of larger well-formed crystals in a matrix or groundmass of finer grain.

Pre-Cambrian: The immense period of geological time prior to the Cambrian; also, the rocks formed in that time. The pre-Cambrian rocks, which are generally much altered (metamorphosed), were formerly thought to be devoid of fossils, and so antedated the beginning of life on earth; but fossil remains of organisms without hard parts have now been found in the pre-Cambrian.

Primary: An obsolete term, at one time used to designate what is now called the pre-Cambrian; then extended to include the Palaeozoic; and then used to refer to what is now called the Palaeozoic. (Thus at one time the Palaeozoic, Mesozoic, and Tertiary eras were called Primary, Secondary, and Tertiary.)

Primitive rocks: An obsolete term, formerly used to refer to rocks that were believed to be those first formed on the earth—such as the ancient gneisses of northwest Scotland.

Principle of superposition: The assumption that in a sequence of sedimentary layers the ones lying above are younger than those lying below.

Pyroxene: A family of generally dark-colored ferromagnesian minerals, the commonest being augite, found in basic igneous rocks. The pyroxenes commonly have two cleavage systems, approximately perpendicular to one another, which feature is useful for recognition in hand specimens. (Cf. amphiboles.)

Quartz: The commonest naturally occurring variety of silica (silicon dioxide). Well-shaped clear crystals are called "rock crystal." Quartz is an essential component of acidic igneous rocks, and sand grains are commonly fragments of the mineral.

Quartz rock: A name formerly used for an important stratigraphic unit in the rocks of the Northwest Highlands: a commonly brilliant white quartzite now regarded as belonging to the Cambrian system.

Quartzite: Either a metamorphic rock in which the quartz grains of a sandstone have recrystallized under heat and pressure to form an interlocking mass or a quartz-rich sandstone that has been cemented by silica that has grown in optical continuity round each fragment.

Recumbent fold: An asymmetrical fold in which the crest is turned over such that it lies on its side—that is, such that its axial plane lies toward the horizon.

Reverse fault: A fault arising from the action of compressive forces, such that the movement of strata on one side of the fault has been upward. (Cf. normal fault.)

Roche moutonée: A glacial landscape form consisting of a resistant rock mass carved so that it has a gentle slope on the side that formerly faced the advancing glacier and a steep slope on the other side, down-valley.

Rutile: A dark mineral composed of titanium oxide, common as an accessory mineral in igneous and metamorphic rocks.

Salterella: See Serpulite.

Schist: Medium- to coarse-grained metamorphic rock with strong foliations, fissile from the preferred orientation of abundant inequidimensional mineral grains such as micas (flakes) or amphiboles (rods).

Secondary rocks: At one time used as an alternative term for Werner's *Floetz* series (the stratified rocks). Also used to refer to Mesozoic rocks. Now obsolete.

Sericite: A fine-grained type of white mica occurring in schists, usually near muscovite in composition.

Serpentine: A hydrous magnesium silicate, usually dull green or white. Found in altered basic and ultrabasic rocks where it is produced by the breakdown of olivines and pyroxenes. The name derives from a supposed resemblance of serpentine rocks to a snake skin.

Serpulite: A form of marine worm that lived in a tortuous calcareous tube. In the Northwest Highlands, these were thought to occur in the unit formerly called "Serpulite grit." In the original description of the unit, Murchison regarded the fossil as *Serpulites maccullochi*: hence the name of the unit. Later, the fossil remains were construed as belonging to a mollusc *Salterella,* rather than a worm; and modern texts prefer the term "Salterella grit."

Shale: A laminated fine-grained sediment.

Sillimanite: A white orthorhombic aluminium silicate produced in aluminous rocks under conditions of high temperature and pressure in metamorphism. (Cf. kyanite.)

Silurian: The Palaeozoic system between the Ordovician and the Devonian, beginning about 438 million years ago.

Slate: A fine-grained, usually dark, metamorphic rock with a cleavage due to the preferred orientation of flaky minerals such as mica and chlorite.

Slickenside: A polished and striated fault surface.

Solid map: A geological map prepared to show the distribution of rocks that may be covered by superficial deposits such as river gravels, moraines, peat, etc.

Spherulite: A growth of radiating needle-shaped crystals.

Spherulitic: A texture found in igneous rocks where spherulites occur.

Spinel: A family of oxide minerals, of aluminium and magnesium, occurring in high-temperature metamorphic rocks. (Some forms are used as gems.)

Stratification: The character of sedimentary rocks produced by the processes of deposition: beds, laminae, lenses, wedges, etc.

Stratigraphy: The branch of geology that deals with the formation, sequence, composition, and correlation of strata.

Stratum: A part of a formation consisting of approximately the same kind of rock material throughout; or a bed.

Strike: The direction or bearing of a horizontal line on an inclined bed or other structural plane (e.g., a fault) at right-angles to the direction of the dip. Equivalent to a contour line on a bed, etc.

Syenite: A coarse-grained igneous rock of intermediate acid/base composition, containing alkalic feldspar such as orthoclase, and some dark minerals such as horneblende or brown mica. Small amounts of quartz and plagioclase may also be present.

Syncline: A fold in strata such that the concave side is uppermost.

Taconic: A geological system proposed in America by Emmons, corresponding approximately to the Cambrian system. Now obsolete.

Talc: A soft, greasy hydrated magnesium silicate found as a result of hydrothermal alteration of basic and ultrabasic rocks.

Tertiary: A name that has survived from the old division of rocks into Primary, Secondary, and Tertiary, now used to refer to the sum of the following epochs: Palaeocene, Eocene, Oligocene, Miocene, and Pliocene. Coming after the Cretaceous, the Tertiary period began about 65 million years ago and lasted about 63 million years.

Thrust fault: A low-angle reverse fault.

Torridonian: As used in this book, an informal term for the red-brown clastic rocks of western Scotland, of the late pre-Cambrian age. More precisely, it is used to refer to the period from about 1,400 to 700 million years ago when the Torridon Sandstone was formed and partly converted to the Moine rocks.

Transition series: In the nomenclature of Werner (see chap. 2) rocks (now designated as Cambrian, Ordovician, and Silurian) made of hard limestones, traps, and greywackes, and constituting the first orderly deposits supposedly formed worldwide from a universal ocean; that is, lying between the crystalline "primitive rocks" such as granite and gneiss and the lowest *Floetz* rock (the Old Red Sandstone).

Trap rock: A term derived from the German *Trappe,* meaning "stair." In some places, successive lava flows may appear as steps or stairs; hence basalts, etc., were formerly referred to as "trap rock."

Triassic: The first period of the Mesozoic era, between the Permian and the Jurassic systems, and beginning about 248 million years ago. In Britain, Triassic rocks form part of the New Red Sandstone.

Twinning of crystals: The arrangement of two or more grains of the same crystalline species, whereby the grains are regularly oriented against each other with respect to a common lattice plane.

Unconformity: A surface of erosion separating younger from older strata, with the older beds generally dipping more steeply than the younger ones.

Uniformitarianism: A term with a variety of meanings including the methodological principle that "the present is the key to the past," the assumption that the laws of nature are uniform, the hypothesis that conditions and processes in the past have been essentially the same as at the present, and that geological changes occur slowly and steadily without sudden discontinuities or "catastrophes."

Vulcanist: An upholder of the doctrines of Hutton regarding the Plutonic origin rocks; and of the view that basalts were igneous in origin, not deposited from solution as envisaged by Werner (see chap. 2).

Window: A patch of rock(s) lying beneath a thrust or recumbent fold, and exposed by erosion. Also called a "fenster."

Xenolith: An inclusion of "foreign" rock in an igneous body, such as a block of "country rock" caught up in an intrusion.

Zone: A group of beds, characterized by a particular assemblage of fossils that together define the biostratigraphical subdivision.

Bibliography

Manuscripts

Bonney, Thomas G. Correspondence. Geological Society of London.
British Geological Survey. General archive at Keyworth, Nottingham; Archive of Scottish Branch, Edinburgh; Public administrative documents at Public Record Office (DSIR).
Cadell, Henry M. Papers. British Geological Survey (Edinburgh).
Cunningham, Robert J. S. Hand-colored map of the geology of Sutherland. Royal Scottish Museum.
Forbes, James. Correspondence. St. Andrews University Library.
Geikie, Archibald. Field notebooks. Haslemere Educational Museum; Edinburgh University Library; British Geological Survey (Edinburgh).
———. Correspondence and papers. British Geological Survey; British Library; Edinburgh University; Geological Society of London; National Library of Scotland; National Museum of Scotland; Royal Society.
Gordon, George. Correspondence. Elgin Museum, Elgin.
Lapworth, Charles. Correspondence, papers, and field maps. Birmingham University.
Mérimée, Prosper. Cartoon of Murchison and Geikie. National Library of Scotland.
Murchison, Roderick I. Correspondence and papers. British Geological Survey; Edinburgh University; Elgin Museum; Geological Society of London.
Nicol, James. Correspondence. Elgin Museum; Geological Society of London.
———. Diary, 1840–41, Edinburgh University Library.
Ramsay, Andrew C. Correpondence and papers. British Geological Survey; Imperial College, London.
Scrope, George P. Correspondence, Geological Society of London.

Printed Sources

Aristotle. 1961. *The metaphysics* . . . With an English translation by Hugh Tredennick. London and Cambridge: William Heinemann and Harvard University Press.
Bacon, Francis. [1620] 1960. *The new organon and related writings.* Edited by F. H. Anderson. Indianapolis: Bobbs-Merrill.
Barber, A. J., et al. 1978. *The Lewisian and Torridonian rocks of north-west Scotland.* London: The Geologists' Association.
Barrande, Joachim. 1852–1911. *Système Silurien du centre de la Bohème.* 8 vols. (in 29 books). Prague and Paris: The Author.

————. 1859–60. Colonies dans le bassin Silurien de la Bohème. *Bulletin de la Société Géologique de France* 17:602–66.

Barrow, George. 1893. On an intrusion of muscovite-biotite-gneiss in the south-eastern Highlands of Scotland and its accompanying metamorphism. *Quarterly Journal of the Geological Society of London* 49:330–53.

————. 1895. On the origin of the crystalline schists, with special reference to the southern Highlands. *Proceedings of the Geologists' Association* 13:48–49.

Bertrand, Marcel A. 1883–84. Rapports de structure des Alpes de Glaris et du bassin houiller du nord. *Bulletin de la Société Géologique de France,* series 3, 12:318–30.

————. 1893. The mountains of Scotland. *Geological Magazine,* decade 3, 10:118–29.

Blake, John F. 1883–84. The North-west Highlands and their teachings. *Proceedings of the Geologists' Association* 8:419–37.

————. 1893. On the still possible Cambrian age of the Torridon Sandstone. *Report of the Sixty-second Meeting of the British Association for the Advancement of Science; held at Edinburgh in August 1892,* 713. London: John Murray.

Bonney, Thomas G. 1879. The pre-Cambrian rocks of Great Britain. *Proceedings of the Philosophical Society of Birmingham* 1:140–59.

————. 1880a. Petrological notes on the vicinity of the upper part of Loch Maree. *Quarterly Journal of the Geological Society of London* 36:93–108.

————. 1880b. The "pre-Cambrian" rocks of Ross-shire. *Geological Magazine,* decade 2, 7:329–30.

————. 1885. The anniversary address of the president. *Proceedings of the Geological Society of London* 41:37–96.

Bonney, Thomas G., and Edwin Hill. 1877–80. On the pre-carboniferous rocks of Charnwood Forest. *Quarterly Journal of the Geological Society of London* 33:754–89; 34:199–239; 36:337–50.

Boué, Aimé. 1820. *Éssai géologique sur l'Écosse.* Paris: Courcier.

Brøgger, Waldemar C. 1882. *Die Silurischen Etagen 2 und 3 im Kristianagebiet und auf Eker, ihre Gliederung, Fossilien, Schichtenstörungen und Contactmetamorphosen: Universitätsprogramm für 2, Sem. 1882.* Kristiana [Oslo]: A. W. Brøgger.

Burnet, Thomas. 1684. *The theory of the earth: Containing an account of the origin of the earth and all of the general changes which it hath already undergone, or is to undergo till the consummation of all things. The two first books concerning the Deluge, and concerning Paradise.* London: R. Norton for W. Kettilby.

Cadell, Henry M. 1888. Experimental researches in mountain building. *Transactions of the Royal Society of Edinburgh* 35:337–57.

————. 1895. The scenery of Sutherland. *Scottish Geographical Magazine* 11:385–93.

————. 1896. *The geology and scenery of Scotland.* Edinburgh: David Douglas.

Callaway, Charles. 1879–82. The pre-Cambrian rocks of Shropshire. *Quarterly Journal of the Geological Society of London* 35:643–62; 38:119–23.

————. 1881. The limestone of Durness and Assynt. *Quarterly Journal of the Geological Society of London* 37:239–45.

————. 1882. The Torridon Sandstone in relation to the Ordovician rocks of the northern Highlands. *Quarterly Journal of the Geological Society of London* 38:114–18.

————. 1883a. The Highland problem. *Geological Magazine,* decade 2, 10:139–40.

————. 1883b. The age of the newer gneissic rocks of the northern Highlands . . . With notes on the lithology by Prof. T. G. Bonney . . . *Quarterly Journal of the Geological Society of London* 39:355–422.

————. 1884. The Archaean and Lower Palaeozoic rocks of Anglesey. *Quarterly Journal of the Geological Society of London* 40:567–83.

————. 1885. A plea for comparative lithology. *Geological Magazine*, decade 3, 2 : 258–64.

————. 1889. The present state of the Archaean controversy in Britain. *Geological Magazine*, decade 3, 6 : 319–25.

Craig, Gordon Y. 1983. *Geology of Scotland*. 2d ed. Edinburgh: Scottish Academic Press.

Craig, Gordon Y., and E. K. Walton. 1959. Sequence and structure in the Silurian rocks of Kirkcudbrightshire. *Geological Magazine* 96 : 209–20.

Croll, James. 1875. *Climate and time in their geological relations: A theory of secular changes of the earth's climate*. London: Daldy, Isbister and Co.

Crosskey, Henry W. 1865. On a section near Inch-na-damff. *Transactions of the Geological Society of Glasgow* 2 : 19.

Cunningham, Robert J. H. 1841. Geognostic account of the County of Sutherland. *Transactions of the Highland and Agricultural Society of Scotland*, n.s. 7 : 73–114 and map.

Darwin, Charles R. [1890.] *On the structure and distribution of coral reefs: also geological observations on the volcanic islands and parts of South America visited during the voyage of H.M.S. Beagle* . . . London: Ward, Lock and Bowden.

De Lapparent, Auguste-A. C. 1883. *Traité de géologie*. Paris: F. Savy.

Dinner to Sir Archibald Geikie. 1901. *Nature* 64 : 34–36.

Dixon, John H. 1886. *Gairloch in north-west Ross-shire: Its records, traditions, inhabitants, and natural history with a guide to Gairloch and Loch Maree and a map and illustrations*. Edinburgh: Co-operative Printing Co.

Élie de Beaumont, J. B. A. L. Léonce. 1852. *Notice sur les système des montagnes*. 3 vols. Paris: Bertrand.

Emmons, Ebenezer. 1844. *The Taconic system: Based on observations in New York, Massachusetts, Maine, Vermont and Rhode Island*. Albany: Carroll and Cook.

[Escher von der Linth, Arnold.] 1841. Geologische und mineralogische Section. Erste Sitzung. Dinstags den 3. August. *Verhandlungen der Schweizerischen Naturforschend Gesellschaft*, 53–73.

[Geikie, Archibald.] 1861. Art. 6: 1. *Geological map of Scotland*, by John Macculloch, M.D. London (1832) [*sic*]. 2. *Geological map of Scotland*, by James Nicol, F.R.S.E. (Edinburgh: W. and A. K. Johnston). 3. *Geological map of Scotland*, by J. A. Knipe, F.G.S. (London: Stanford). 4. *Palaeontological map of the British Isles*, by Edward Forbes, F.R.S. (Edinburgh: W. and A. K. Johnston). 5. *Geological Survey of Scotland* (Sheets 32 and 33). 6. *Memoirs of the Geological Survey: Geology of the neighbourhood of Edinburgh* (1861). 7. *First sketch of a new geological map of Scotland*, by Sir R. I. Murchison, F.R.S., and Archibald Geikie, F.R.S.E. (Edinburgh: W. and A. K. Johnston). 8. *Quarterly Journal of the Geological Society of London*, vols. from 1850 to 1861. *North British Review* 35 : 125–56.

————. 1867. Siluria. *Pall Mall Gazette*, Dec. 9, p. 11.

————. 1868a. Siluria. *The Times*, Jan. 17.

————. 1868b. Art. 7: *Siluria, a history of the oldest rocks in the British Isles and other countries*, by Sir Roderick I. Murchison, Bart, K.C.B., 4th [i.e., 3d] ed. (London, 1867). *Quarterly Review* 125 : 188–217.

Geikie, Archibald. 1855. [Remarks accompanying an] Exhibition of a collection of Liassic fossils from Pabba and Skye. *Edinburgh New Philosophical Journal*, n.s. 1 : 366–68.

————. 1858a. The geology of Strath, Skye. *Quarterly Journal of the Geological Society of London* 14 : 1–23.

————. 1858b. *The story of a boulder, or, Gleanings from the note-book of a field geologist*. Edinburgh and London: Constable and Hamilton, Adams and Co.

————. 1861a. My first geological excursion: A chapter of geology for boys. *Good Words* 2 : 8–11.

————. 1861b. Lessons in the lime quarry. *Good Words* 2 : 73–76.

————. 1861c. The old sea-country. *Good Words* 2 : 211–13.

————. 1863a. *The geology of eastern Berwickshire (Map 34).* London: Her Majesty's Stationery Office.

————. 1863b. *On the phenomena of the glacial drift of Scotland.* Glasgow: John Gray.

————. 1864. *Outlines of the geology of the British Isles to accompany the geological map.* Edinburgh: W. and A. K. Johnston.

————. 1865. *The scenery of Scotland viewed in connection with its physical geology . . . With a geological map by Sir Roderick I. Murchison and Archibald Geikie.* Macmillan: London and Cambridge.

————. 1868. On the order of succession among the Silurian rocks of Scotland. *Transactions of the Geological Society of Glasgow* 3 : 74–95.

————. 1869–70. Opening address [to the Thirty-sixth Anniversary meeting of the Edinburgh Geological Society]. *Transactions of the Edinburgh Geological Society* 2 : 1–13.

————. 1871a. The Tertiary volcanic rocks of the British Isles. *Quarterly Journal of the Geological Society of London* 27 : 279–310.

————. 1871b. *Memoirs of the Geological Survey, Scotland. Explanation of Sheet 15. Dumfriesshire (north-west part); Lanarkshire (south part); Ayrshire (south-east part).* Edinburgh: Her Majesty's Stationery Office.

————. 1873. *Memoirs of the Geological Survey, Scotland. Explanation of Sheet 3. Western Wigtownshire.* Edinburgh: Her Majesty's Stationery Office.

————. 1877. *Mountain architecture: A lecture delivered in the City Hall, Glasgow: under the auspices of the Glasgow Science Lectures Association, on Thursday, 27th January 1876.* London and Glasgow: William Collins.

————. 1879. Geology. *Encyclopaedia Britannica,* 9th ed., 10 : 212–375. Edinburgh: Adam and Charles Black.

————. 1880a. The lava-fields of north-western Europe. *Nature* 23 : 3–5.

————. 1880b. A fragment of primeval Europe. *Nature* 22 : 400–403.

————. 1880c. Rock weathering, as illustrated in Edinburgh church yards. *Proceedings of the Royal Society of Edinburgh* 10 : 518–32.

————. 1882a. *Text-book of geology.* London: Macmillan.

————. 1882b. *Geological sketches at home and abroad.* London: Macmillan.

————. 1883. The supposed pre-Cambrian rocks of St. David's *Quarterly Journal of the Geological Society of London* 39 : 261–326.

————. 1884a. Report of the director-general of (1) the Geological Survey of the United Kingdom, (2) the Museum of Practical Geology, and (3) the Mining Records Office. *Thirty-first Report of the Science and Art Department of the Committee of Council on Education, with Appendices,* 282–96. London: Eyre and Spottiswoode.

————. 1884b. The origin of the crystalline schists. *Nature* 30 : 121–23.

————. 1884c. The crystalline schists of the Scottish Highlands. *Nature* 31 : 29–31.

————. 1885a. Report of the director-general of the Geological Survey of the United Kingdon, and of the Museum of Practical Geology. *Thirty-second Report of the Science and Art Department of the Committee of Council on Education, with Appendices,* 240–54. London: Eyre and Spottiswoode.

————. 1885b. *Text-book of geology.* 2d ed. London: Macmillan.

————. 1886. Report of the director-general of the Geological Survey of the United Kingdom, and of the Museum of Practical Geology. *Thirty-third Report of the Science* and Art Department of the Committee of Council on Education, with Appendices, 324–39 London: Eyre and Spottiswoode.

———. 1890. The history of volcanic action during the Tertiary period in the British Isles. *Transactions of the Royal Society of Edinburgh* 35:21–184.

———. 1892a. Discovery of the Olenellus-zone in the North-west Highlands. *Report of the Sixty-first Meeting of the British Association for the Advancement of Science; held at Cardiff in August 1891*, 633–34. London: John Murray.

———. 1892b. Annual report of the Geological Survey and Museum of Practical Geology for the year ending December 31, 1891. *Thirty-ninth Report of the Department of Science and Art of the Committee of Council on Education, with Appendices*, 381–94. London: Eyre and Spottiswoode for Her Majesty's Stationery Office.

———. 1892c. *Geological sketches at home and abroad.* 2d ed. New York: Macmillan.

———. 1893. The geology of the North-west Highlands. *Nature* 47:292–93.

———. 1894. On the relations of the basic and acid rocks of the Tertiary volcanic series of the Inner Hebrides. *Quarterly Journal of the Geological Society of London* 50:212–31.

———. 1897. *The ancient volcanoes of Great Britain.* 2 vols. London and New York: Macmillan.

———. 1898. *Summary of progress of the Geological Survey of the United Kingdom for 1897. With an introduction regarding the history, organization, and work of the Survey.* London: Her Majesty's Stationery Office.

———. 1900. *Outlines of field-geology.* 5th ed. London and New York: Macmillan.

———. 1902. *The geology of eastern Fife: Being a description of Sheet 41 and parts of Sheets 40, 48, and 49 of the Geological Map.* Glasgow: His Majesty's Stationery Office.

———. 1903. *Text-book of geology.* 4th ed. 2 vols. London and New York: Macmillan.

———. 1904. *Scottish reminiscences.* Glasgow: James Maclehose.

———. 1905. *Landscape in history and other essays.* London and New York: Macmillan.

———. 1912. *The love of nature among the Romans during the latter decades of the Republic and the first century of the Empire.* London: John Murray.

———. 1916. *The birds of Shakespeare.* Glasgow: James Maclehose.

———. 1924. *A long life's work: An autobiography.* London: Macmillan.

Geikie, Archibald, William Gunn, Benjamin N. Peach, and Alfred Harker. 1903. *The geology of North Arran, South Bute, and the Cumbraes, with parts of Ayrshire and Kintyre (Sheet 21, Scotland); the description of North Arran, South Bute, and the Cumbraes by W. Gunn; part of Ayrshire by Sir A. Geikie; part of Kintyre by B. N. Peach; with chapters on the petrography of the Tertiary igneous rocks of Arran, South Bute and the Cumbrae Islands by A. Harker.* Glasgow: His Majesty's Stationery Office.

Geikie, Archibald, and Jethro J. H. Teall. 1894. On the banded structure of some Tertiary gabbros in the Isle of Skye. *Quarterly Journal of the Geological Society of London* 50:645–60.

The Geological Survey. 1881. *Geological Magazine*, decade 2, 8:39–43.

Geologists' Association. 1885. Societies and academies London: Geologists' Association, Jan. 2. *Nature* 31:258–59.

Gosselet, Jules. 1880. Sur la structure générale du bassin houiller franco-belge. *Bulletin de la Société Géologique de France* 8:505–11.

Green, John F. N. 1935. The Moines. *Proceedings of the Geological Society of London* 91:lxiv–lxxxiv.

Harker, Alfred. 1895. *Petrology for students: An introduction to the study of rocks under the microscope.* Cambridge: Cambridge University Press.

———. 1909. *The natural history of igneous rocks.* London: Methuen.

————. 1932. *Metamorphism: A study of the transformation of rock masses*. London: Methuen.

Harker, Alfred, and Charles T. Clough. 1904. *The Tertiary igneous rocks of Skye*. Glasgow: His Majesty's Stationery Office.

Harkness, Robert. 1851. On the Silurian rocks of Dumfriesshire and Kirkcudbrightshire. *Quarterly Journal of the Geological Society of London* 7:46–58.

————. 1856. On the lowest sedimentary rocks of the south of Scotland. *Quarterly Journal of the Geological Society of London* 12:238–45.

————. 1861. On the rocks of portions of the Highlands of Scotland south of the Caledonian Canal; and their equivalents in the north of Ireland. *Quarterly Journal of the Geological Society of London* 17:256–71.

————. 1873. The Southern Uplands of Scotland [being a review of] Memoirs of the Geological Survey of Scotland, Sheets 1, 2, 3 and 15, &c. Explanations of, 1871, 1872, 1873. *Nature* 9:22–24, 57–59.

Heddle, M. Forster. 1879–84. The county geognosy and mineralogy of Scotland. *Mineralogical Magazine and Journal of the Mineralogical Society* 2:9–35, 106–33, 155–90; 3:18–56, 147–77, 219–51; 4:21–35, 135–80, 197–254; 5:71–106, 133–89, 217–63, 271–324.

————. 1882. Description of the geological map of Sutherland. *Mineralogical Magazine and Journal of the Mineralogical Society* 5:41–48.

————. 1901. *The mineralogy of Scotland*. Edited by John G. Goodchild. 2 vols. Edinburgh: David Douglas.

Heim, Albert. 1878. *Untersuchungen über den Mechanismus der Gebirgsbildung: Im Anschluss an die geologische Monographie der Tödi-Windgällen-Gruppe*. 2 vols. and atlas. Basle: Benno Schwabe.

————. 1919–21. *Geologie der Schweiz*. 3 vols. Leipzig: C. H. Tauchnitz.

Hicks, Henry. 1872. On the classification of Cambrian and Silurian rocks. *Geological Magazine*, decade 1, 9:383–84.

————. 1873–86. The "pre-Cambrian" and Lower Palaeozoic rocks of Pembrokeshire, particularly those in the neighbourhood of St. David's [collective title]. *Quarterly Journal of the Geological Society of London* 29:39–52; 31:167–95; 33:229–41; 34:153–69; 35:285–94; 40:507–60; 42:351–56.

————. 1878. On the metamorphic and overlying rocks in the neighbourhood of Loch Maree, Ross-shire. *Quarterly Journal of the Geological Society of London* 34:811–18.

————. 1880a. On the pre-Cambrian rocks of west and central Ross-shire. With petrological notes by T. Davies, F.G.S., of the British Museum. *Geological Magazine*, decade 2, 7:103–9, 155–66, 222–26, 266–71.

————. 1880b. On the pre-Cambrian rocks of the north-western and central Highlands of Scotland. *Proceedings of the Geological Society of London* 36:101–3.

————. 1880c. Pre-Cambrian volcanoes and glaciers. *Geological Magazine*, decade 2, 7:488–91.

————. 1881a. The pre-Cambrian rocks of Britain and Bohemia. *Geological Magazine*, decade 2, 8:142–44.

————. 1881b. On some recent researches among pre-Cambrian rocks in the British Isles. *Proceedings of the Geologists' Association* 7:59–87.

————. 1885–86. On some recent views concerning the geology of the north-west of Scotland. *Proceedings of the Geologists' Association* 9:43–65.

Hicks, Henry, and Robert Harkness. 1871. On the ancient rocks of St. David's promontory, South Wales and their fossil contents. *Quarterly Journal of the Geological Society of London* 27:384–404.

Hicks, Henry, and John W. Salter. 1867. Second report on the "Menevian Group" and the other formations of St. David's Pembrokeshire. *Report of the Thirty-sixth Meeting of the British Association for the Advancement of Science; held at Nottingham in August 1866*, 182–86. London: John Murray.

Horne, John. 1886. The origin of the andalusite-schists of Aberdeenshire. *Mineralogical Magazine and Journal of the Mineralogical Society* 6:98–100.

———. 1907. Previous literature relating to the history of the region described in this memoir. In *The geological structure of the North-west Highlands of Scotland*, edited by A. Geikie, 11–32. Glasgow: His Majesty's Stationery Office.

———. 1915. Report of an excursion to the Assynt district of the North-west Highlands. *Proceedings of the Geologists' Association* 26:127–37.

Horne, John, and Jethro J. H. Teall. 1895. On borolanite—an igneous rock intrusive in the Cambrian limestone of Assynt, Sutherlandshire, and the Torridon Sandstone of Ross-shire. *Transactions of the Royal Society of Edinburgh* 37:163–78 and plates.

Howell, Herbert H., and Archibald Geikie. 1861. *The geology of the neighbourhood of Edinburgh*. London: Her Majesty's Stationery Office.

Hudleston, Wilfred H. 1879–80. On the controversy respecting the gneiss rocks of the North-west Highlands. *Proceedings of the Geologists' Association* 6:47–79.

———. 1882. First impressions of Assynt. *Geological Magazine*, decade 2, 9:390–99.

Hull, Edward. 1881. The geological age of the North Highlands of Scotland. *Nature* 23:289–90.

———. 1882–84. On the geological structure of the northern Highlands of Scotland; being notes of a recent tour. *Journal of the Royal Geological Society of Ireland*, n.s. 6:56–68.

Hunt, T. Sterry. 1875. *Chemical and geological essays*. Boston and London: James Osgood and Trübner and Co.

———. 1883. A historical account of the Taconic question in geology, with a discussion of the relations of the Taconic Series to the Older Crystalline and to the Cambrian rocks. *Transactions of the Royal Society of Canada* 1:217–70.

Jameson, Robert. 1805. *A mineralogical description of the County of Dumfries*. Edinburgh: Bell and Bradfute, and Blackwood.

Johnson, M. R. W. 1983. Torridonian-Moine. In *Geology of Scotland*. edited by Gordon Y. Craig, 49–75. 2d ed. Edinburgh: Scottish Academic Press.

Judd, John W. 1874. The Secondary rocks of Scotland. Second paper. On the ancient volcanoes of the Highlands and the relations of their products to the Mesozoic strata. *Quarterly Journal of the Geological Society of London* 30:220–302.

———. 1886. Presidential address to the Geology Section of the British Association. *Report of the Fifty-fifth Meeting of the British Association for the Advancement of Science; held at Aberdeen in September 1885*, 994–1013. London: John Murray.

———. 1889. The Tertiary volcanoes of the Western Isles of Scotland. *Quarterly Journal of the Geological Society of London* 45:175–86.

———. 1890a. The propylites of the Western Isles of Scotland and their relation to the andesites and diorites. *Quarterly Journal of the Geological Society of London* 46:341–85.

———. 1890b. Critical introduction [to *Geological observations on South America*]. In Charles Darwin, *On the structure and distribution of coral reefs: Also geological observations on the volcanic islands and parts of South America visited during the voyage of H.M.S. Beagle . . . with maps, plates, and numerous illustrations. And a critical introduction to each work by Prof. John W. Judd F.R.S.*, 276. London: Ward, Lock and Bowden.

———. 1893. On inclusions of Tertiary granite in the gabbros of the Cuillin Hills,

Skye, and on the products resulting from the partial fusion of the acid by the basic rock. *Quarterly Journal of the Geological Society of London* 49: 175–95.

Kayser, E. 1893. *Text-book of comparative geology.* Translated and edited by Philip Lake. London and New York: Swan Sonnenschein and Macmillan.

King, W., and T. H. Rowney. 1881. *An old chapter of the geological record with a new interpretation: Or, rock metamorphism (especially the methylosed kind) and its resultant imitations of organisms. With an introduction giving an annotated history of the controversy on the socalled "Eozoon Canadense." And an appendix.* London: John von Hoorst.

Lapworth, Charles. 1870. On the Lower Silurian rocks of Galashiels. *Geological Magazine,* decade 1, 7: 204–9, 279–84.

———. 1878. The Moffat Series. *Quarterly Journal of the Geological Society of London* 34: 240–346.

———. 1879. On the tripartite classification of the Lower Palaeozoic rocks. *Geological Magazine,* decade 2, 6: 1–15.

———. 1882a. Recent discoveries among the Silurians of South Scotland. *Transactions of the Geological Society of Glasgow* 6: 78–84.

———. 1882b. The Girvan succession. *Quarterly Journal of the Geological Society of London* 38: 537–666.

———. 1882c. History of the discovery of Cambrian rocks in the neighborhood of Birmingham. *Geological Magazine,* decade 2, 9: 563–66.

———. 1883a. History of the discovery of Cambrian rocks in the neighbourhood of Birmingham. *Proceedings of the Birmingham Philosophical Society* 3: 234–38.

———. 1883b. Review of *Text-book of geology,* by Archibald Geikie. *Geological Magazine,* decade 2, 10: 39–42, 80–86.

———. 1883c. The secret of the Highlands. *Geological Magazine,* decade 2, 10: 120–28, 193–99, 337–44.

———. 1883–84. On the structure and metamorphism of the rocks of the Durness-Eriboll district. *Proceedings of the Geologists' Association* 8: 438–42.

———. 1885. On the close of the Highlands controversy. *Geological Magazine,* decade 3, 2: 97–106.

———. 1886. The Highlands controversy in British geology: Its causes, course, and consequences. *Report of the Fifty-fifth Meeting of the British Association for the Advancement of Science: held at Aberdeen in September 1885.* 1025–27. London: John Murray.

———. 1888. On the discovery of the Olenellus fauna in the Lower Cambrian rocks of Britain. *Geological Magazine,* decade 3, 5: 484–87.

———. 1889. On the Ballantrae rocks of South Scotland and their place in the Upland sequence. *Geological Magazine* 26: 20–24, 59–69.

———. 1899. The Survey memoir on the Scottish Uplands. *Geological Magazine,* decade 4, 6: 472–79, 510–20.

———. 1902. The place of geology in economics and education. *Transactions of the Liverpool Biological Society* 16: 485–504.

Lapworth, Charles, and James Wilson. 1871. On the Silurian rocks of the counties of Roxburgh and Selkirk. *Geological Magazine,* decade 1, 8: 456–64.

———. 1872. Note on the results of some recent researches among the graptolitic black shales of the south of Scotland. *Geological Magazine,* decade 1, 9: 533–35.

Lehmann, Johannes Georg. 1884. *Untersuchungen über die Entstehung der Altkrystallinischen Schiefergestein mit besonderer Bezugnahme auf das Sächsische Granulitgebirge, Erzgebirge, Fichtelgebirge und Bairisch-Böhmisch Grenzgebirge.* Bonn: M. Hochgürtel.

Logan, William E. 1852. On the foot-prints occurring in the Potsdam Sandstone of Canada. *Quarterly Journal of the Geological Society of London* 8: 199–213.

Lugeon, Maurice. 1901. Les grandes nappes de recouvrement des Alpes du Chablis et de la Suisse. *Bulletin de la Société Géologique de France*, series 4, 1:723–825.

Lyell, Charles. 1866. *Elements of geology: Or, the ancient changes of the earth and its inhabitants as illustrated by geological monuments.* 5th ed. New York: Appleton.

Macconochie, Arthur. 1881–84. Review of the southern Silurian question. *Transactions of the Geological Society of Glasgow* 7:370–72, 431.

Macculloch, John. 1814. Remarks on several parts of Scotland which exhibit quartz rock, and on the nature and connexions of this rock in general. *Transactions of the Geological Society of London* 2:450–87.

———. 1819. *A description of the Western Islands of Scotland including the Isle of Man: With remarks on their agriculture, scenery and antiquities.* 3 vols. London: Constable.

———. 1821. *A geological classification of rocks with descriptive synopses of the species and varieties comprising the elements of practical geology.* London: Longman, Hurst, Rees, Orme and Brown.

———. 1822. Supplementary remarks on quartz rock. *Transactions of the Geological Society of London*, n.s. 1:53–60.

———. 1831. *A system of geology, with a theory of the earth and an explanation of its connection with ancient records.* 2 vols. London: Longman, Rees, Orme, Brown and Green.

———. 1836. *Memoirs to His Majesty's Treasury respecting a geological survey of Scotland.* London: S. Arrowsmith.

McIntyre, Donald B. 1954. The Moine Thrust: Its discovery, age, and tectonic significance. *Proceedings of the Geologists' Association* 65:203–23.

Miller, Hugh. 1841. *The Old Red Sandstone; or, New walks in an old field.* Edinburgh: J. Johnstone.

———. 1854. *My schools and schoolmasters; or, The story of my education.* 3d ed. Edinburgh: Johnstone and Hunter.

———. 1857. *The testimony of the rocks; or, Geology in its bearings on the two theologies, natural and revealed.* Edinburgh: Constable.

———. 1861. *Foot-prints of the Creator; or, The Asterolepis of Stromness.* Edinburgh and London: Adam and Charles Black; and Hamilton, Adams and Co.

———. 1886. *Sketch-book of popular geology.* Edinburgh: W. P. Nimmo.

———. 1906. On the red sandstone, marble, and quartz deposits of Assynt; with their supposed organisms and probable analogues. In *The Old Red Sandstone*, 309–26. London and New York: Dent and Dutton.

Miller, John. 1859. On the succession of rocks on the north coast of Scotland. *Quarterly Journal of the Geological Society of London* 11:544–49.

Murchison, Roderick I. 1829. Geological sketch of the north-western extremity of Sussex, and the adjoining parts of Hants and Surrey. *Transactions of the Geological Society of London*, n.s. 2:97–108.

———. 1833. Address to the Geological Society, delivered on the evening of the 17th of February 1832 by the president. *Proceedings of the Geological Society of London* 1:362–86.

———. 1839. *The Silurian system, founded on researches in the counties of Salop, Hereford, Radnor, Montgomery, Caermathen, Brecon, Pembroke, Monmouth, Gloucester, Worcester, and Stafford, with descriptions of the overlying formations . . .* London: John Murray.

———. 1848. On the geological structure of the Alps, Appenines, and Carpathians, more especially to prove a transition from Secondary to Tertiary rocks, and the development of Eocene deposits in Southern Europe. *Quarterly Journal of the Geological Society of London* 5:157–312.

———. 1851. On the Silurian rocks of the south of Scotland. *Quarterly Journal of the Geological Society of London* 7:139–78.

———. 1854. *Siluria: A history of the oldest known rocks containing organic remains, with a brief sketch of the distribution of gold over the earth.* London: John Murray.

———. 1856. On the relations of the crystalline rocks of the North Highlands to the Old Red Sandstone of that region, and on the recent discoveries of fossils in the former by Mr. Charles Peach. *Report of the Twenty-fifth Meeting of the British Association for the Advancement of Science; held at Glasgow in September 1855*, 85–88. London: John Murray.

———. 1858a. The quartz rocks, crystalline limestones, and micaceous schists of the north-western Highlands of Scotland, proved to be of Silurian age, through the recent discoveries of Mr. C. Peach. *Report of the Twenty-seventh Meeting of the British Association for the Advancement of Science; held at Dublin in August and September 1857*, 82–83. London: John Murray.

———. 1858b. On the crystalline rocks of the North Highlands of Scotland (in the form of a letter addressed to Sir William Logan). *Proceedings of the American Association for the Advancement of Science; Eleventh Meeting, held at Montreal, Canada East, August, 1857*, part 2, 57–61. Cambridge: Joseph Lovering.

———. 1857–58. Sur une nouvelle classification des terrains de l'Écosse. *Bulletin de la Société Géologique de France* 15:367–68.

———. 1859a. Speech of thanks on the occasion of the award of the Wollaston Donation Fund. *Proceedings of the Geological Society of London* 15:xxv.

———. 1859b. On the succession of the older rocks in the northernmost counties of Scotland; with some observations on the Orkney and Shetland Islands. *Quarterly Journal of the Geological Society of London* 15:353–418.

———. 1859c. Some results of recent researches among the older rocks of the Highlands of Scotland. *Report of the Twenty-eighth Meeting of the British Association for the Advancement of Science; held at Leeds in September 1858*, 94–96. London: John Murray.

———. 1860. Supplemental observations on the order of the ancient stratified rocks of the north of Scotland, and their associated eruptive rocks. *Quarterly Journal of the Geological Society of London* 16:215–40.

———. 1861. Appendix. *Quarterly Journal of the Geological Society of London* 17:228–32.

———. 1867. *Siluria: A history of the oldest rocks in the British Isles and other countries; with sketches of the origin and distribution of native gold, the general succession of geological formations, and changes in the earth's surface.* "4th" ed. London; John Murray.

———. 1870. On the succession of the Laurentian, Cambrian and Lower Silurian rocks on the shores of Loch Broom. *Geological Magazine* 7:134–36.

Murchison, Roderick I., Edouard de Verneuil, and Alexander von Keyserling. 1845. *The geology of Russia in Europe and the Ural Mountain*, vol. 1. London: John Murray. *Géologie de la Russie d'Europe et des montagnes d'Oural*, vol. 2, Paris: Bertrand.

Murchison, Roderick I., and Archibald Geikie. 1861a. On the altered rocks of the western islands of Scotland, and the north-western and central Highlands. *Quarterly Journal of the Geological Society of London* 17:171–229.

———. 1861b. On the coincidence between stratification and foliation in the crystalline rocks of the Scottish Highlands. *Quarterly Journal of the Geological Society of London* 17:232–40.

Naumann, Carl F. 1847. Über die wahrscheinlich eruptive Natur mancher Gneisse und Gneiss-Granite. In *Neues Jahrbuch für Mineralogie, Geognosie, Geologie und Petrefakten-Kunde*, edited by K. C. von Leonhard and H. G. Bronn, 297–310. Stuttgart: E. Schweizerbart'sche Verlagshandlung und Druckerei.

————. 1848. On the problable eruptive origin of several kinds of gneiss and of gneiss-granite. *Quarterly Journal of the Geological Society of London,* part 2, 4 : 1–9.

————. 1856. Ueber die Bildung der sächsischen Granulit-Formation. *Geologisches Jahrbuch* 7 : 766–71.

Nicholson, Henry Alleyne. 1872. *A monograph of the British graptolitidae.* Edinburgh and London: Blackwood.

Nicol, James. 1840. *An historical and descriptive account of Iceland, Greenland and the Faroe Islands; with illustrations of their natural history. Maps by Wright and engravings by Jackson and Bruce.* Edinburgh: Oliver and Boyd.

————. 1842. *A catechism of geology; or, Natural history of the earth.* Edinburgh: Oliver and Boyd.

————. 1844a. *Guide to the geology of Scotland: Containing an account of the character, distribution, and more interesting appearances of its rocks and minerals. With a geological map and plates.* Edinburgh and London: Oliver and Boyd and Simkin Marshall and Co.

————. 1844b. *Introductory book of the sciences.* Edinburgh: Oliver and Boyd.

————. 1848. On the geology of the Silurian rocks in the valley of the Tweed. *Quarterly Journal of the Geological Society of London* 4 : 195–209.

————. 1849. *Manual of mineralogy; or, The natural history of the mineral kingdom, containing a general introduction to the science, and descriptions of the separate species including the recent discoveries and chemical analyses.* Edinburgh: Adam and Charles Black.

————. 1850. Observations on the Silurian rocks of the south-east of Scotland. *Quarterly Journal of the Geological Society of London* 6 : 53–65.

————. 1852. On the geology of the southern portion of the Peninsula of Cantyre. *Quarterly Journal of the Geological Society of London* 7 : 406–25.

————. 1853a. On the structure of the south Silurian mountains of Scotland. *Report of the Twenty-second Meeting of the British Association for the Advancement of Science; held at Belfast in September 1852,* 55 (title only). London: John Murray.

————. 1853b. *On the study of natural history as a branch of general education: An inaugural lecture, at Marischal College [Aberdeen].* Edinburgh: Oliver and Boyd.

————. 1856. On the striated rocks and other evidences of ice-action observed in the north of Scotland. *Report of the Twenty-fifth Meeting of the British Association for the Advancement of Science; held at Glasgow in September 1855,* 88–89. London: John Murray.

————. 1857. On the red sandstone and conglomerate, and the superposed quartz-rocks, limestones, and gneiss of the north-west coast of Scotland. *Quarterly Journal of the Geological Society of London* 13 : 17–39.

————. 1858. *Elements of mineralogy; containing a general introduction to the science, with descriptions of the species.* Reprinted from the eighth edition of the *Encyclopaedia Britannica.* Edinburgh: Adam and Charles Black.

————. 1859. On the age and relations of the gneiss rocks in the north of Scotland. *Report of the Twenty-eighth Meeting of the British Association for the Advancement of Science; held at Leeds in September 1858,* 94–96. London: John Murray.

————. 1860. On the relations of the gneiss, red sandstone, and quartzite in the Northwest Highlands. *Report of the Twenty-ninth Meeting of the British Association for the Advancement of Science; held at Aberdeen in September 1859,* 119–20. London: John Murray.

————. 1861a. On the structure of the north-western Highlands, and the relationship of the gneiss, red sandstone, and quartzite of Sutherland and Ross-shire. *Quarterly Journal of the Geological Society of London* 17 : 85–113.

———. 1861b. On the geological structure of the southern Grampians. *Quarterly Journal of the Geological Society of London* 18:443.

———. 1862. On the geological succession of the southern Grampians. *Quarterly Journal of the Geological Society of London* 19:180–209.

———. 1866. *The geology and scenery of the north of Scotland: Being two lectures given at the Philosophical Institution, Edinburgh. With notes and an appendix.* Edinburgh and London: Oliver and Boyd and Simkin Marshall and Co.

Peach, Benjamin N. 1894. Additions to the fauna of the *Olenellus* zone of the North-west Highlands. *Quarterly Journal of the Geological Society of London* 50:661–76.

———. 1908. *Monograph on the higher crustacea of the Carboniferous rocks of Scotland.* Glasgow: His Majesty's Stationery Office.

Peach, Benjamin N., and John Horne. 1879. The glaciation of the Shetland Isles. *Quarterly Journal of the Geological Society of London* 35:778–811.

———. 1880. The glaciation of the Orkney Islands. *Quarterly Journal of the Geological Society of London* 36:648–63.

———. 1881. The glaciation of Caithness. *Proceedings of the Royal Physical Society of Edinburgh* 6:316–52.

———. 1884. Report on the geology of the north-west of Scotland. *Nature* 31:31–35.

———. 1892. The *Olenellus* zone in the north-west Highlands of Scotland. *Quarterly Journal of the Geological Society of London* 48:227–42.

———. 1899. *The Silurian rocks of Britain, Vol. 1 Scotland . . . with petrological chapters and notes by J. J. H. Teall.* Glasgow: Her Majesty's Stationery Office.

———. 1914. *Guide to the geological model of the Assynt mountains.* Edinburgh: His Majesty's Stationery Office.

———. 1930. *Chapters on the geology of Scotland.* Edited by M. MacGregor et al. London: Oxford University Press.

Peach, Benjamin N., John Horne, William Gunn, Charles T. Clough, and Lionel Hinxman. 1907. *The geological structure of the North-west Highlands of Scotland . . . with petrological notes by J. J. H. Teall . . .* Edited by Archibald Geikie. Glasgow: His Majesty's Stationery Office.

Peach, Benjamin N., John Horne, William Gunn, Charles T. Clough, Lionel Hinxman, and Henry M. Cadell. 1888. Report on the recent work of the Geological Survey in the North-west Highlands of Scotland, based on the field notes and maps of Mssrs. B. N. Peach, J. Horne, W. Gunn, C. T. Clough, L. Hinxman, and H. M. Cadell. *Quarterly Journal of the Geological Society of London* 44:378–441.

Peach, Charles W. 1858. Notice of the discovery of fossils in the limestones of Durness, in the County of Sutherland. *Proceedings of the Royal Physical Society of Edinburgh* 1:23–24.

Pennant, Thomas. 1775. *A tour in Scotland, and voyage to the Hebrides.* 2 vols. Dublin: A. Leathley.

Pringle, J. 1948. *British regional geology: the south of Scotland.* 2d ed. Edinburgh: His Majesty's Stationery Office.

Ramsay, Andrew C. 1841. *The geology of the Island of Arran, from original survey.* Glasgow: Richard Griffin and Co.; London: Thomas Tegg.

———. 1860. *The old glaciers of Switzerland and North Wales.* London: Longman, Green, Longman and Roberts.

———. 1863. *The physical geology and geography of Great Britain.* London: Stanford.

———. 1866. *The geology of North Wales, . . . with an appendix on the fossils, with plates by J. W. Salter . . .* London: Longmans, Green, Reader, and Dyer.

———. 1878. *The physical geology and geography of Great Britain: A manual of British geology.* 5th ed. London: Stanford.

———. 1881. Address by Andrew Crombie Ramsay . . . President (On the recurrence of certain phenomena in geological time). *Report of the Fiftieth Meeting of the British Association for the Advancement of Science; held at Swansea in September 1880,* 1–22. London: John Murray.

Read, Herbert H. 1934. Age-problems of the Moine Series of Scotland. *Geological Magazine* 71:302–17.

Rogers, Henry D. 1857. On the laws of structure of the more disturbed zones of the earth's crust. *Transactions of the Royal Society of Edinburgh* 21:431–71.

Rogers, Henry D., and William B. Rogers. 1843a. On the physical structure of the Appalachian chain, as exemplifying the laws which have regulated the elevation of great mountain chains generally. *Reports of the First, Second, and Third Meetings of the Association of American Geologists and Naturalists at Philadelphia in 1840 and 1841, and at Boston in 1842,* 1:474–531. Boston: Gould, Kendall and Lincoln.

———. 1843b. Summary of 1843a. *Report of the Twelfth Meeting of the British Association for the Advancement of Science; held at Manchester in June 1842,* 40–42. London: John Murray.

Salter, John W. 1858. Note on the fossils of Durness. *Report of the Twenty-seventh Meeting of the British Association for the Advancement of Science; held at Dublin in August and September 1857,* 83–84. London: John Murray.

———. 1859. Durness Limestone fossils described. *Quarterly Journal of the Geological Society of London* 15:374–81.

———. 1865a. On some new forms of Olenoid trilobites from the lowest fossiliferous rocks of Wales. *Report of the Thirty-fourth Meeting of the British Association for the Advancement of Science; held at Bath in September 1864,* 67. London: John Murray.

———. 1865b. On the old pre-Cambrian (Laurentian) island of St. David's, Pembrokeshire. *Report of the Thirty-fourth Meeting of the British Association for the Advancement of Science; held at Bath in September 1864,* 67–68. London: John Murray.

Sedgwick, Adam, and Roderick I. Murchison. 1827–28. On the old conglomerates, and other Secondary deposits on the north coasts of Scotland. *Proceedings of the Geological Society of London* 1:77–80.

———. 1829. On the structure and relations of the deposits contained between the Primary rocks and the Oolitic Series in the north of Scotland. *Transactions of the Geological Society of London,* n.s. 3:125–60.

Sharpe, Daniel, 1847 and 1849. The relation between distortion and slaty cleavage. *Quarterly Journal of the Geological Society of London* 3:74–105; 5:111–29.

———. 1850. On the Secondary district of Portugal which lies on the north of the Tagus. *Quarterly Journal of the Geological Society of London* 6:135–201.

———. 1852a. On the arrangement of the foliation and cleavage of the rocks of the north of Scotland. *Philosophical Transactions of the Royal Society, London* 142:445–61.

———. 1852b. On the "Quartz Rock" of Macculloch's map of Scotland. *Quarterly Journal of the Geological Society of London* 8:120–26.

———. 1856. On the structure of Mont Blanc and its environs. *Quarterly Journal of the Geological Society of London* 11:11–26.

Suess, Eduard. 1875. *Die Enstehung der Alpen.* Vienna: Wilhelm Braumüller.

———. [1883] 1904–1909. *The face of the earth (Das Antlitz der Erde).* Translated by Hertha B. C. Sollas under the direction of William J. Sollas. 4 vols. Oxford; Clarendon Press.

Symonds, W. S. 1872. *Record of the rocks; or, notes on the geology, natural history, and antiquities of North & South Wales, Devon and Cornwall.* London: John Murray.

Teall, Jethro J. H. 1885. The metamorphosis of dolerite into hornblende-schist. *Quarterly Journal of the Geological Society of London* 41 : 133–45.

———. 1886. Notes on some hornblende-bearing rocks from Inchnadamph. *Geological Magazine,* decade 3, 3 : 346–53.

———. 1888. *British petrograph* London; Dulau.

———. 1900. On nepheline syenite and its associates in the north-west of Scotland. *Geological Magazine,* decade 4, 7 : 385–92.

Tilley, Charles E. 1925. A preliminary survey of metamorphic zones in the southern Highlands of Scotland. *Quarterly Journal of the Geological Society of London* 81 : 100–12.

Von Penck, Albrecht. 1897. Geomorphologische Probleme aus Nordwest-Schottland. *Zeitschrift der Gesellschaft für Erdkunde zu Berlin* 32 : 146–91.

———. 1898. The Highland controversy. *Geographical Journal* 11 : 163–65.

Wegener, Alfred. 1966. *The origins of continents and oceans.* Translated by John Biram. New York: Dover Publications.

Wharton Committee. 1900. *Report of the Committee on the Geological survey and the Musuem of Practical Geology* [Wharton Inquiry]. No place or publisher stated.

Williams, John. 1789. *The natural history of the mineral kingdom. In three parts . . .* 2 vols. Edinburgh: T. Ruddiman.

Woodward, John. 1695. *An essay towards a natural history of the earth; and terrestrial bodies, especially minerals: As also of the sea, rivers, and springs. With an account of the Universal Deluge; and of the effects that it had upon the earth.* London: R. Wilken.

Maps

British Geological Survey. 1976. 1 : 50,000 series. *Broadford, Scotland, Sheet 71W.* Southampton.

British Geological Survey Map Library (Edinburgh). 1913. *Sutherland O.S. 1 : 10, 560 Sheet 15.* Reg. no. 044149.

Geikie, Archibald. 1876. *Geological map of Scotland . . . the topography by T. B. Johnston.* Edinburgh and London: W. and A. K. Johnston.

Geological Survey of Great Britain (Scotland). 1948. Four miles to one inch. *Sheet 5* [*Sutherland*]. Chessington.

Geological Survey of Scotland. 1891. *Sutherland, Sheet 71, Assynt Sutherland 6″ to mile. Geologically surveyed in 1885–6 by B. N. Peach, J. Horne and L. W. Hinxman.* London: Geological Survey Office.

Heddle, M. Forster. 1881. *Geological and mineralogical map of Sutherland . . . to accompany his papers on the geognosy of Sutherland published in the Mineralogical Magazine.* Edinburgh; W. and A. K. Johnston.

Institute of Geological Sciences. 1952. One inch series. *Scotland Sheet 70: Minginish.* Chessington.

———. 1962. One inch series. *Scotland. Sheet 92: Inverbroom.* Chessington.

———. 1965. One inch series. *Assynt district: Scotland special sheet. Parts of Sheets 101, 102, 107 & 108.* Chessington.

———. 1975. 1 : 50,000 Series. *Scotland. Sheet 81(E). Loch Torridon.* Southampton.

Kitchen, Thomas. N.d. *North Britain or Scotland divided into its counties: Corrected from the best surveys & astronomical observations by Thos. Kitchen. Hydrographer to His Majesty.* London: W. Faden.

Macculloch, John. 1836. *A geological map of Scotland by Dr. MacCulloch, F.R.S. &c. &c. &c. published by order of the Lords of the Treasury by S. [sic] Arrowsmith. Hydrographer to the King.* London. (This is the title of the second issue, as pasted over the title of the topographical map prepared by Arrowsmith. See Eyles 1939.)

————. 1843. *Map of Scotland constructed from original materials obtained under the authority of the parliamentary commissioners for making roads and building bridges in the Highlands of Scotland. With their permission it is now published by their much obliged and obedient servant A. [sic] Arrowsmith. Hydrographer to his Majesty . . . Additions to 1843.* London: G. F. Crunchley. (This map, first issued in 1840, was used as the topographical basis for the fourth and final version of Macculloch's map. See Eyles 1939.)

Murchison, Roderick I., and Archibald Geikie. 1861. *First sketch of a new geological map of Scotland with explanatory notes.* Edinburgh and London: W. and A. K. Johnston and W. Blackwood and Stanford.

Murchison, Roderick I., and James Nicol. 1856. *Geological map of Europe exhibiting the different systems of rocks according to the most recent research and inedited materials . . . constructed by A. Keith Johnston.* Edinburgh: Blackwood and Johnston.

Nicol, James. 1858. *Geological map of Scotland from the most recent authorities & personal observations.* Edinburgh and London: K. Johnston.

Secondary References

Albritton, Claude C. 1963. *The fabric of geology.* Stanford: Freeman, Cooper and Co.

Anderson, Angela. 1980. *Ben Peach's Scotland: Landscape sketches by a Victorian geologist.* Edinburgh: Institute of Geological Sciences.

Anderson, J. G. C. 1979. The concept of Precambrian geology and the recognition of Precambrian rocks in Scotland and Ireland. In *History of concepts in Precambrian geology,* edited by W. O. Kupsch and W. A. S. Sarjeant, 1–11. Toronto: Geological Association of Canada.

————. 1983. *Field geology in the British Isles: A guide to regional excursions.* Oxford: Pergamon.

A[nderson], P. J. 1899. *Aurora borealis academica: Aberdeen University appreciations.* Aberdeen: University Printers.

Agassiz, Louis. 1861. Hugh Miller. In Hugh Miller, *Foot-prints of the creator, or, The Asterolepis of Stromness,* iii–xxxvii. Edinburgh and London: Adam and Charles Black and Hamilton Adams and Co.

Arbenz, P. 1937. Albert Heim, 1849–1937. *Verhandlungen der Schweizerischen naturforschenden Gesellschaft* 118:330–53.

Athenaeum. 1886. Mr. C. W. Peach. *Athenaeum,* no. 3046:362–63.

Bailey, Edward B. 1926. Benjamin Neeve Peach. *Geological Magazine* 63:187–90.

————. 1935. *Alpine essays mainly tectonic.* Oxford: Clarendon Press.

————. 1939. Professor Albert Heim, 1849–1937. *Obituary Notices of Fellows of the Royal Society of London* 7:471–74.

————. 1952. *Geological Survey of Great Britain.* London: Murby.

————. 1967. *James Hutton: The founder of modern geology.* Amsterdam: Elsevier.

Bayne, P. 1871. *The life and letters of Hugh Miller.* 2 vols. London: Strachan and Co.

Bonney, Thomas G. 1893. John Macculloch, M.D. (1773–1835). *Dictionary of National Biography* 35:17–19.

————. 1894. Sir Roderick Impey Murchison (1792–1871). *Dictionary of National Biography* 39:317–20.

————. 1895a. James Nicol (1810–79). *Dictionary of National Biography* 41 : 38–39.

————. 1895b. Charles William Peach (1800–86). *Dictionary of National Biography* 44 : 131.

————. 1896. Sir Andrew Crombie Ramsay (1814–91). *Dictionary of National Biography* 47 : 236–38.

————. 1897. Daniel Sharpe (1806–56). *Dictionary of National Biography* 51 : 421–22.

————. 1905. Henry Hicks (1837–99). *Proceedings of the Royal Society of London* 75 : 106–109.

————. 1916. Professor John Wesley Judd. *Geological Magazine*, decade 4, 3 : 190–92.

————. 1921. *Memories of a long life*. Cambridge: Metcalfe and Co.

Boulton, William S., G. L. Elles, W. G. Fearnsides, and Murray Macgregor. 1951. The work of Charles Lapworth. *Advancement of Science* 7 : 433–22.

Bourdieu, Pierre. 1971. Intellectual field and creative project. In *Knowledge and control: New directions for the sociology of education*, edited by M. F. D. Young, 161–68. London and New York: Collier Macmillan.

————. 1975. The specificity of the scientific field and the social conditions for the progress of learning. *Social Science Information* 14 : 19–47.

Bradley, Francis H. 1914. *Essays on truth and reality*. Oxford: Clarendon Press.

Brock, William H. 1979. Chemical geology or geological chemistry? In *Images of the earth: Essays in the history of the environmental sciences*, edited by Ludmilla J. Jordanova and Roy S. Porter, 147–70. Chalfont St. Giles: British Society for the History of Science.

Brown, J. R. ed. 1984. *Scientific rationality: The sociological turn*. Dordrecht: Reidel.

Browne, E. Janet. 1981. The making of the *Memoir* of Edward Forbes, F.R.S. *Archives of Natural History* 10 : 205–19.

Chambers, T. G. 1896. *Register of the associates and old students of the Royal College of Chemistry, the Royal School of Mines and the Royal College of Science; with historical introduction and biographical notices and portraits of past and present professors*. London: Hazell, Watson and Viney.

Cole, Grenville A. J. 1914–22. *In memoriam* John Wesley Judd, C.B., LL.D., F.R.S., F.G.S. (1840–1916). *Proceedings of the Yorkshire Geological Society*, n.s. 19 : 327–29.

Collingwood, R. G. 1939. *An autobiography*. Oxford: Clarendon Press.

Collins, Harry M. 1981. *Changing order: Replication and induction in scientific practice*. London: Sage Publications.

Cribb, H. G. S. 1976. History of the Geological Survey of Queensland. In *History and role of governmental geological surveys in Australia*, edited by R. K. Johns, 46–55. South Australia: Government Printer.

Cumming, David A. 1981. Geological maps in preparation: John Macculloch on western islands. *Archives of Natural History* 10 : 255–71.

————. 1983. John Macculloch: Pioneer of "Precambrian" geology. Ph.D. dissertation, University of Glasgow.

————. 1984. John Maculloch's "Millstone survey" and its consequences. *Annals of Science* 41 : 567–91.

————. 1985. John Macculloch: Blackguard, thief and high priest. In *From Linnaeus to Darwin: Commentaries on the history of biology and geology*, edited by A. Wheeler and J. H. Price, 77–88. London: Society for the History of Natural History.

Cutter, Eric. 1974. Sir Archibald Geikie: A bibliography. *Journal of the Society for the Bibliography of Natural History* 7 : 1–18.

De Lapparent, Auguste-A. C. 1893. Scientific worthies. No. 28, Sir Archibald Geikie. *Nature* 47 : 217–20.

De Margerie, Emmanuel. 1946. Three stages in the evolution of Alpine geology: De Saussure–Studer–Heim. *Proceedings of the Geological Society of London* 102 : xcvii–cxiv.

De Margerie, Emmanuel, and Albert Heim. 1888. *Les dislocations de l'écorce terrestre. Die Dislocationen der Erdinde. Essai de définition et de nomenclature. Versuch einer Definition und Bezeichung.* Zürich: J. Wurster.

Dott, R. H. 1969. James Hutton and the concept of a dynamic earth. In *Toward a history of geology*, edited by Cecil J. Schneer, 122–41. Cambridge, Mass.: MIT Press.

Ellenberger, François. 1972. La métaphysique de James Hutton (1726–1797) et le drame écologique du XXe siècle. *Revue de Synthèse* 93 : 267–83.

———. 1978. The First International Geological Congress: Paris, 1878. *Episodes: Geological Newsletter, International Union of Geological Sciences*, no. 2 : 20–24.

Engelhardt, H. Tristram, and Arthur L. Caplan, eds. 1987. *Scientific controversies: Case studies in the resolution and closure of disputes in science and technology.* Cambridge: Cambridge University Press.

Eyles, Victor A. 1937. John Macculloch, F.R.S., and his geological map: An account of the first geological survey of Scotland. *Annals of Science* 2 : 114–29.

———. 1939. Macculloch's geological map of Scotland: An additional note. *Annals of Science* 4 : 107.

———. 1948. Louis Albert Necker of Geneva, and his geological map of Scotland. *Transactions of the Geological Society of Edinburgh* 14 : 93–127.

Feyerabend, Paul K. 1975. *Against method: Outlines of an anarchistic theory of knowledge.* London and Atlantic Highlands: New Left Books and Humanities Press.

Flett, John S. 1924–25. Sir Jethro Teall, 1849–1924. *Proceedings of the Royal Society of London*, series B, 97 : xv–xvii.

———. 1925. Sir Jethro Justinian Harris Teall. *Proceedings of the Geological Society of London* 81 : lxiii–lxv.

———. 1928–29. John Horne, 1848–1928. *Proceedings of the Royal Society of London*, series B, 104 : i–viii.

———. 1937. *The first hundred years of the Geological Survey of Great Britain* London: His Majesty's Stationery Office.

Flinn, Derek. 1981. John Macculloch, M.D., F.R.S., and his geological map of Scotland: His years in the Ordnance; 1795–1826. *Notes and Records of the Royal Society of London* 36 : 83–101.

Ford, Trevor D. 1979. The history of the study of the Precambrian rocks of Charnwood Forest. In *History of concepts in Precambrian geology*, edited by W. O. Kupsch and W. A. S. Sarjeant, 65–80. Toronto: Geological Association of Canada.

Geikie, Archibald. 1875. *Life of Sir Roderick I. Murchison Bart., K.C.B., F.R.S., sometime director of the Geological Survey of the United Kingdom. Based on his journals and letters with notices of his scientific contemporaries and a sketch of the rise and growth of palaeozoic geology in Britain . . .* 2 vols. London: John Murray.

———. 1878. Robert Harkness, F.R.S. *Nature*, Oct. 10, 628.

———. 1892. Sir Andrew Crosbie Ramsay. *Proceedings of the Geological Society of London* 48 : xxxviii–xlvii.

———. 1895. *Memoir of Sir Andrew Crosbie Ramsay.* London and New York: Macmillan.

———. 1908. Marcel Bertrand. *Proceedings of the Geological Society of London* 64 : l–li.

———. 1917. Professor Edward Hull, F.R.S. *Geological Magazine*, decade 6, 4 : 553–55.

———. [1905] 1962. *The founders of geology.* Reprint. New York: Dover. (Originally published 1897 by Macmillan.)

Geological Magazine. 1878. Professor Robert Harkness, F.R.S., F.G.S. *Geological Magazine*, decade 2, 5 : 574–76.

————. 1882. Eminent living geologists: Sir Andrew C. Ramsay. LL.D., F.R.S., V.P.G.S., etc., etc. *Geological Magazine,* decade 2, 9:289–93.

————. 1890. Eminent living geologists: Professor Archibald Geikie, LL.D., F.R.S. *Geological Magazine,* decade 3. 7:49–51.

————. 1899a. Eminent living geologists: Henry H. Howell, F.G.S., formerly Director of the Geological Survey of Great Britain. *Geological Magazine,* decade 4, 6:433–37.

————. 1899b. Dr. Henry Hicks, F.R.S., F.G.S. *Geological Magazine,* decade 4, 6:574–75.

————. 1901. Eminent living geologists: Professor Charles Lapworth. *Geological Magazine,* decade 4, 38:289–303.

————. 1904. Eminent living geologists: Wilfred Hudleston Hudleston, J.P., M.A., F.R.S., F.L.S., F.G.S., F.C.S., etc. *Geological Magazine,* decade 5, 1:431–38.

————. 1905. Eminent living geologists: John Wesley Judd, C.B., LL.D., F.R.S., F.G.S. *Geological Magazine,* decade 5, 2:385–97.

————. 1909a. Eminent living geologists: Jethro Justinian Teall, M.A., D.Sc., F.R.S., F.G.S., Director of the Geological Survey of Great Britain and the Museum of Practical Geology. *Geological Magazine,* decade 5, 6:1–8.

————. 1909b. Wilfred Hudleston Hudleston, J.P., M.A., F.R.S., F.L.S., F.G.S., F.C.S., etc. *Geological Magazine,* decade 5, 6:143–44.

————. 1920. The Nicol Memorial. *Geological Magazine* 57:387–92.

Geological Society of London. 1938. [Albert Heim.] *Proceedings of the Geological Society of London* 94:cxi–cxiii.

Gerstner, Patsy A. 1968. James Hutton's theory of the earth and his theory of matter. *Isis* 59:25–29.

————. 1975. A dynamic theory of mountain building: Henry Darwin Rogers. *Isis* 66:26–37.

Glasgow Herald. 1924. A great scientist: Death of Sir Archibald Geikie—the reorganiser of the Geological Survey. *Glasgow Herald,* Nov. 12.

Gohau, Gabriel. 1983. Idées anciennes sur la formation des montagnes. *Cahiers d'Histoire et de Philosophie des Sciences* 7:1–86.

Goodchild, John G. 1882–83. Professor Robert Harkness, F.R.S., F.G.S. *Transactions of the Cumberland Association for the Advancement of Literature and Science* 8:145–88.

————. 1898. Emeritus Professor M. Forster Heddle, M.D., F.R.S.E. *Mineralogical Magazine and Journal of the Mineralogical Society* 12:38–42.

Greene, Mott T. 1982. *Geology in the nineteenth century: Changing views of a changing world.* Ithaca and London: Cornell University Press.

Greenly, Edward. 1928–32. Benjamin Neeve Peach: A study. *Transactions of the Edinburgh Geological Society* 12:1–12.

————. 1938. *A hand through time: Memories—romantic and geological; studies in the arts and religion; and the grounds of confidence in immortality.* 2 vols. London: Murby.

Gregory, John W. 1928. Dr. John Horne, F.R.S. *Nature* 121:991–92.

————. 1929. John Horne. *Proceedings of the Geological Society of London* 85:lx–lxii.

Hamilton, Beryl M. 1979. The development of hard rock geology in Britain. Ph.D. dissertation, University of Lancaster.

————. 1982. The influence of the polarizing microscope on late nineteenth century geology. *Janus* 69:51–68.

————. 1984. The contributions of Gustav Linnarsson to British stratigraphic geology. *Geologiska Föreningens i Stockholm Förhandlingar* 106:185–92.

Hanson, N R. 1961. *Patterns of discovery: An inquiry into the conceptual foundations of science.* Cambridge: Cambridge University Press.

Harker, Alfred. 1917. Anniversary address of the president. *Proceedings of the Geological Society of London* 73:lv–xcvi.

———. 1918. Edward Hull. *Proceedings of the Geological Society of London* 74:liv.

Harrison, J. M. 1963. Nature and significance of geological maps. In *The fabric of geology*, edited by Claude C. Albritton, 225–32. Stanford: Freeman, Cooper and Co.

Heritsch, F. 1929. *The nappe theory in the Alps (Alpine tectonics, 1905–1928).* Translated by P. G. H. Boswell. London: Methuen.

Herries-Davies, Gordon L. 1983. *Sheets of many colours: The mapping of Irish rocks 1750–1890.* Dublin: Royal Dublin Society.

Hooykaas, Reijer. 1959. *Natural law and divine miracle: An historical-critical study of the principle of uniformity in geology, biology, and theology.* Leiden: Brill.

Horne, John. 1926a. B. N. Peach, 1842–1926. *Proceedings of the Royal Society of London*, series B, 100:xi–xiii.

———. 1926b. Sir Archibald Geikie, 1835–1924. *Proceedings of the Royal Society of London*, series A, 3:xxiv–xxxix.

Horny, R. 1980. Joachim Barrande (1799–1883): Life, work, collections. *Journal of the Society for the Bibliography of Natural History* 9:365–68.

Howell, J. V. 1957. *Glossary of geology and related sciences: A cooperative project of the American Geological Institute.* Washington, D.C.: American Geological Institute/National Academy of Sciences–National Research Council.

Houghton, Walter E., et al. 1966 *The Wellesley index to Victorian periodicals, 1824–1900: Tables of contents and identification of contributors with bibliographies of their articles and stories.* Vol. 1. Toronto and Buffalo: University of Toronto Press; London: Routledge and Kegan Paul.

Houston, Robert S. 1925. Sir Archibald Geikie, geologist. *Paisley Express*, June 1, 2, and 3.

Hull, David L. 1988. *Science as a process: An evolutionary account of the social and conceptual development of science.* Chicago and London: University of Chicago Press.

Hull, Edward. 1910. *Reminiscences of a strenuous life.* London: Hugh Rees.

Illustrated London News. 1871. Sir Roderick Murchison. *Illustrated London News*, Oct. 28, 414.

Jehu, T. J. 1934. Dr. H. M. Cadell. *Nature* 133:822–23.

Johns, R. K., ed. 1976. *History and role of government geological surveys in Australia.* South Australia: Government Printer.

Johnson, M. R. W., and I. Parsons. 1979. *MacGregor & Phemister's geological excursion guide to the Assynt district of Sutherland.* Edinburgh: Edinburgh Geological Society.

Jones, Jean. 1985. James Hutton's agricultural research and his life as a farmer. *Annals of Science* 42:573–601.

Kendall, Percy F., and Herbert E. Wroot. 1924. *Geology of Yorkshire: An illustration of the evolution of northern England.* Vienna: The Authors.

Kuhn, Thomas S. 1962. *The structure of scientific revolutions.* Chicago and London: University of Chicago Press.

Kupsch, W. O., and W. A. S. Sarjeant, eds. 1979. *History of concepts in Precambrian geology.* Toronto: Geological Association of Canada.

Lapworth, Charles. 1882. The life and work of Linnarsson. *Geological Magazine*, decade 2, 9:1–7, 119–22, 171–76.

———. 1895. Dr. Crosskey's scientific researches and publications. In *The life and work of Henry William Crosskey, LL.D., F.G.S.*, edited by R. A. Armstrong, 307–448. Birmingham: Cornish and Co.

Latour, Bruno. 1987. *Science in action: How to follow scientists and engineers through society.* Milton Keynes: Open University Press.

Latour, Bruno, and Steven Woolgar. 1979. *Laboratory life: The social construction of scientific facts.* Beverly Hills and London: Sage Publications.

Laudan, Rachel. 1976. William Smith: Stratigraphy without palaeontology. *Centaurus* 20:210–26.

———. 1987. *From mineralogy to geology: The foundations of a science, 1650–1830.* Chicago: University of Chicago Press.

Linnean Society. 1857. Daniel Sharpe, Esq. *Proceedings of the Linnean Society of London* 13:275–79.

Literary Gazette. 1856. Daniel Sharpe, F.R.S. *The Literary Gazette, and Journal of Archaeology, Science, and Art,* June 7, 351–52.

MacGregor, M. 1928. John Horne. *Geological Magazine* 65:381–84.

McMullin, Ernan. 1984. The rational and the social in the history of science. In *Scientific rationality: The sociological turn,* edited by J. R. Brown, 127–63. Dordrecht: Reidel.

———. 1987. Scientific controversy and its termination. In *Scientific controversies: Case studies in the resolution and closure of disputes in science and technology,* edited by H. Tristram Engelhardt and Arthur L. Caplan, 49–91. Cambridge: Cambridge University Press.

Macnair, P., and F. Mort, eds. 1908. *History of the Geological Society of Glasgow.* Glasgow: Geological Society of Glasgow.

Macpherson, H. G., and A. Livingstone. 1982. *Glossary of Scottish mineral species.* Edinburgh and Glasgow: Scottish Journal of Geology.

Marr, John E. 1924. Sir Jethro Justinian Harris Teall. *Geological Magazine* 61:382–84.

Marsden, W. E. 1979. Archibald Geikie, 1835–1924. *Geographers: Biographical Studies* 3:39–52.

———. 1980. Sir Archibald Geikie (1835–1924) as geographical educationist. In *Historical perspectives on geographical education,* edited by W. E. Marsden, 54–65. London: University of London Institute of Education.

Mather, Kirtley F., and Shirley L. Mason. 1939. *A source book in geology.* New York and London: McGraw-Hill.

Menard, H. William. 1971. *Science: Growth and Change.* Cambridge, Mass.: Harvard University Press.

Merrill, George P. 1924. *The first hundred years of American geology.* New Haven: Yale University Press.

Merton, Robert K. 1968. The Matthew effect in science: The reward and communication system in science. *Science* 199:55–63.

Mineralogical Society. 1909. Wilfred Hudleston Hudleston (1828–1909). *Mineralogical Magazine and Journal of the Mineralogical Society* 15:265.

Nature (memorials). 1885–86. Charles William Peach. *Nature* 33:446–47.

———. 1901–2. Sir John Donnelly, K.C.B. *Nature* 65:538–39.

Nature Conservancy Council. 1986. *Geological trail Knockan Cliff: Inverpolly National Nature Reserve.* Inverness: Nature Conservancy Council.

———. n.d.(a). *Inverpolly: National Nature Reserve.* Edinburgh: Nature Conservancy Council.

———. n.d.(b). *Knockan Cliff Nature Trail: Inverpolly Nature Trail.* Edinburgh: Nature Conservancy Council.

O'Brien, Charles F. 1970. *Eozoön Canadense:* The dawn animal of Canada. *Isis* 61:206–23.

O'Connor, J. G., and A. J. Meadows. 1976. Specialization and professionalization in British geology. *Social Studies of Science* 6 : 77–89.

Oldham, R. D. 1922. Robert Logan Jack. *Proceedings of the Geological Society of London* 78 : xlix.

Oldroyd, David R. 1980. Sir Archibald Geikie (1835–1924), geologist, romantic aesthete, and historian of geology: The problem of Whig historiography of science. *Annals of Science* 37 : 441–62.

———. 1987a. Punctuated equilibrium theory and time: A case study in problems of coherence in the measurement of geological time (the "KBS" Tuff controversy and the dating of rocks in the Turkana Basin, Kenya).). In *Measurement, realism and objectivity*, edited by John C. Forge, 89–152. Dordrecht: Reidel.

———. 1987b. Science action packed. *Social Epistemology* 1 : 341–50.

O'Rourke, James E. 1978. A comparison of James Hutton's *Principles of knowledge* and *Theory of the Earth*. *Isis* 69 : 4–20.

Ospovat, Alexander M. 1967. The place of the *Kurze Klassifikation* in the work of A. G. Werner. *Isis* 58 : 90–95.

———. 1969. Reflections on A. G. Werner's "Kurze Klassifikation." In *Toward a history of geology*, edited by C. J. Schneer, 242–56. Cambridge, Mass.: MIT Press.

———. 1971. Introduction and notes to A. G. Werner's *Short classification and description of the various rocks*, translated by A. M. Ospovat, 1–35. New York: Hafner.

———. 1980. The importance of regional geology in the geological theories of Abraham Gottlob Werner: A contrary opinion. *Annals of Science* 37 : 433–40.

Peach, Benjamin N., and John Horne. 1925. The scientific career of Sir Archibald Geikie, O.M., K.C.B., F.R.S. *Proceedings of the Royal Society of Edinburgh* 45 : 346–61.

Perrin, C. A. 1989. Document, text and myth: Lavoisier's crucial year revisited. *British Journal for the History of Science* 22 : 3–25.

Pingree, J. 1986. *Sir Andrew Crosbie Ramsay, 1814–1891: List of papers*. London: Imperial College Archives, University of London.

Porter, Roy S. 1978. Gentlemen and geology: The emergence of a scientific career, 1660–1920. *Historical Journal* 21 : 809–36.

Portlock, Joseph E. 1857. Daniel Sharpe. *Proceedings of the Geological Society of London* 13 : xlv–lxiv.

Rastall, R. H. 1937. Thomas George Bonney (1833–1923). *Dictionary of National Biography, 1922–1930*, 89–91.

Reeks, Margaret. 1920. *Register of the associates and old students of the Royal School of Mines and history of the Royal School of Mines*. London: Royal School of Mines (Old Students') Association.

Renevier, E. 1879. Notice of Prof. A. Heim's work on the mechanism of the formation of mountains. *Geological Magazine*, decade 1, 6 : 131–35.

Richardson, L. 1915. Charles Callaway, M.A., D.Sc. *Geological Magazine*, decade 6, 2 : 525–28.

Rosie, George. 1981. *Hugh Miller, outrage and order: A biography and selected writings*. Edinburgh: Mainstream Publishing.

Royal Society of London. 1971–19. Edward Hull, 1829–1917. *Proceedings of the Royal Society of London*, series B, 90 : xxviii–xxxi.

Rudwick, Martin J. S. 1985. *The great Devonian controversy: The shaping of scientific knowledge among gentlemanly specialists*. Chicago: University of Chicago Press.

Sarjeant, William A. S. 1980. *Geologists and the history of geology: An international bibliography from the origins to 1978*. 4 vols. New York: Arno.

———. 1987. *Supplement, 1979 to 1984*. 2 vols. Melbourne, Florida: Krieger.

Sarjeant, William A. S., and Anthony P. Harvey. 1979. Uriconian and Longmyndian: A history of the study of the Precambrian rocks of the Welsh Borderland. In *History of concepts in Precambrian geology*, edited by W. O. Kupsch and W. A. S. Sarjeant, 181–224. Toronto: Geological Association of Canada.

Schneer, Cecil J. 1969. Ebenezer Emmons and the foundation of American geology. *Isis* 60:439–50.

Scottish Geographical Magazine (memorials). 1934, Dr. H. M. Cadell. *Scottish Geographical Magazine* 50:169–71.

Secord, James A. 1981–82. King of Siluria: Roderick Murchison and the imperial theme in nineteenth-century British geology. *Victorian Studies* 25:413–42.

———. 1985. John W. Salter: The rise and fall of a Victorian palaeontological career. In *From Linnaeus to Darwin: Commentaries on the history of geology and biology*, edited by A. Wheeler and J. H. Price, 61–75. London: Society for the History of Natural History.

———. 1986a. *Controversy in Victorian geology: The Cambrian-Silurian dispute.* Princeton: Princeton University Press.

———. 1986b. The Geological Survey of Great Britain as a research school, 1839–1855. *History of Science* 24:223–75.

Shapin, Stephen, and Simon Schaffer. 1985. *Leviathan and air-pump: Hobbes, Boyle, and the experimental life.* Princeton: Princeton University Press.

Smiles, Samuel. 1878. *Robert Dick, baker of Thurso: Geologist and botanist.* London: John Murray.

Smith Woodward, Arthur. 1916a. Charles Callaway. *Proceedings of the Geological Society of London* 72:lvii.

———. 1916b. Henry Hyatt Howell. *Proceedings of the Geological Society of London* 72:lx.

Sollas, William J. 1909. Wilfred Hudleston Hudleston (born Simpson), 1828–1909. *Proceedings of the Geological Society of London* 65:lxi–lxiii.

Sorby, Henry C. 1879. Robert Harkness. *Proceedings of the Geological Society of London* 35:xli–xliv.

———. 1880. James Nicol, F.R.S.E., F.G.S., &c. *Proceedings of the Geological Society of London* 36:xxxiii–xxxvi.

Speakman, Colin. 1982. *Adam Sedgwick, geologist and dalesman: A biography in twelve themes.* Heathfield: Broad Oak Press.

Stafford, Robert A. 1984. Geological surveys, mineral discoveries, and British expansion, 1835–71. *Journal of Imperial and Commonwealth History* 12:5–32.

———. 1988a. Roderick Murchison and the structure of Africa: A geological prediction and its consequences for British expansion. *Annals of science* 45:1–40.

———. 1988b. The long arm of London: Sir Roderick Murchison and imperial science in Australia. In *Australian science in the making*, edited by Roderick W. Home, 69–101. Cambridge: Australian Academy of Science and Cambridge University Press.

———. 1989. *Scientist and empire: Sir Roderick Murchison's scientific exploration and Victorian imperialism.* Cambridge: Cambridge University Press.

Stephen, L., and S. Lee, eds. 1885–1903. *Dictionary of National Biography.* 63 vols., Index and Epitome. London: Smith, Elder and Co.

Strachan I., and P. Toghill. 1975. Dobb's Linn, Moffat. In *The geology of the Lothians and South-east Scotland: An excursion guide.* edited by G. Y. Craig and P. McL. D. Duff, 167–76. Edinburgh: Scottish Academic Press.

Strahan, Aubrey. 1913. Award of the Murchison Medal [to George Barrow]. *Proceedings of the Geological Society of London* 69:xlvi–xlvii.

———. 1924. Sir Archibald Geikie, O.M., K.C.B., F.R.S. *Nature* 114:758–60.

———. 1925. Sir Archibald Geikie. *Proceedings of the Geological Society of London* 81:lii–lx.

———. 1926. Sir Archibald Geikie. *Annual report of the regents of the Smithsonian Institution . . . 1925*, 591–98. Washington: Government Printing Office (reprint of Strahan 1925).

Sweeting, George S. 1958. *The Geologists' Association, 1858–1958: A history of the first hundred years.* Colchester: Benham and Co.

Teall, Jethro J. H., and William W. Watts. 1921. Charles Lapworth, 1842–1920. *Proceedings of the Royal Society of London* 92:xxxi–xl.

Thackray, John C. 1972. Essential source-material of Roderick Murchison. *Journal of the Society for the Bibliography of Natural History* 6:162–70.

———. 1978a. R. I. Murchison's *Geology of Russia* (1845). *Journal of the Society for the Bibliography of Natural History* 8:421–33.

———. 1978b. R. I. Murchison's *Silurian System* (1839). *Journal of the Society for the Bibliography of Natural History* 9:61–73.

———. 1981. R. I. Murchison's *Siluria* (1854 and later). *Archives of Natural History*. 10:37–43.

Thomas, Herbert H. 1933. George Barrow. *Proceedings of the Geological Society of London* 89:lxxxvii–lxxxix.

———. 1937. Sir Jethro Justinian Harris Teall (1849–1924). *Dictionary of National Biography, 1922–1930,* 826–27.

Thoms, A. 1901. Memoir of Dr. Heddle. In M. Forster Heddle's *The mineralogy of Scotland,* edited by John G. Goodchild, 1:xi–xiii. Edinburgh: David Douglas.

Tomkeieff, Sergei I. 1962. Unconformity: An historical study. *Proceedings of the Geologists' Association* 73:383–417.

Tyrrell, G. W. 1926. Sir Archibald Geikie. *Zeitschrift für Vulkanologie* 9:149–55.

Vallance, Thomas G. 1984. The span of metamorphism. *Proceedings of the Twenty-seventh International Geological Congress* 21:67–84.

Von Franks, S., and B. Glaus. 1987. Albert Heim (1849–1937). *Gesnerus* 44:85–98.

Von Kobell, F. 1874. Dr. Karl Friedrich Naumann. Geb 1797 am 30. Mai zu Dresden, Gest. 1873 am 26. November ebenda. *Sitzungsberichte der Bayerischen Akademie der Wissenschaften zu Munchen (1. Math.: Phys. C1.)* 4:81–84.

Walton, E. K., B. A. O. Randall, M. H. Battey, and O. Tomkeieff, eds. 1983. *Dictionary of petrology: S. I. Tomkeieff.* New York: John Wiley.

Waterston, Charles D. 1957. Robert James Hay Cunningham (1815–1842). *Transactions of the Edinburgh Geological Society* 17:260–72.

———. 1966. *Hugh Miller: The Cromarty Stonemason.* Edinburgh: National Trust for Scotland.

Watts, William W. 1921a. Charles Lapworth. *Proceedings of the Geological Society of London* 77:lv–lxi.

———. 1921b. The geological work of Charles Lapworth, M.Sc., LL.D., F.R.S., F.G.S. *Proceedings of the Birmingham Natural History and Philosophical Society* 14, special supplement.

———. 1925–26. Thomas George Bonney, 1833–1923. *Proceedings of the Royal Society,* series B, 99:xvii–xxvii.

———. 1939. The author of the Ordovician system: Charles Lapworth. *Proceedings of the Geologists' Association* 50:235–86.

Watts, William W., and Jethro J. H. Teall. 1921. Charles Lapworth, 1842–1920. *Proceedings of the Royal Society of London* 92:xxxi–xl.

Weaver, J. R. H. ed. 1937. *Dictionary of National Biography, 1922–1930*. London: Oxford University Press.

Wheeler, A. and J. H. Price, eds. 1985. *From Linnaeus to Darwin: Commentaries on the history of geology and biology*. London: Society for the History of Natural History.

Whitaker, William. 1900. Dr. Henry Hicks, F.R.S. *Proceedings of the Geological Society of London* 56 : lviii–lix.

Wilson, H. E. 1985. *Down to earth: One hundred and fifty years of the British Geological Survey*. Edinburgh and London: Scottish Academic Press.

Wilson, R. B. 1977. *A history of the Geological Survey in Scotland*. Edinburgh: National Environment Research Council and Institute of Geological Sciences.

Woodward, Henry B. 1899–1900. Dr. Henry Hicks, F.R.S. *Nature* 61 : 109–10.

———. 1908. *The history of the Geological Society of London*. London: Longmans, Green and Co.

Woolgar, Steven. 1988. *Science: The very idea*. Chichester: Horwood and Tavistock.

Wrottesley, John (Baron). 1856–57. Daniel Sharpe. *Proceedings of the Royal Society of London* 8 : 275–79.

Young, John, 1893. The late Sir A. C. Ramsay, F.G.S., Director-General of the Geological Survey of Great Britain. *Transactions of the Geological Society of Glasgow*. 9 : 256–63.

Zhang Zi Ping. 1935. *Biography of Archibald Geikie*. Peking: Business House Press (in Chinese).

Index